The Imidazolinone Herbicides

T0258814

Editors

Dale L. Shaner, Ph.D.

Research Fellow
Agriculture Research Center
American Cyanamid Company
Princeton, New Jersey

Susan L. O'Connor

Technical Writer/Editor
Agriculture Research Center
American Cyanamid Company
Princeton, New Jersey

CRC Press
Taylor & Francis Group
Boca Raton London New York

CRC Press is an imprint of the
Taylor & Francis Group, an **informa** business

PREFACE

The imidazolinone herbicides were discovered in the 1970s by scientists at American Cyanamid Company. At the writing of this book, the discovery of the imidazolinone herbicides has led to the development of four world-class herbicides. Because of their versatility, low toxicity, and environmental safety, these herbicides are used in many different crops and play a vital role in the production of food and fiber throughout the world. The imidazolinones represent one of the new classes of herbicides that will lead us into the 21st century.

Due to the importance of the imidazolinones in weed control in agriculture, a great deal of research has been conducted. Some research has been performed in the public domain, but much of it has been conducted by the research scientists at Cyanamid during the past 10 years. Unfortunately, the work performed by industrial scientists is not widely disseminated nor readily available to researchers in the public sector. This book was written to help rectify this situation.

The purpose of this book is to provide as complete a picture as possible of the imidazolinones. The chapters have been written by scientists and product development managers who work for Cyanamid.

The volume is divided into five sections: Chemistry, Biology, Metabolism and Residues, Environmental Fate, and Product Performance. The Chemistry section includes two chapters written by Dr. Los on the history of the discovery of the imidazolinones and on the chemical synthesis of these compounds. The next chapter, written by Dr. Wepplo, summarizes the current knowledge of the chemical and physical properties of the imidazolinone herbicides. In the fourth chapter, Dr. Ladner has written about the structure-activity relationships of the herbicidal activity of the imidazolinones.

In the Biology section, Chapter 5, written by Dr. Little and Dr. Shaner, describes the absorption and translocation properties of these herbicides. Next, Drs. Stidham and Singh discuss acetohydroxyacid synthase and the interaction of the imidazolinones. Chapter 7 is a review of the mechanism of selectivity of the imidazolinones, while Chapter 8 highlights how different environmental factors affect the herbicidal activity of these chemicals. The physiological effects of these herbicides at the whole plant level is covered in Chapter 9, written by Dr. Shaner. In the last chapter in this section, Drs. Newhouse, Wang, and Anderson describe the discovery and development of imidazolinone-resistant crops.

The Metabolism and Residues section includes chapters on the metabolic transformation of the imidazolinones in plants and animals and the analytical methodologies for detection of the herbicides in various agricultural commodities and environmental samples. The chapter written by Cyanamid toxicologists summarizes the data obtained from the extensive research that has been performed to demonstrate the safety of these herbicides.

Dr. Mangels has written two chapters on the environmental fate of the imidazolinones in both water and soil. This section clearly describes how the herbicides dissipate after application.

In the final section, the product performance of the imidazolinones is summarized. Each chapter was written by the product development managers who were responsible for registering the herbicides for use throughout the world. These chapters offer concise descriptions of how these herbicides are used under practical, ''real world'' conditions.

This book will provide the research scientist with vital information on the imidazolinone herbicides. It will also be a useful tool for those who teach undergraduate and graduate students about the mode of action and chemical properties of herbicides, as well as for anyone who wants to learn more about the imidazolinones.

The Imidazolinone Herbicides would not have been completed without the support of the management at American Cyanamid or the authors who took the time and effort to write the individual chapters. We would like to thank An-horng Lee, Edward Lignowski, Marinus Los, and Mary Anna Reeves, who, as members of the Imidazolinone Book Committee, provided technical expertise, direction, assistance, and encouragement when needed. We would also like to thank Gail Schmitt for her expert editorial assistance. We are especially grateful to Dr. Jim Gramlich, who originally conceived the idea of this book and provided us the resources and time to complete it. This book is dedicated to all the researchers who have made the imidazolinone herbicides what they are today for weed control in world agriculture.

TRADEMARK STATEMENT

EDITORS

Dale L. Shaner, Ph.D., is a Research Fellow at American Cyanamid Company. He is a member of the Plant Directed Basic Research Department at the Agricultural Research Center in Princeton, where he has directed the research of scientists in the areas of plant biochemistry, plant physiology, and plant-chemical-environment interactions since 1988. Dr. Shaner received his B.A. in Botany in 1970 from DePauw University, an M.A. in Plant Ecology in 1972 from the University of Colorado, and a Ph.D. in Plant Physiology in 1976 from the University of Illinois.

After obtaining his Ph.D., Dr. Shaner served as assistant professor of Weed Science/Plant Physiology at the University of California, Riverside until 1979. At that time he joined the Herbicide Discovery Group at American Cyanamid, where he helped conduct the early research on the greenhouse and field activity of the imidazolinones. He then moved into the laboratory and was instrumental in discovering the mode of action of the imidazolinone herbicides. In 1984 he received Cyanamid's Scientific Achievement Award for his work on imazamethabenz-methyl. He has also won several awards for scientific publication given at Cyanamid. Dr. Shaner started the Plant Biotechnology Group at Cyanamid.

Dr. Shaner is a member of the Weed Science Society of America and the American Chemical Society. He is an associate editor for the *Weed Science journal.*

Dr. Shaner is author of more than 35 papers and has given more than 25 invited lectures at symposia and seminars. His research interests include all aspects of the mode of action of herbicides, the effect of environmental factors on herbicidal activity, and on herbicide resistance in crops and weeds.

Susan L. O'Connor is a Technical Writer/Editor at American Cyanamid Company. She is a member of the International/LAG Plant Industry Development Department at the Agricultural Research Center where she has written, edited, and designed the department's Technical Bulletins and Technical Information Reports since 1988. Ms. O'Connor received her B.A. in Communication in 1984 from Seton Hall University. Since 1984, Ms. O'Connor has worked in various technical editing positions in the medical and research and development fields.

Ms. O'Connor is a member of the New York Chapter of the Society for Technical Communications where she acts as Public Service Advertising Campaign Manager.

CONTRIBUTORS

Paul C. Anderson, Ph.D.
Research Director
Plant Science Research, Inc.
Minnetonka, Minnesota

Richard A. Beardmore, Ph.D.
Senior Sales Representative
American Cyanamid Company
Salisbury, Maryland

Barnett Bernstein, Ph.D.
Product Development Manager
Agricultural Research Center
American Cyanamid Company
Princeton, New Jersey

Prithivi R. Bhalla, Ph.D.
Senior Turf Specialist
Agricultural Research Center
American Cyanamid Company
Princeton, New Jersey

Timmy Y. Chiu, Ph.D.
Senior Research Chemist
Agricultural Research Center
American Cyanamid Company
Princeton, New Jersey

James M. Devine
Manager, Residue Chemistry
Agricultural Research Center
American Cyanamid Company
Princeton, New Jersey

Scott Diehl
Toxicology Clerk
Agricultural Research Center
American Cyanamid Company
Princeton, New Jersey

Ruth Fiala
Agricultural Research Center
American Cyanamid Company
Princeton, New Jersey

Joel E. Fischer
Senior Research Toxicologist
Agricultural Research Center
American Cyanamid Company
Princeton, New Jersey

Chien H. Fung, Ph.D.
Senior Research Chemist
Agricultural Research Center
American Cyanamid Company
Princeton, New Jersey

James A. Gagne, Ph.D.
Associate Research Fellow
Agricultural Research Center
American Cyanamid Company
Princeton, New Jersey

Paul E. Gatterdam, Ph.D.
Senior Research Chemist
Agricultural Research Center
American Cyanamid Company
Princeton, New Jersey

Barbara Gingher, Ph.D.
Product Registration Manager
Agricultural Research Center
American Cyanamid Company
Princeton, New Jersey

Neil M. Hackett, Ph.D.
Product Development Manager
Agricultural Research Center
American Cyanamid Company
Princeton, New Jersey

Jane E. Harris, Ph.D.
Director of Toxicology
Agricultural Research Center
American Cyanamid Company
Princeton, New Jersey

Ronaldo G. Hart
Product Development Manager
Agricultural Research Center
American Cyanamid Company
Princeton, New Jersey

Frederick G. Hess, Ph.D.
Program Manager, Toxicology
Agricultural Research Center
American Cyanamid Company
Princeton, New Jersey

Richard D. Iverson
Product Development Manager
Agricultural Research Center
American Cyanamid Company
Princeton, New Jersey

Kenneth Kirkland, Ph.D.
Manager, International Plant Product
  Development
Agricultural Research Center
American Cyanamid Company
Princeton, New Jersey

David W. Ladner, Ph.D.
Group Leader, Herbicide Synthesis I
Agricultural Research Center
American Cyanamid Company
Princeton, New Jersey

An-horng Lee, Ph.D.
Group Leader, Metabolism II
Agricultural Research Center
American Cyanamid Company
Princeton, New Jersey

Edward M. Lignowski, Ph.D.
Senior Product Development Manager
Agricultural Research Center
American Cyanamid Company
Princeton, New Jersey

Desiree L. Little, Ph.D.
Product Development Coordinator
Agricultural Research Center
American Cyanamid Company
Princeton, New Jersey

Marinus Los, Ph.D.
Associate Director, Plant Sciences
Agricultural Research Center
American Cyanamid Company
Princeton, New Jersey

Timothy Malefyt, Ph.D.
Group Leader, Herbicide Discovery
Agricultural Research Center
American Cyanamid Company
Princeton, New Jersey

N. Moorthy Mallipudi, Ph.D.
Senior Research Chemist
Agricultural Research Center
American Cyanamid Company
Princeton, New Jersey

Gary Mangels, Ph.D.
Group Leader, Environmental Chemistry
Agricultural Research Center
American Cyanamid Company
Princeton, New Jersey

Phillip Miller, Ph.D.
Group Leader, Metabolism Research
Agricultural Research Center
American Cyanamid Company
Princeton, New Jersey

Keith E. Newhouse, Ph.D.
Senior Research Biologist
Agricultural Research Center
American Cyanamid Company
Princeton, New Jersey

Laura Sue Quakenbush, Ph.D.
Senior Research Weed Scientist
Agricultural Research Center
American Cyanamid Company
Princeton, New Jersey

Mark A. Risley, Ph.D.
Product Development Manager
Agricultural Research Center
American Cyanamid Company
Princeton, New Jersey

Dale L. Shaner, Ph.D.
Research Fellow
Agricultural Research Center
American Cyanamid Company
Princeton, New Jersey

Rajendar K. Sharma, Ph.D.
Principal Research Toxicologist
Agricultural Research Center
American Cyanamid Company
Princeton, New Jersey

S. M. Shehata, Ph.D.
Product Development Manager
Agricultural Research Center
American Cyanamid Company
Princeton, New Jersey

Bijay K. Singh, Ph.D.
Senior Research Biochemist
Agricultural Research Center
American Cyanamid Company
Princeton, New Jersey

Mark A. Stidham, Ph.D.
Group Leader
Agricultural Research Center
American Cyanamid Company
Princeton, New Jersey

Fred R. Taylor
Product Development Manager
Agricultural Research Center
American Cyanamid Company
Princeton, New Jersey

Karl A. Traul, Ph.D.
Senior Group Leader, Toxicology
Agricultural Research Center
American Cyanamid Company
Princeton, New Jersey

Mark C. Trimmer, Ph.D.
Product Development Manager
Agricultural Research Center
American Cyanamid Company
Princeton, New Jersey

Theodora Wang, Ph.D.
Toxicology Administrator
Agricultural Research Center
American Cyanamid Company
Princeton, New Jersey

Peter Wepplo, Ph.D.
Principal Research Chemist
Agricultural Research Center
American Cyanamid Company
Princeton, New Jersey

Cletus Youmans
Field Agriculturalist
American Cyanamid Company
Dyersburg, Tennessee

TABLE OF CONTENTS

Chapter 1

DISCOVERY OF THE IMIDAZOLINONE HERBICIDES

Marinus Los

TABLE OF CONTENTS

I. BACKGROUND

To direct the synthesis efforts of chemists in their search for a new herbicide, three approaches are possible. The first, and most likely to produce active compounds, is to use the structure of a known herbicide as a template. Modification of such a structure often results in compounds with desirable properties not shared by the template. The dangers in this approach are that the selected structures may already have been examined and discarded, that structure modifications may lead to a product of marginal advantage, and finally, that the resulting compound may encounter serious patent problems.

This approach has produced highly successful commercial products that share many properties but have unique characteristics that set them apart from the rest. A particularly good example of this approach is pendimethalin (Prowl herbicide). Pendimethalin shares many of the same properties with trifluralin, yet has an important additional attribute — it does not have to be incorporated in the soil to maintain activity.

The second approach, often called the biorational method, derives from the knowledge of the structure of the target enzyme and/or the detailed mechanism of the reaction involved. This method has been successfully applied in pharmaceutical research but has not, as far as is known, led to any new agrochemical product. This area is still in its infancy but will, in time, play a more prominent role in the search for new agricultural compounds.

The third method is that of random screening of chemicals. This method, though lacking the sophistication of the second, has been highly successful and has resulted in many new compounds with diverse structures and mechanisms of action. Random screening is possible since the supply of the target — weeds — is not a limiting factor and the results obtained can frequently be translated immediately into practical uses. This is in sharp contrast with the biorational approach, where a great deal has to be done to translate an active inhibitor of a target site into a practical product. Although this method is relatively simple, some parameters in the screening process need to be established. For example, the rate of application of the compounds will obviously determine the percentage of herbicidally active compounds, whereas the weed species included in the test will define the types of activities discovered. Difenzoquat (Avenge herbicide) would not have been discovered had *Avena fatua* not been in this primary random screen, since wild oats is the only weed difenzoquat controls.

It would be inappropriate for any company engaged in agrochemical research to disregard any of these three methods, for it is vitally important that new structures with new modes of action be discovered. Random screening is the simplest method by which this can be achieved. As modes of action are unraveled and the target sites are identified, enzyme screens can frequently augment other screens utilizing whole plants. Then, as the details of the enzyme structure and its active site become available, the role of "rational" synthesis can increase.

II. DISCOVERY OF THE IMIDAZOLINONE HERBICIDES

The discovery of the imidazolinone herbicides started with a random screening test in which the phthalimide, **1**, was found to have sufficient activity at 4 kg/ha to warrant the initiation of a synthesis program around this structure.

1 R=H
2 R=Cl

3 AC 94,377

GA$_3$

A systematic evaluation of the effects of changes in the acid function, phenyl substitution, and alkyl group of the lead phthalimide 1 was undertaken. The primary amide function appeared to be optimal, since any change to this function (to acid, ester, or substituted amide) resulted in reduced herbicidal activity. The effect of substitution in the phenyl ring was quite surprising. It was found, for example, that the chloro- derivative, 2, had extremely poor herbicidal properties but did have pronounced growth-stimulating effects similar to those of gibberellic acid (GA). This exciting diversion resulted in a considerable effort to optimize the growth-stimulating effects. As a result of this research, a new compound, 3 (AC 94,377), was prepared which essentially had no herbicidal activity but showed re-markable growth-stimulating activity.[1] In a number of gibberellic acid bioassays, AC 94,377 had approximately one tenth the activity of GA_3, a remarkable result when the rather complex structure of GA_3 is compared to the relatively simple structure of AC 94,377.[2] Preliminary data indicate that AC 94,377 does, in fact, compete for the same binding site as GA_3.[3] The structural modifications to molecules 2 and 3, which result in GA-like activity, are fully described in the patent literature.[4-6]

The unusual activity associated with 3 resulted in extensive field trials on a number of important crops. Because this required the preparation of several hundred kilograms of material, various synthetic methods were evaluated. One of them involved the cyclization of the phthalamic acid 4 using a variety of reagents. One of the reagents was cold trifluo-roacetic anhydride (TFA), which gave in low yield the easily isolated imidazoisoindole 5. When this compound was tested, it showed very similar properties to those of the noncyclized chemical 3, although it appeared to be slower acting. This effect was rationalized by assuming that 5 was hydrolyzing slowly to AC 94,377. Because of the novel structure of 5 and its attendant biological activity, a synthesis effort was directed to explore this area of chemistry.

4   5

The first requirement was to find a practical method to synthesize the imidazoisoindoles. Several acidic and basic reagents were found to effect this transformation.[7,8] In the laboratory, the most useful reagent was sodium hydride in toluene, while large scale preparations utilized NaOH pels in toluene or xylene.

One of the earliest compounds prepared in this series was derived from 1, the chemical that started this program. Although the product, 6, did not have GA-like activity, its herbicidal activity was promising compared to that of the original phthalimide 1. Because of its lack of crop selectivity, this compound was tested as a total vegetation control agent. Although compound 6 proved to be the most active member in this series, particularly on perennial weeds such as *Cyperus* spp. and *Convolvulus* spp., it was not sufficiently active to warrant further development.

1   6

The chemistry of these novel heterocycles was studied further. The compounds were readily reduced with sodium borohydride to give a mixture of diastereomers, for example **7a** and **7b** from **6**. The mixture of **7a** and **7b** could be separated by fractional crystallization. The mixture, as well as each isomer, was also shown to have very good nonselective herbicidal activity.[8-10]

It was further found that these imidazoisoindoles such as **6** would add a variety of alcohols, thiols, and amines to the imine double bond to yield a mixture (**8**) which was also highly active herbicidally. The addition occurs usually at room temperature and is reversed by heating.[8-10]

Another reaction of the imidazoisoindoles such as **6** was of paramount significance in that it allowed the first entry into the imidazolinone group of herbicides. Specifically, it was the reaction of sodium methoxide in methanol with compound **6** that resulted in the first example in the imidazolinone series of herbicides.[11-15] In contrast to methanol under nonbasic conditions, this reaction with methoxide occurs at the benzoyl carbonyl group, rather than at the imine double bond, to give the benzoate **9**. Similarly, reaction with an amine at an elevated temperature results in the formation of benzamides.

Thus, the study of the chemistry of the imidazoisoindoles led to the discovery of the imidazolinones.

REFERENCES

1. **Los, M., Kust, C. A., Lamb, G., and Diehl, R. E.,** Phthalimides as plant growth regulators, *Hortic. Sci.,* 15, 22, 1980.

2. **Suttle, J. C. and Schreiner, D. R.,** Biological activity of AC 94,377[1-(3-chlorophthalimido)-cyclohexanecarboxamide], *J. Plant Growth Regul.,* 1, 139, 1982.

3. **Yalpani, N., Suttle, J. C., Hultstrand, J. F., and Rodaway, S. J.,** Competition for *in vitro* [^{3}H] gibberellin A$_4$ binding in cucumber by substituted phthalimides: comparison with *in vivo* gibberellin-like activity (in preparation).

4. **Diehl, R. E. and Walworth, B. L.,** Phthalimide Derivatives and Their Use as Plant-Growth Regulants, U.S. Patent 3,940,419, 1976.

5. **Diehl, R. E. and Walworth, B. L.,** Phthalimide Derivatives as Plant-Growth Regulants, U.S. Patent 4,017,299, 1977.

6. **Los, M. and Walworth, B. L.,** Dithiindicarboximide, Dithiolanedicarboximide, Thiapyrandicarboximide and Pyrancarboximide Derivatives as Plant-Growth Regulants, U.S. Patent 4,164,404, 1979.

7. **Los, M.,** Imidazoisoindolediones and the Use thereof as Herbicidal Agents, U.S. Patent 4,017,510, 1977.

8. **Los, M.,** Method of Preparing Imidazoisoindolediones, U.S. Patent 4,125,727, 1978.

9. **Los, M.,** Dihydroimidazoisoindolediones and the Use thereof as Herbicidal Agents, U.S. Patent 4,041,045, 1977.

10. **Los, M.,** Dihydroimidazopyrrolopyridine, -quinoline, Thieno- and Furo[2,3-*b*]pyridine, Dihydrothieno- and -furo[2,3-*b*]pyridine, Thieno- and Furo[3,2-*b*]pyridine and Dihydrothieno- and -furo[3,2-*b*]pyridine Herbicides and Method for Preparation thereof, U.S. Patent 4,701,208, 1987.

11. **Los, M.,** Imidazolinyl Benzoic Acids, Esters, and Salts and Their Use as Herbicidal Agents, U.S. Patent 4,188,487, 1980.

12. **Los, M.,** Imidazolinyl Benzamides as Herbicidal Agents, U.S. Patent 4,122,275, 1978.

13. **Los, M.,** Preparation of Imidazolinyl Benzoic Acids, U.S. Patent 4,608,437, 1986.

14. **Los, M.,** Herbicidal Imidazolinyl Naphthoic Acids, U.S. Patent 4,554,013, 1985.

15. **Los, M.,** Preparation of Imidazolinyl Benzoic Acids, Esters, and Salts, U.S. Patent 4,544,754, 1985.

Chapter 2

SYNTHESIS OF THE IMIDAZOLINONE HERBICIDES

Marinus Los

TABLE OF CONTENTS

I. SYNTHESIS VIA TRICYCLIC INTERMEDIATES

As described in Chapter 1, the first imidazolinone, **2,** was prepared by the reaction of the imidazoisoindole **1** with methoxide in methanol. This method was employed for several years to synthesize a large number of analogs.

During that time, the product now known as imazamethabenz-methyl (Assert herbicide) was synthesized and is now a commercial product used to control *Avena fatua, Alopecurus myosuroides,* and certain broadleaved weeds in wheat and barley. This product differs from the other imidazolinones in that it is a mixture of isomers, a result of the method of synthesis. The compound is unique in that one isomer is particularly active on *Avena* and *Alopecurus,* while the other has excellent activity on *Sinapis arvensis.*[1]

Assert Herbicide

A useful and versatile method for the synthesis of imidazolinones is illustrated in the above sequence. In the case of Assert, the formation of the mixture of products proved advantageous, but this generally is not the case, because one of the isomers is virtually inactive yet wastes a significant portion of starting materials and requires removal at the end of the synthesis. This synthetic sequence leads to an ester which can be converted to the corresponding acids by simple base hydrolysis.[2,3]

A more appropriate strategy may be to prepare the benzyl ester from the imidazoisoindole and then remove this group by hydrogenolysis.

II. SYNTHESIS VIA ARYL-1,2-DICARBOXYLIC ACIDS

The simplest syntheses of the imidazolinones start with an o-dicarboxylic acid or corresponding anhydride. Many of these compounds are described in the chemical literature. Of particular importance is the ability to produce these anhydrides in large quantities by a commercially acceptable process and at an economical price. Frequently, new methods have to be devised, as was done by Doehner[4] for quinoline-2,3-dicarboxylic acid.

With an appropriate o-dicarboxylic acid available, a number of strategies can be followed to produce the desired imidazolinone. These will be exemplified with pyridine-2,3-dicarboxylic acid, which must first be converted to the corresponding anhydride. The simplest available coreactant is the α-aminonitrile derived from the appropriate ketone via a Strecker synthesis. It should be pointed out that this is also one of the most convenient steps for introducing a ^{14}C or ^{13}C marker into the molecule by means of labeled cyanide. The cyano group, 3, is then hydrolyzed with sulfuric acid and the resulting amide cyclized to the desired imidazolinone 4. Although 4 is the major product, the isomeric picolinic acid is also formed since the anhydride reacts at both carbonyl groups. Thus, all similar syntheses of arylimidazolinyl carboxylates utilizing an asymmetrical anhydride as intermediate will result in a mixture of products.

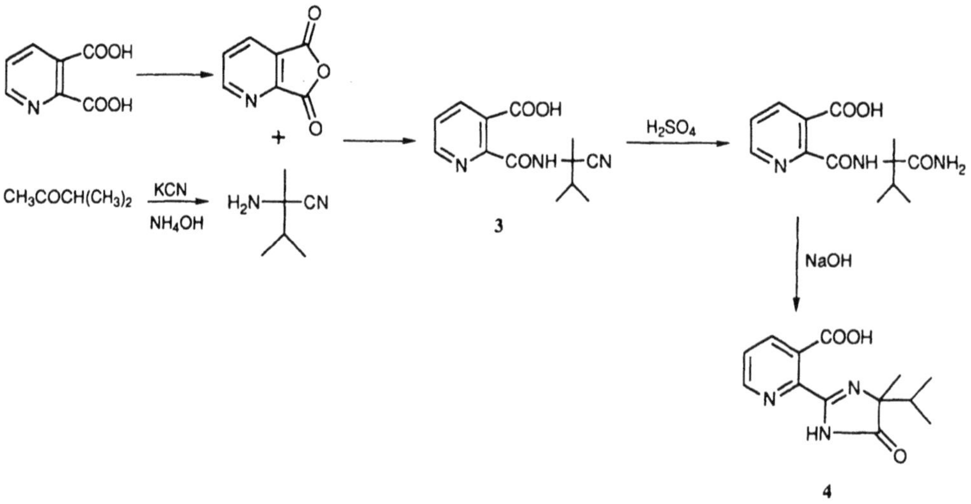

A variant on the preceding route to the imidazolinones utilizes the α-aminocarboxylic acid amide rather than the nitrile. This substantially simplifies the synthesis. The initial condensation is carried out in methylene chloride, and the product is extracted into sodium hydroxide solution and then heated to effect cyclization to the imidazolinone.[5]

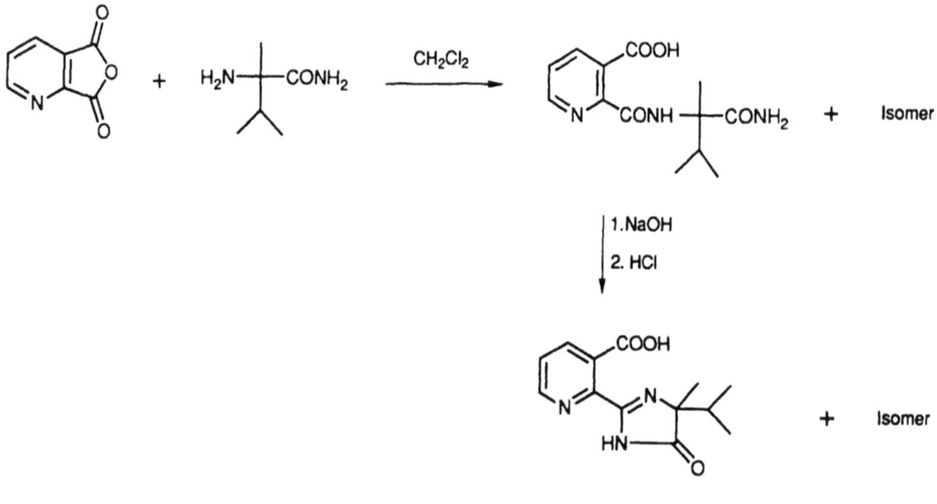

III. SYNTHESIS VIA ARYL-1,2-DICARBOXYLIC ACID ESTERS

A new method reported recently (shown below) is a valuable addition to the methods described above in that it is essentially a one-step synthesis.

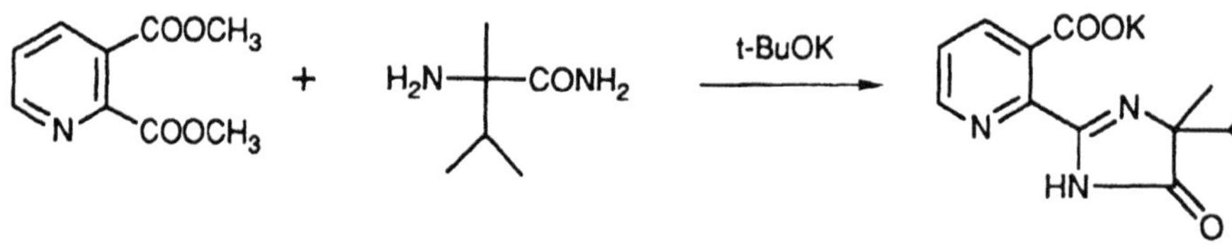

The value of this process is that synthetic methods for *o*-dicarboxylic acids frequently proceed through the diester. This reaction is usually carried out in toluene at 80°C.[6]

In order to advance the program and probe the requirements for activity and selectivity in the imidazolinone series, the development of either new routes to *o*-dicarboxylic acids or other alternative syntheses was required.

A good route to the intermediate required for the synthesis of imazethapyr (Pursuit herbicide) was developed, based on the known process for the preparation of pyridine-2,3-dicarboxylic acid.[7,8]

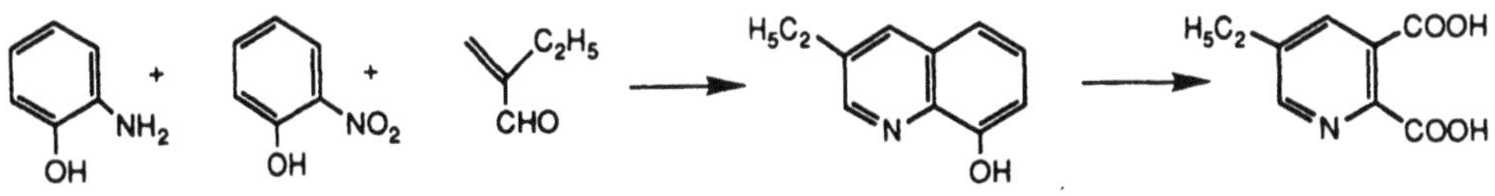

IV. SYNTHESIS VIA METALLATION OF 2-ARYL-SUBSTITUTED-2-IMIDAZOLIN-4-ONES

A radically different route to these products was developed which made possible the synthesis of a large number of analogs. Many of these would have been extremely difficult to prepare by the traditional *o*-dicarboxylic acid route. The key step involved the *o*-directed and stabilized metallation of 2-aryl-substituted-2-imidazolin-4-one.[3,9] This is shown below, using the pyridine derivative as an example.

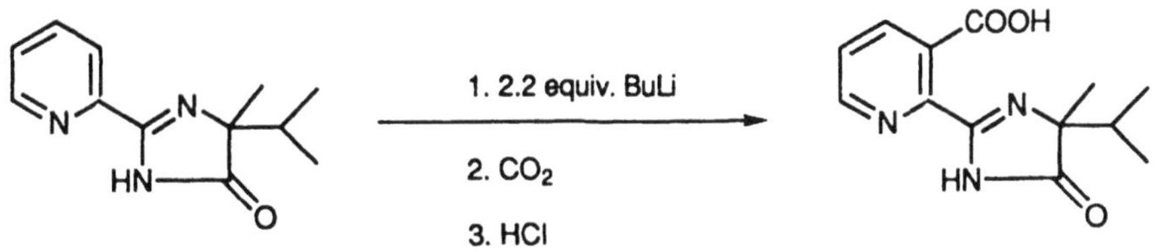

1. 2.2 equiv. BuLi
2. CO₂
3. HCl

The requirements for this synthetic route are obviously less stringent; the intermediate must be a carboxylic acid and must only bear substituents that will be stable toward an alkyllithium reagent. The acid can be activated through its acyl halide or mixed anhydride, reacted with the α-aminocarboxamide, and the product cyclized by base to the desired imidazolinone.

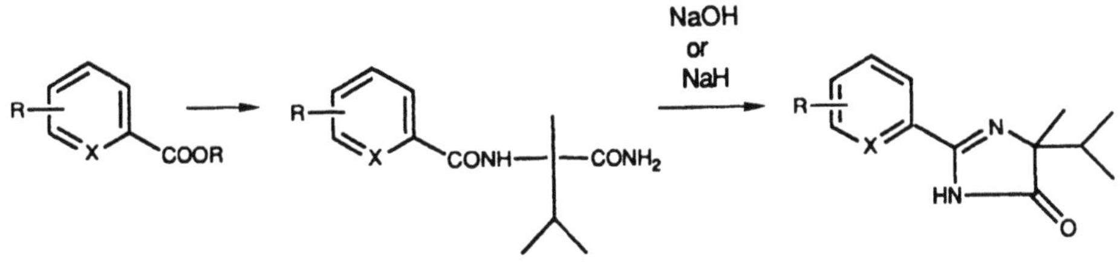

NaOH
or
NaH

It should be pointed out that metallation will occur at both positions ortho to the imidazolinone group if these are available. Thus, the products derived from *m*-toluic acid are illustrative.

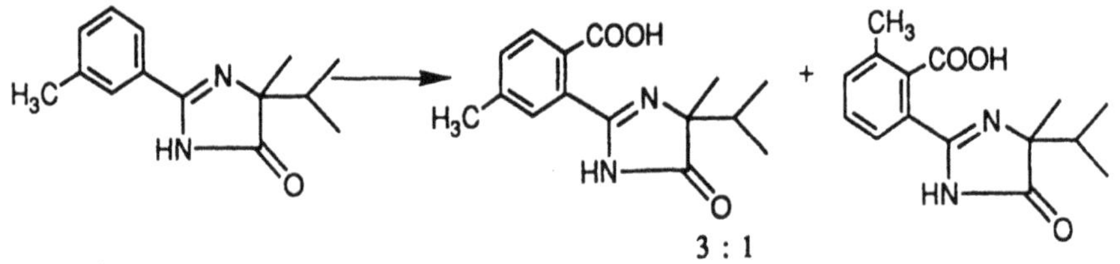

3 : 1

V. SYNTHESIS VIA WILLGERODT REACTION

In certain cases the arylimidazolinone can be prepared in an even simpler manner than that described above for the corresponding carboxylic acid. It is well known that an activated methyl group (as in α-picoline) can be converted by a Willgerodt reaction in a single reaction to a thioamide.[10] Wepplo[11] reasoned that it should be possible to replace the aniline by 2-amino-2,3-dimethylbutyramide to obtain the valuable intermediate **5** en route to the required imidazolinone **6**. In fact, the reaction yielded directly product **6**, carboxylation of which then gives the herbicide imazethapyr (Pursuit herbicide).

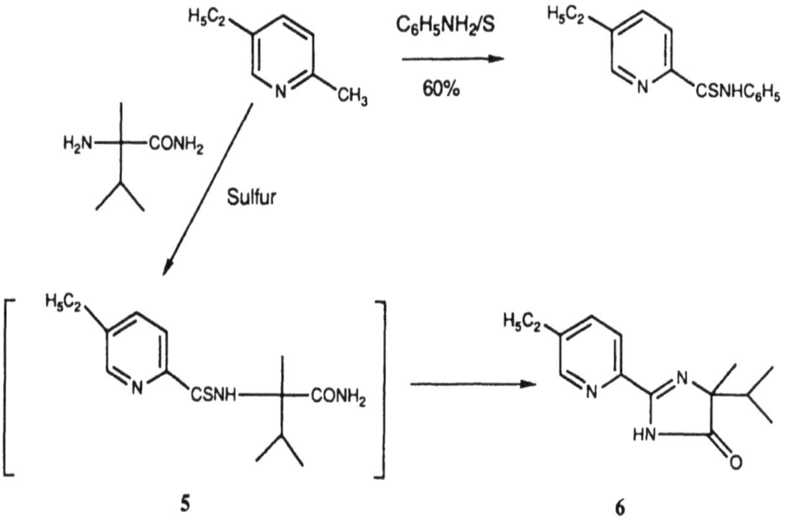

5

6

VI. SYNTHESIS OF OPTICALLY ACTIVE IMIDAZOLINONES

Another feature of the imidazolinones requires elaboration. The compounds of greatest interest are substituted at C-4 by a methyl and an isopropyl group, which results in a chiral center. In order to evaluate the biological activity of each enantiomer, a method was developed to prepare appropriate chiral intermediates. This was achieved by the enzymatic resolution of the N-acetyl derivative of α-methylvaline by porcine kidney acylase.[12]

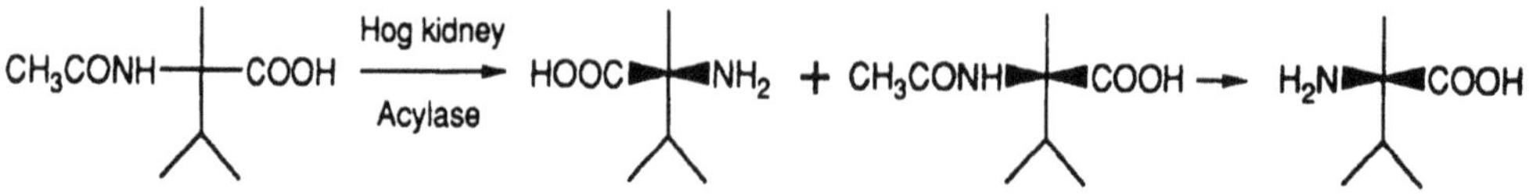

The synthesis of the imidazolinone proceeded by condensation of the amino acid with the appropriate *o*-dicarboxylic acid anhydride to give an imide and was completed by conthe imidazolinone. U.S. Patent 4,683,324[13] describes the kinetic resolution of the nitrile **7** in which the racemic **7** is converted in essentially quantitative yield to **R-7** by means of (2*S*,3*S*)-tartaric acid. The **R-7** proved to be somewhat unstable but, by hydrolysis with concentrated sulfuric acid, could be readily converted to the stable amide **R-8**. The absolute configuration of these compounds was determined by the crystallographic analysis of the more active enantiomer of the imidazolinone **9**.

Each enantiomer of a number of active herbicides was prepared and their biological activities were compared. In general, the activity of the *R*-isomer is approximately twice that of the racemic compound. The *S*-isomer, although not completely inactive, is nevertheless much less active than the *R*-isomer.

VII. SUMMARY

The methodologies described in this chapter have been applied to the preparation of a large number of imidazolinone derivatives. Details of these methodologies can be found in the patent literature.[3,14-17]

REFERENCES

1. **Los, M.**, *o*-(5-Oxo-2-imidazolin-2-yl)arylcarboxylates: a new class of herbicides, in *Pesticide Synthesis Through Rational Approaches*, Magee, P. S., Kohn, G. K., and Menn, J. J., Eds., ACS Symp. Ser. No. 255, American Chemical Society, Washington, D.C., 1984, 29.
2. **Los, M.**, Preparation of Imidazolinyl Benzoic Acids, U.S. Patent 4,608,437, 1986.
3. **Los, M.**, 2-(2-Imidazolin-2-yl)pyridines and -quinolines and Use of Said Compounds as Herbicidal Agents, U.S. Patent 4,798,619, 1989.
4. **Doehner, R. F.**, Alkyl Esters of Substituted 2-Methyl-3-quinolinecarboxylic Acid and Quinoline-2,3-dicarboxylic Acid, U.S. Patent 4,656,283, 1987.
5. **Barton, J. M., Long, D. W., and Lotts, K. D.**, Process for the Preparation of 2-(5,5-Disubstituted-4-oxo-2-imidazolin-2-yl)nicotinic Acids, -quinoline-3-carboxylic Acids and -benzoic Acids, U.S. Patent 4,658,030, 1987.
6. **Ciba-Geigy**, 2-Imidazolinyl-pyridine- and -quinoline-3-carboxylic Acid Production by Reaction of Pyridine- or Quinoline-2,3-dicarboxylic Acid Esters with a 2-Amino-alkanoic Acid Amide, EP 233-150-A, 1986.
7. **Reiker, W. F. and Daniels, W. A.**, Method for the Preparation of Pyridine-2,3-dicarboxylic Acids, U.S. Patent 4,816,588, 1989.
8. **Sturrock, M. G., Cline, E. L., Robinson, K. R., and Zercher, K. A.**, Pyridine Carboxylic Acids, U.S. Patent 2,964,529, 1960.
9. **Gaschwend, H. W. and Rodriquez, H. R.**, Heteroatom-facilitated lithiations, *Org. React.*, 26, 1, 1979.
10. **Brown, E. V.**, The Willgerodt reaction, *Synthesis*, 358, 1975.
11. **Wepplo, P. J.**, Process for the Preparation of Pyridyl and Quinolyl Imidazolinones, U.S. Patent 4,474,962, 1984.
12. **Los, M.**, Imidazoisoindolediones, the Use thereof as Herbicidal Agents, U.S. Patent 4,017,510, 1977.

13. **Gastrock, W. H. and Wepplo, P. J.,** Process for the Resolution of Certain Racemic Amino Nitriles, U.S. Patent 4,683,324, 1987.

14. **Los, M.,** Herbicidal 2-(2-Imidazolin-2-yl)fluoroalkoxy-, alkenyloxy- and -alkynyloxypyridines, U.S. Patent 4,647,301, 1987.

15. **Los, M., Ladner, D. W., and Cross, B.,** (2-Imidazolin-2-yl)thieno- and -furo[2,3-*b*] and -[3,2-*b*]pyridines and Intermediates for the Preparation thereof, and Use of Said Compounds as Herbicidal Agents, U.S. Patent 4,650,514, 1987.

16. **Los, M.,** Herbicidal 2-(2-Imidazolin-2-yl)fluoroalkoxy-, alkenyloxy- and alkynyloxyquinolines, U.S. Patent 4,772,311, 1988.

17. **Los, M., Ladner, D. W., and Cross, B.,** (2-Imidazolin-2-yl)thieno- and furo[2,3-*b*] and [3,2-*b*]pyridines and Use of Said Compounds as Herbicidal Agents, U.S. Patent 4,752,323, 1988.

Chapter 3

CHEMICAL AND PHYSICAL PROPERTIES OF THE IMIDAZOLINONES

Peter J. Wepplo

TABLE OF CONTENTS

I. INTRODUCTION

This chapter will discuss the solubility properties, partition coefficients, pK_as, hydrolytic and chemical stability, spectroscopic data, and chromatographic data of the five imidazolinone herbicides — imazapyr, imazamethabenz-methyl, imazethapyr, imazaquin, and AC 263,222 — the structures of which are shown below.

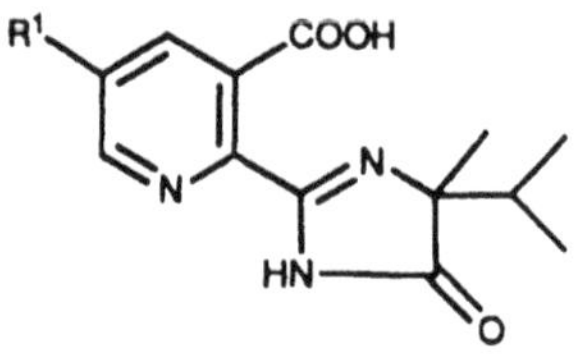

R^1 = H   Imazapyr
R^1 = CH$_3$ AC 263,222
R^1 = C$_2$H$_5$ Imazethapyr

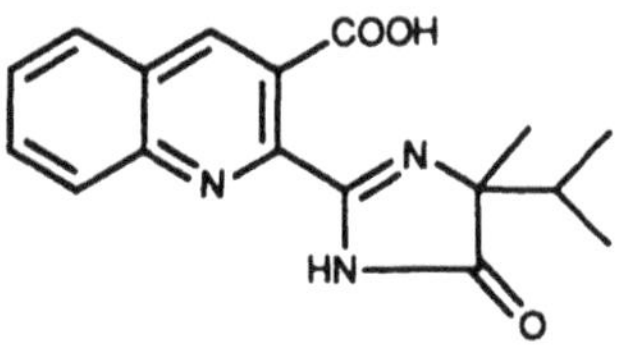

Imazaquin

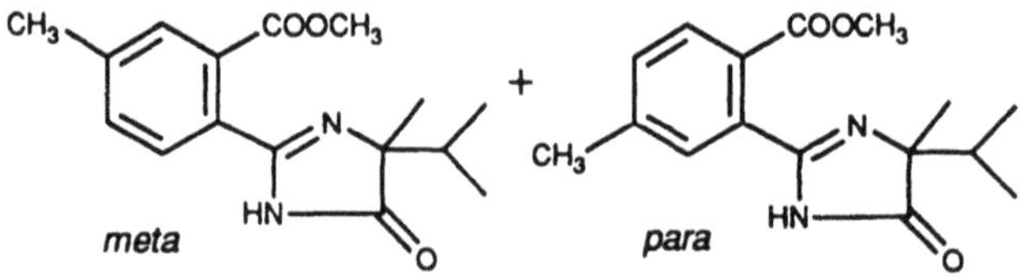

meta   para

Imazamethabenz-methyl

II. SOLUBILITY STUDIES

The basic physical properties of the imidazolinone herbicides are listed in Appendix A. The water solubility and vapor pressure of the imidazolinone carboxylic acids are low, whereas the octanol-water partition coefficients are pH dependent. At pH 4, the imidazolinone carboxylic acids are neutral and therefore lipophilic, whereas at pH 7 the carboxylic acids are ionized. Ionization increases the water solubility and decreases the octanol solubility of the compounds. Table 1 lists the water solubility of selected imidazolinone salts which are used in the formulation of imazapyr, imazethapyr, imazaquin, and AC 263,222.

Imazamethabenz-methyl, which is selective for wheat, derives its selectivity from its ester functionality and is formulated as a mixture of toluate esters (55% *para*, 45% *meta*). Since the solubility of imazamethabenz-methyl in water is low, it cannot be formulated as an aqueous solution. Table 2 lists the solubility of imazamethabenz-methyl in various organic solvents. Imazamethabenz-methyl is lipophilic, with the *meta* isomer having a higher octanol-water partition coefficient than the *para* isomer. From the mixture of isomers, a pure sample of the less soluble *para* isomer can be crystallized from an acetonitrile or ethyl acetate solution.

A comparison of the 2-imidazolinylnicotinic acid, imazapyr (**1**), with the isomeric, 3-imidazolinylpicolinic acid, **2a** or **2b**, shows imazapyr to be less soluble in water and more soluble in organic solvents, such as acetone, acetonitrile, ethyl acetate, and DMF. The picolinic acid prefers to exist as the zwitterionic tautomer **2b**, i.e., internally ionized, over the uncharged species **2a** and is consequently more soluble in water and less soluble in organic solvents. This solubility difference is significant in the preparation of the imidazolinone herbicides since it helps to avoid contamination of the product with the isomeric picolinic acid.

TABLE 1	
Water Solubility of Imidazolinone Salts	
Compound	Solubility
Imazapyr	>57% (isopropylammonium salt) at 5°C
AC 263,222	>24% (ammonium salt) at 0°C
Imazethapyr	24% (ammonium salt) at 15°C
Imazaquin	17% (ammonium salt) at 18°C

TABLE 2	
Solubility of Imazamethabenz-Methyl in Various Solvents at 25°C (g/100 g solvent)	
Solvent	Solubility
Heptane	0.06
Water	0.24
Toluene	4.5
Dichloromethane	17.2
Isopropyl alcohol	18.3
Dimethyl sulfoxide	21.6
Acetone	23
Methanol	30.9

III. IONIZATION CONSTANTS

The pK_as of several imidazolinone carboxylic acids have been measured (see Table 3). Up to three inflection points have been observed in the titration of imidazolinone carboxylic acids (see Scheme 1). The imidazolinone ring is amphoteric and behaves as a weak base or a weak acid. The first inflection point corresponds to protonation of the imidazolinone ring at pK_1, ca. 1.9 to 3.3, while the second reflects ionization of the carboxylic acid at pK_2, ca. 3.6 to 3.9. The third inflection, determined only for AC 263,222, corresponds to deprotonation of the imidazolinone ring at pK_3, ca. 11.4. The pK_a for pyridine ring protonation was not determined.

Table 3 shows that the pK_a of the protonated imidazolinone ring is close to that of the carboxylic acid. While the neutral species in Scheme 1 can be written as **1b** or zwitterion **1c**, our data suggest that compounds exist as **1b** rather than **1c**.

IV. STABILITY STUDIES

A. HYDROLYTIC STABILITY

The hydrolytic stability of the imidazolinone ring is pH dependent. At environmentally relevant pHs and temperatures, hydrolysis is very slow. When the hydrolysis of imazaquin, imazapyr, and imazethapyr was studied at pH 5, 7, and 9 for 30 d at 25°C, there was no detectable degradation in the pH 5 and 7 buffers over the 30-day period. At pH 9, the hydrolytic half-lives of imazaquin, imazapyr, and imazethapyr were calculated as 169, 325, and 288 d, respectively. The hydrolysis product in each case was a diamide. At pHs less than 7, the imidazolinone ring is stable. Imazapyr was subjected to digestion at pH 4 for 5 h at 70°C and at pH 3 for 15 h at 90°C without degradation. The imidazolinone ring is stable even to refluxing 6 N hydrochloric acid.

Diamide **3** is the immediate precursor of imazapyr **1** in the current commercial process. For imazapyr in aqueous base, an equilibrium between the diamide **3a**, imidazolinone **1d**, and the doubly ionized imidazolinone **1e** is reached. Between pHs 8 and 12, the diamide

TABLE 3
Ionization Constants[a]

Compound	pK_1	pK_2	pK_3	Structure
AC 213,493	3.3			
Imazamethabenz-methyl	2.9			
Imazapyr	1.9	3.6		
Imazaquin		3.8		

AC 263,222

3.1[b] 3.9 11.4[b]

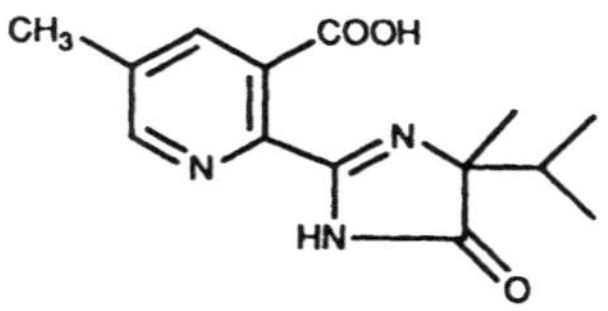

Imazethapyr

2.1 3.9

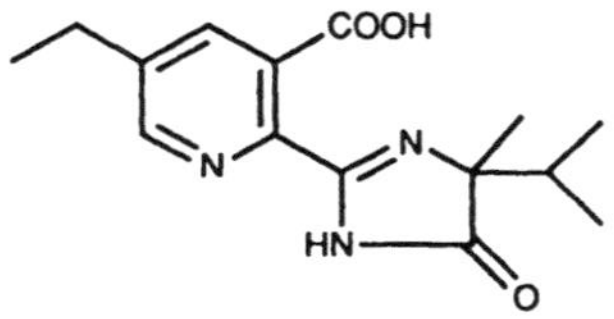

[a] Ionization constants measured by potentiometric titration, except where indicated.
[b] Ionization constants measured by spectrophotometric titration.

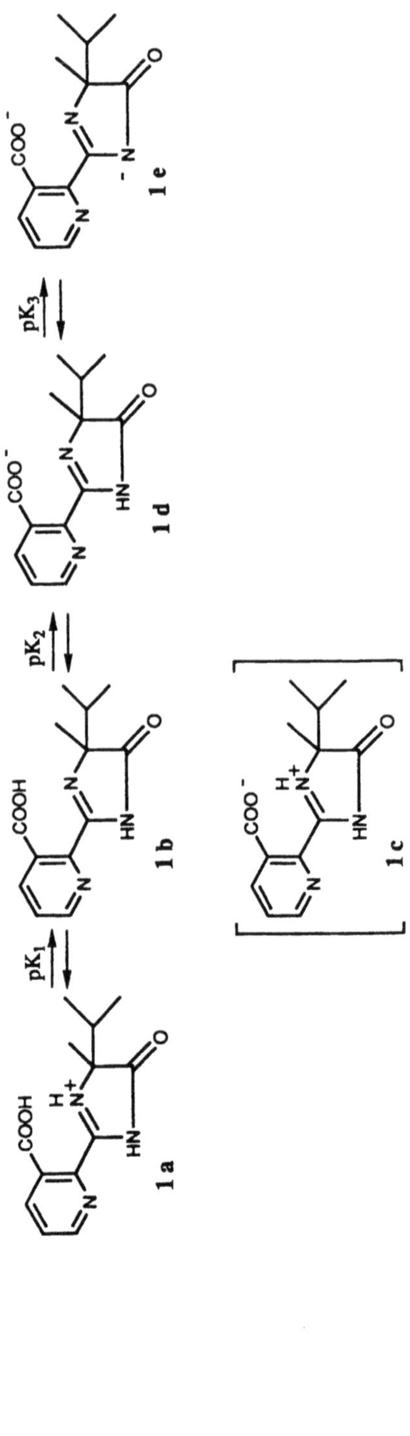

SCHEME 1.   Imazapyr tautomers.

TABLE 4			
Effect of Sodium Hydroxide Concentration and Stoichiometry on Amount of 1 and 3			
Moles NaOH/moles 1	NaOH molarity	%1	%3
2	2.5	92.4	7.3
2	3.75	94.6	4.9
2	5.0	95.9	3.6
3	2.5	98.0	1.7
3	3.75	98.7	0.9
3	5.0	98.1	0.5
4	2.5	98.5	1.1
4	3.75	98.9	0.6
4	5.0	99.3	0.3

TABLE 5			
Stability of Imazapyr during Acidification at 80°C			
pH	%1	%3	Elapsed time (min)
12.1	97.9	2.3	9
11.5	96.4	3.6	30
10.9	91.4	8.6	60
9.4	77.8	22.2	75
6.0	65.5	34.5	93
5.0	64.3	35.7	97
4.0	66.7	33.3	102

3a and imidazolinone 1d predominate. At pHs greater than 12, 1d ionizes to 1e and shifts the equilibrium toward the cyclic forms.

In an attempt to optimize the imidazolinone ring-forming reaction, the levels of diamide 3 and imidazolinone 1, beginning from pure 1 (imazapyr), were examined. Each reaction was allowed to equilibrate at 60°C for 3 h. The percentage of 1 and 3 is reported in Table 4 for two, three, and four equivalents of sodium hydroxide at different molar concentrations.

Examination was made of the stability of imazapyr to the pHs encountered during its preparation and isolation. A sample was dissolved in base at 80°C and hydrochloric acid was added at a constant rate. Aliquots were analyzed for the relative amounts of imazapyr and the free acid 3 at varying stages of acidification. These data are presented in Table 5.

Tables 4 and 5 indicate that the imidazolinones can be unstable in the pH range of approximately 8 to 12. Although appreciable ring opening occurred during acidification at 80°C, such opening could be controlled by carrying out the acidification at lower temperatures.

A variety of imidazolinone esters have been prepared and saponified in order to produce analogs. The saponification process exposed these compounds to 1.1 to 2.2 equivalents of 1.0 to 2.0 normal sodium hydroxide solutions at temperatures up to 50°C and the nicotinic acids could be isolated without significant loss. In a few experiments, complete hydrolysis of the imidazolinone ring occurred when saponification times of 16 h or more were used.

TABLE 6
Chemical Stability of the Imidazolinone Herbicides

Stable to	Labile to
Concentrated hydrochloric acid	Mild base (pH ~8—12)
Catalytic hydrogenation	Alkylation
(5% Pd/carbon at 1 atm)	(DMF acetal, iodomethane, excess
	diazomethane)
Strong base (OH⁻ or RO⁻)	*m*-Chloroperoxybenzoic acid
Sodium borohydride reduction	*N*-Bromosuccinimide
(except imazamethabenz-methyl)	
Pyridinium chlorochromate oxidation	Refluxing phosphorus oxychloride
	(slow degradation)
	Sodium cyanoborohydride/pH 3
	(reduces imidazolinone ring)
	Phosphorus pentasulfide
	(reacts with imidazolinone ring)

TABLE 7
Optical Rotation of Selected Imidazolinones

Compound	R-$[\alpha]_D$	S-$[\alpha]_D$	Solvent
Imazapyr	+ 18.37°	− 18.14°	THF
Imazethapyr	+ 13.41°	− 12.92°	THF
Imazaquin	+ 28.3°	− 29.2°	CH_2Cl_2
Imazamethabenz-methyl	+ 0.81°	− 0.89°	THF
m-isomer	+ 3.29°	—	THF
p-isomer	+ 1.06°	—	THF

B. CHEMICAL STABILITY

During the analog synthesis program referred to in Chapter 1, the imidazolinylnicotinic acids and esters were treated under a variety of reaction conditions. Table 6 lists conditions to which the imidazolinone ring has been subjected.

V. SPECTROSCOPIC PROPERTIES

A. THREE-DIMENSIONAL STRUCTURE AND PROPERTIES

The optical rotations for the chiral isomers of a number of imidazolinones are shown in Table 7.

An X-ray crystal structure obtained on the *R* isomer of the *p*-bromobenzyl ester of imazapyr confirmed the absolute configuration of the stereogenic atom as *R* and showed that the imidazolinone rings were hydrogen bonded to each other in the unit cell (see below).

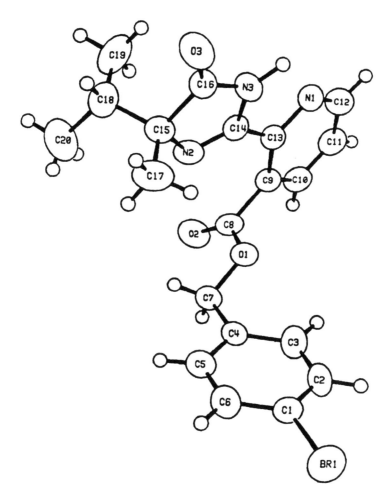

FIGURE 1. X-ray crystal structure of the *p*-bromobenzyl ester of imazapyr.

The crystal structure confirmed our representation of the imidazolinone ring as the nonconjugated tautomer in the solid state and showed the imidazolinone ring to be 19° and the carboxylate carbonyl group to be 75° out of the plane of the pyridine ring (see Figure 1).[2]

B. NUCLEAR MAGNETIC RESONANCE (NMR) SPECTROSCOPY

For reference purposes, the proton (Tables 8 and 9) and carbon (Tables 10 and 11) NMR data are provided for imazapyr and imazaquin. The imidazolinone methyl signals are useful for identification of an imidazolinone ring. In general, the isopropyl methyl signals are well separated, and the signal for the methyl group attached to the ring is shifted to a higher field by approximately 0.3 ppm compared to the corresponding diamide precursor.

A key feature of the carbon spectrum has been the three-bond coupling of H_4 with the carboxyl carbon (carbon 3a, $^3J_{CH} = 4.4$). This coupling is diagnostic for determination of a nicotinic vs. picolinic acid.

C. INFRARED (IR) SPECTROSCOPY

The solid state infrared spectra of the imidazolinones are sensitive to the electronic environment of the imidazolinone ring. For example, an imidazolinone ring can be written at tautomers, **1f, 1b,** or **1c** (see below). Schipper and Chinery[1] studied the infrared spectra

TABLE 8
Imazapyr Proton NMR Chemical Shifts

Chemical shift (ppm)	Coupling (Hz)	Hydrogen
8.25	$J_{46} = 1.6$, $J_{45} = 7.9$	H_4 (1)
7.70	$J_{54} = 7.9$, $J_{56} = 4.8$	H_5 (1)
8.81	$J_{64} = 1.6$, $J_{65} = 4.8$	H_6 (1)
1.25		(3, s)
1.94	$J = 6.7$	(1, h)
1.00	$J = 6.7$	(3, d)
0.83	$J = 6.7$	(3, d)

TABLE 9
Imazaquin Chemical Shifts (DMSO-d₆)

Chemical shift (ppm)	Coupling (Hz)	Hydrogen
8.91		H_4 (1,s)
8.22	$J = 8.4$	H_8 (1,d)
8.16	$J = 8.4$	H_5 (1, d)
7.97	$J = 7.6, 1.1$	H_7 (1, t)
7.8	$J = 7.6, 1.1$	H_6 (1, t)
1.95	$J = 6.9$	(1, h)
1.27		(3, s)
1.01	$J = 6.7$	(3, d)
0.85	$J = 6.7$	(3, d)

of some model imidazolinones in which the tautomers are fixed by alkylation of each of the nitrogen atoms. From these models, the infrared spectra suggest that the imidazolinone carboxylic acids are the nonconjugated isomer **1b** rather than the conjugated isomer **1f**.

TABLE 10
Carbon Chemical Shifts for Imazapyr (CDCl₃)

Carbon	Chemical shift (ppm)	Coupling constants (Hz)	Carbon	Chemical shift (ppm)	Coupling constants (Hz)
2	142.06	$^3J_{CH} = 7$	2a	162.54	—
		$^3J_{CNCH} = 7$	2b	182.12	br m
3	131	$^3J_{CH} = 7$	2c	75.93	br
4	144.26	$^1J_{CH} = 168.9$	2d	20.93	$^1J_{CH} = 130$
		$^3J_{CH} = 5.7$			$^3J_{CH} = 2.2$
		$^2J_{CH} = 2.4$	2e	34.98	$^1J_{CH} = 129$
5	127.14	$^1J_{CH} = 166.6$			$^3J_{CH} = 3.5$
		$^2J_{CH} = 8.8$	2f	16.64	$^1J_{CH} = 126$
		$^2J_{CH} = 1.8$	2g	17.04	$^1J_{CH} = 126$
6	151.36	$^1J_{CH} = 165$	3a	164.5	$^3J_{CH} = 4.4$

TABLE 11
Carbon Chemical Shifts for Imazaquin (DMSO-d₆)

Carbon	Chemical shift (ppm)	Coupling constant (Hz)	Carbon	Chemical shift (ppm)	Coupling constant (Hz)
2	146.73	$^3J_{CH} = 6$	2b	186.91	$^3J_{CH} = 4$
3	126.48	—	2c	74.1	$^3J_{CH} = 3$
4	138.8	$^1J_{CH} = 161$			$^3J_{CH} = 5$
		$^3J_{CH} = 9$	2d	20.27	$^1J_{CH} = 126$
5	132.03	$^1J_{CH} = 162$			$^3J_{CH} = 5$
		$^3J_{CH} = 4.8$	2e	34.06	$^1J_{CH} = 130$
6	132.03	$^1J_{CH} = 162$			$^3J_{CH} = 2.6$
		$^3J_{CH} = 4.8$	2f	16.92	$^1J_{CH} = 128$
7	128.81	$^1J_{CH} = 162$	2g	16.66	$^1J_{CH} = 128$
		$^3J_{CH} = 6$	3a	166.89	$^3J_{CH} = 5$
8	128.81	$^1J_{CH} = 162$	4a	166.89	$^3J_{CH} = 7$
		$^3J_{CH} = 6$			$^3J_{CH} = 5.3$
2a	161.01	—	8a	146.48	$^3J_{CH} = 3.8$

TABLE 12
Infrared Carbonyl Frequencies
(cm^{-1})

Structure	Imidazolinone CO	Acid CO
1g	1775	1725
1b	1752	1690
1h	1740	1600

In the case of the imidazolinone carboxylic acids, the compounds can also be charged or uncharged molecules. Table 12 shows how the carbonyl frequency changes as the imidazolinone ring is changed from charged to uncharged and the carboxyl group from uncharged to charged, **1g**, **1b**, and **1h**. It should also be noted that a positively charged imidazolinone, as in **1g**, causes the carboxyl carbonyl to shift to a higher frequency than expected for an aryl carboxylic acid.

1 g   **1 b**   **1 h**

D. MASS SPECTROSCOPY

Under electron impact conditions, the imidazolinone carboxylic acids fragment with patterns that are characteristic of this family of compounds. The parent may lose carbon dioxide and the resulting molecule or parent carboxylic acid fragments following paths similar to those shown in Scheme 2. For imazapyr, the imidazolinone ring *i* or *ii* can undergo a McLafferty rearrangement with loss of propene to give *iii*. The more prevalent path is an α-cleavage reaction with loss of one of the alkyl groups [–(CH$_3$)$_2$CH·(major) or –CH$_3$·] to give *iv* or *v*. Fragments *iv* or *v* are major ions observed in the mass spectra. Fragment *v* can further degrade with simultaneous loss of CO and an RCN to give *ix*. Also observed are the losses of HNCO or CO to give *vi* or *vii* and *viii*. Further fragmentation of daughter ions can occur in expected patterns typical of the resulting pyridine, quinoline, and benzene parents.

VI. CHROMATOGRAPHIC PROPERTIES

A. SILICA GEL CHROMATOGRAPHY

The imidazolinone carboxylic acids can be chromatographed on silica gel thin-layer chromatography (TLC) plates as their ammonium salts (see Table 13). When chromatographed as free acids, the spots streak, as is typical of carboxylic acids. Imazaquin can be chromatographed without extensive streaking. Imazamethabenz-methyl can be chromatographed on silica gel with ethyl ether elution, but the *meta* and *para* isomers are not separable (see Table 14).

Nicotinic acid esters move faster than picolinic acid esters upon chromatography on silica gel TLC plates. This behavior is in agreement with reduced interaction with the silica gel due to intramolecular hydrogen bonding between the imidazolinone hydrogen and the pyridine nitrogen. Such internal hydrogen bonding is not possible with the picolinate esters.

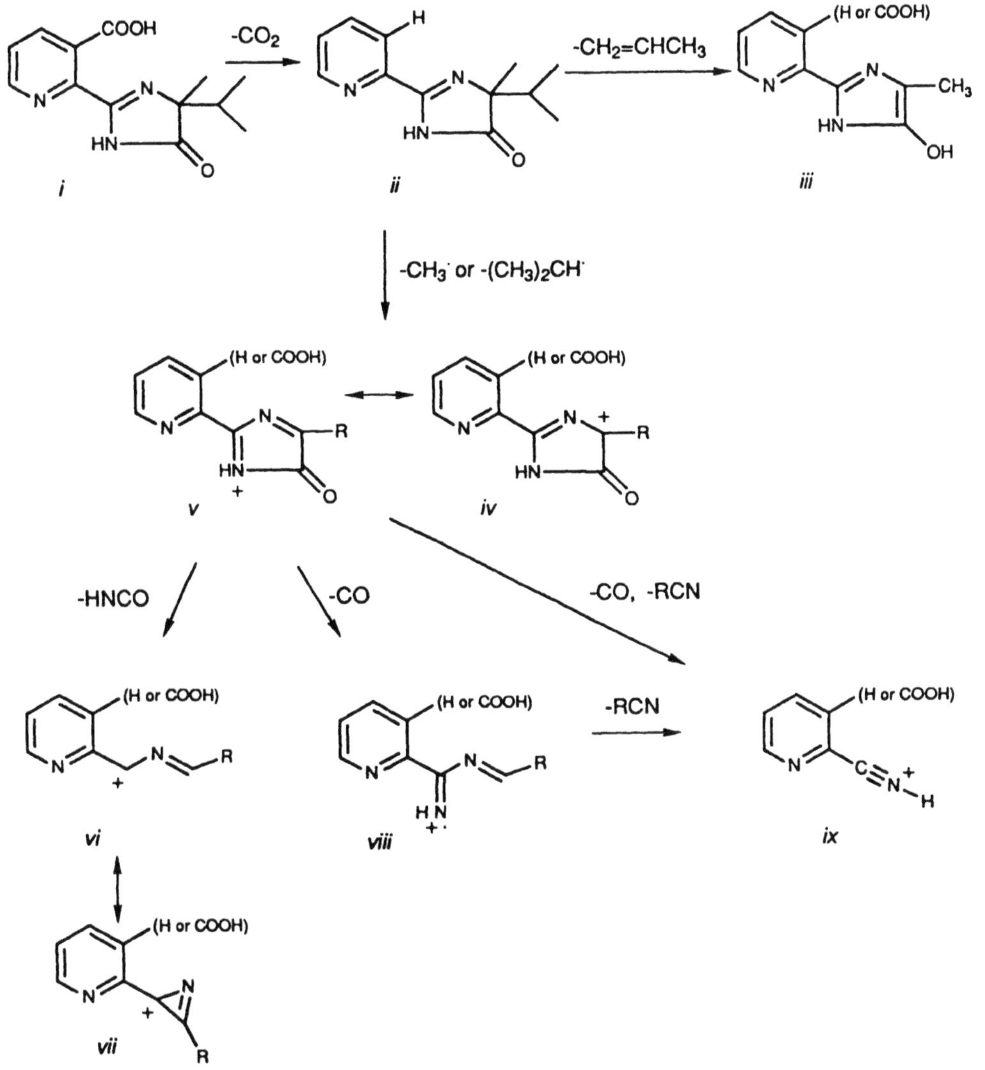

SCHEME 2. Mass spectral fragmentation patterns.

TABLE 13
Thin-Layer Chromatography of
Imidazolinone Carboxylic Acids

Compound	R_f^a	R_f^b
Imazapyr	0.39	ca. 0—0.31 (streak to origin)
Imazethapyr	0.49	ca. 0—0.31 (streak to origin)
Imazaquin	0.55	0.35
AC 263,222	0.42	ca. 0—0.31 (streak to origin)

Note: Conditions: 0.1 mm silica gel plate with $CaSO_4$ binder.

[a] Solvent: 1-propanol (50), ethyl acetate (30), water (15), ammonium hydroxide (5).
[b] Solvent: dichloromethane (95), acetic acid (5).

TABLE 14
Thin-Layer Chromatography of
Imazamethabenz-methyl

Compound	R_f
Imazamethabenz-methyl (*para* and *meta* isomers)	0.45

Note: Conditions: 0.1 mm silica gel plate with $CaSO_4$ binder.
Solvent: anhydrous ethyl ether.

TABLE 15
HPLC of Imidazolinone Carboxylic Acids

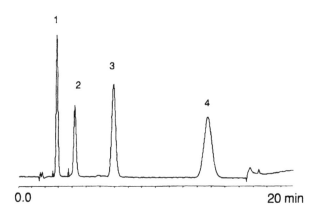

Peak no.	Time (min.)	Peak
1	2.815	Imazapyr
2	4.135	AC 263,222
3	6.875	Imazethapyr
4	13.695	Imazaquin

Note: Conditions: Whatman 5 μ C8 Partisil 4.6 mm i.d. × 12.5
mm. Mobile phase: A, H_2O, 0.1% TFA; B, CH_3CN, 0.1%
TFA. Flow: 90% A, 1.0 ml/min.

B. REVERSED-PHASE HIGH-PERFORMANCE LIQUID CHROMATOGRAPHY (HPLC) CHROMATOGRAPHIC BEHAVIOR

Using HPLC, imazaquin, imazapyr, imazethapyr, and AC 263,222 can be readily separated according to increasing molecular weight (Table 15). The isomers of imazamethabenz-methyl can also be separated under the same conditions (Table 16). Conditions were chosen for this example to show their relative chromatographic elution' properties. Other conditions have been developed for the specific separation of a desired product from its impurities.

ACKNOWLEDGMENTS

The author wishes to acknowledge the contributions made by Cyanamid's Agricultural Research Division scientists in the Analytical, Physical, and Biochemical Research: Chemical Discovery; Directed Basic Research; Metabolism, Residue, and Environmental Research; and Process Development Departments, without whose cooperation this chapter would not have been possible.

TABLE 16
HPLC of Imazamethabenz-methyl

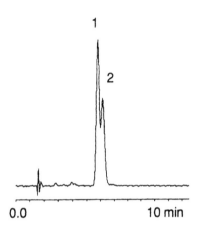

Peak no.	Time (min.)	Peak
1	5.76	Imazamethabenz-methyl (*para* isomer)
2	6.12	Imazamethabenz-methyl (*meta* isomer)

Note: Conditions: Whatman 5 μ C8 Partisil 4.6 mm i.d. × 12.5 mm. Mobile phase: A, H_2O, 0.1% TFA; B, CH_3CN, 0.1% TFA. Flow: 90% A, 1.0 ml/min.

REFERENCES

1. **Schipper, E. and Chinery, E.,** The synthesis and structure of spiroimidazolones, *J. Org. Chem.,* 26, 4480, 1961.
2. **Wepplo, P.,** Imidazolinone herbicides: synthesis and novel chemistry, *Pestic. Sci.,* 39, 293, 1990.

Chapter 4

STRUCTURE-ACTIVITY RELATIONSHIPS AMONG THE IMIDAZOLINONE HERBICIDES*

David W. Ladner

TABLE OF CONTENTS

* This chapter is based upon "Structure-Activity Relationships Among the Imidazolinone Herbicides", by David W. Ladner, which was published in *Pesticide Science*, 29, 317, 1990. ©1990 Society of Chemical Industry.

I. INTRODUCTION

The imidazolinones have been studied for more than 10 years,[1] culminating in the commercial development of four compounds.* Throughout this time, a gradual understanding of the effect of chemical structure upon the biological response has developed. As the weed scientists were discovering which weeds were controlled and which crops were protected by this class of compounds, the synthetic organic chemists were trying to discover what parts of the chemical structure were important for activity by preparing a variety of analogs of various electronic, steric (shape), and lipophilic properties. The goal of this work was to increase potency, selectivity, and the spectrum of weeds controlled. This analog approach has generally worked in herbicide research because of the assumption that the initial lead, usually discovered in greenhouse tests, already has satisfied some complex requirements for activity: uptake, translocation, and interference with a vital biological process. It is also assumed that the structure is reasonably close to the optimum and that the probability of finding similar activity in closely related analogs is high. Conversely, more radical changes in the structure have a lower probability of retaining activity, even though higher activity also becomes possible.

To guide the synthesis of the imidazolinones, the pre- and postemergence greenhouse activity of 18 crops and weeds was routinely measured, using positive controls to help normalize the data. The use of greenhouse data was well justified because these compounds generally behaved identically under field conditions at about the same rate of application. The primary goal was to increase the level of activity; other considerations such as selectivity were either discovered randomly or became secondary goals through minor changes in the most active structures. This approach, of course, was not so rigid as to prevent the opportunistic follow-up of unusual activity when it occurred; indeed, this was the origin of the imidazolinone class itself.

When it became clear that these compounds were of significant value, mode-of-action studies were completed, and *in vitro* screens were available to help guide further work.

This chapter summarizes the relationships found in each of these areas and is divided into three sections. The first is a discussion of analog activity which led to the understanding of the moiety responsible for activity; the second is devoted to a discussion of selectivity; and the third deals with the structure-activity information concerning inhibition at the enzyme level.

II. THE TOXOPHORE — STRUCTURAL REQUIREMENTS FOR HIGH HERBICIDAL EFFECTS

A general representation of the imidazolinone class of herbicides is shown in Figure 1.

For discussion purposes, the molecule has been divided into three sections; the acid or acid equivalent, the backbone, and the imidazolinone ring. Synthetic changes to each of these regions and the resulting effect on greenhouse herbicidal activity will be discussed in turn.

As far as is known, these three structural elements are necessary for activity, although it is possible to disguise or mask these elements. Additionally, the acid equivalent and imidazolinone ring are required to be *ortho* to one another: for example, AC 252,935 was not active at 10 kg/ha compared to activity of 10 g/ha for imazapyr (Figure 2). (For comparative purposes throughout this chapter, activity will be expressed as an estimated average control level for the ten week species used in the imidazolinone rundown tests — Table 1).

---

* The four commercial compounds are Arsenal, Scepter, Assert, and Pursuit herbicides, which are registered trademarks of American Cyanamid Co.

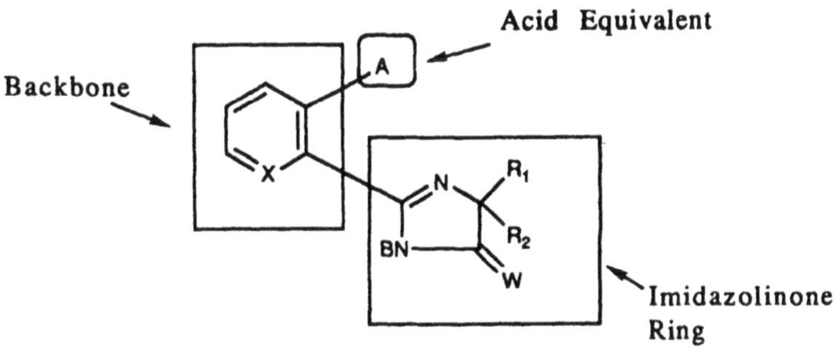

FIGURE 1. The generic imidazolinone.

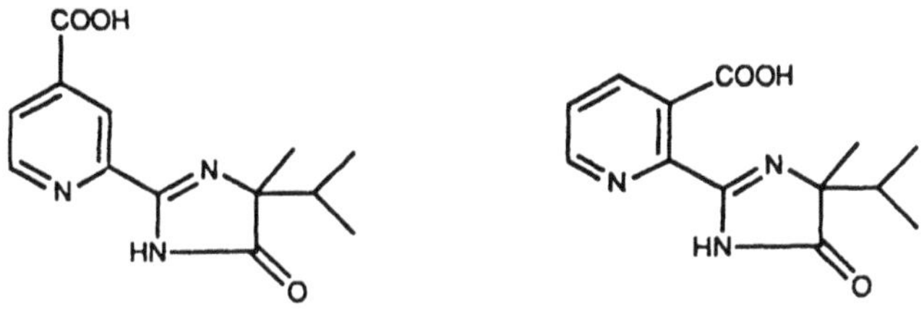

AC 252,935

Inactive at 10 kg/ha

Imazapyr

Active at ≈ 10 g/ha

FIGURE 2. Comparison of positional isomers of imidazolinyl pyridine carboxylic acids.

TABLE 1
Weed Species Used in Greenhouse Tertiary Screen for Imidazolinone Herbicides

Common name	Scientific name
Barnyardgrass	*Echinochloa crus-galli* (L.) Beauv.
Foxtail	*Setaria* spp.
Purple nutsedge	*Cyperus rotundus* L.
Wild oat	*Avena fatua* L.
Quackgrass	*Elytrigia repens* (L.) Nevski
Hedge bindweed	*Calystegia sepium* (L.) Br.
Matricaria	*Matricaria perforata* Merat
Ragweed, common	*Ambrosia artemisiifolia* L.
Velvetleaf	*Abutilon theophrasti* Medicus
Morningglory	*Ipomoea* spp.

A. THE ACID EQUIVALENT

A carboxylic acid group, or a group which can be readily transformed to a carboxylic acid, is necessary for activity. No other group was significantly more active than the carboxyl, but substitution by groups which could be readily converted *in vivo* (via hydrolysis, oxidation, etc.) to an acid were good herbicides. Benzyl, furfuryl, propargyl, and methoxyethyl esters, for example, were more active than the corresponding methyl esters, presumably because of their greater lability. The primary alcohol and the aldehyde followed the esters. Lesser activity was seen for the *o*-methyl analog, although in the pyridine series, activity was observed in several species below 1 kg/ha with the *o*-methyl compounds. Replacement of

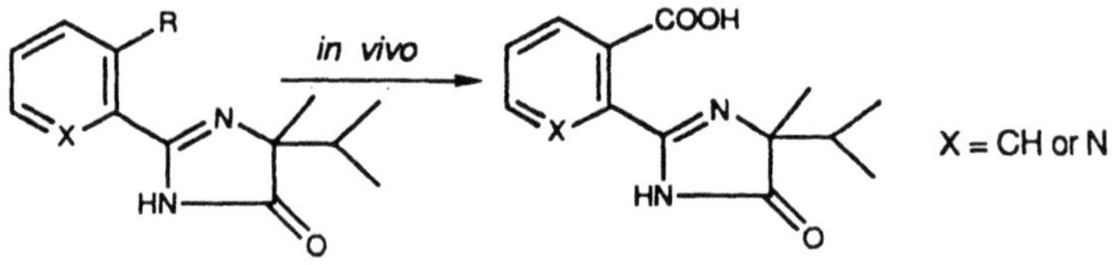

Acid Equivalent   Acid

X = CH or N

R = COO^-M^+ Decreasing
    COOR', R' = CH_3, $CH_2CH_2OCH_3$, $CH_2C{\equiv}CH$, 2-furfuryl   Activity
    CHO, CH_2OH
    CN, $CONH_2$
    CH_3
    H   Inactive

FIGURE 3.   The acid equivalents.

the carboxyl with hydrogen (no substituent) resulted in a compound inactive at 8 kg/ha. Salts of the acid are essentially equivalent to the acids, and this advantage was used in formulations. The cyclized analogs from which the imidazolinones were first discovered may also be considered acid equivalents; these compounds will be discussed later. Figure 3 is a summary of compounds substituted at the acid equivalent site and their comparative activity.

B. THE IMIDAZOLINONE RING

The name given to this class of herbicides underscores the central role played by this ring system. Aside from changes discussed in the next sections, the ring system and substitution pattern discovered early in the program seem to be the optimum for activity. Alterations to the ring include modifications of the R_1,R_2 substituents, of the carbonyl group and of the nitrogen atom substituents, as well as reduction of the double bond, and replacement of the imidazolinone by other hetero rings.

1. R_1,R_2 Substituents

The combination of methyl and isopropyl as substituents on the imidazolinone ring remained far superior in activity, despite the synthesis of many others. Other combinations approaching the same level of activity were methyl cyclopropyl and (2-methyl)spirocyclohexyl. A striking drop in activity for such a simple change as ethyl for methyl demonstrated the sensitivity of the steric requirements in this region of the molecule. Figure 4 contains a summary of the groups at these positions.

When R_1 is methyl and R_2 is isopropyl, the molecule is capable of optical isomerism. It has been shown that the R (+) isomer (configuration determined by X-ray crystallography of the *p*-bromobenzyl ester of (+) imazapyr) has about 1.8 times the postemergence activity of the racemic material. The S (−) isomer is much less active, although it does show some effect at about one tenth the rate of the racemate. Other implications of this phenomenon will be discussed later in this chapter.

2. Replacement of Oxygen

When W (in Figure 5) was replaced by sulfur, the resulting compounds tended to be as herbicidally active as those in which W was oxygen.[2] In contrast to the thiono group, the nitrogen equivalent of the carbonyl group (imino or >C=N–R) caused a severe drop in activity. Complete removal of the carbonyl gave an inactive compound.

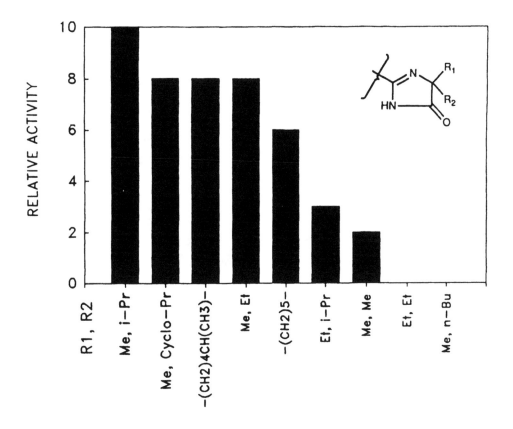

FIGURE 4. Comparison of R_1, R_2 substituents.

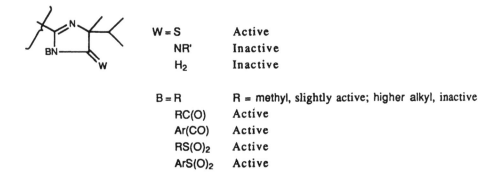

FIGURE 5. Carbonyl replacements and N-substitution.

3. N-Substitution

The activity of N-alkylated compounds (Figure 5, B = alkyl) clearly indicated a det-rimental effect. Easily cleaved groups, however, gave active compounds and therefore the N-acylated compounds were better than N-alkylated analogs. In fact, N-methyl was the only N-alkyl group with appreciable herbicidal effect. Sulfonylation on nitrogen produced com-pounds similar to the N-acyl materials.

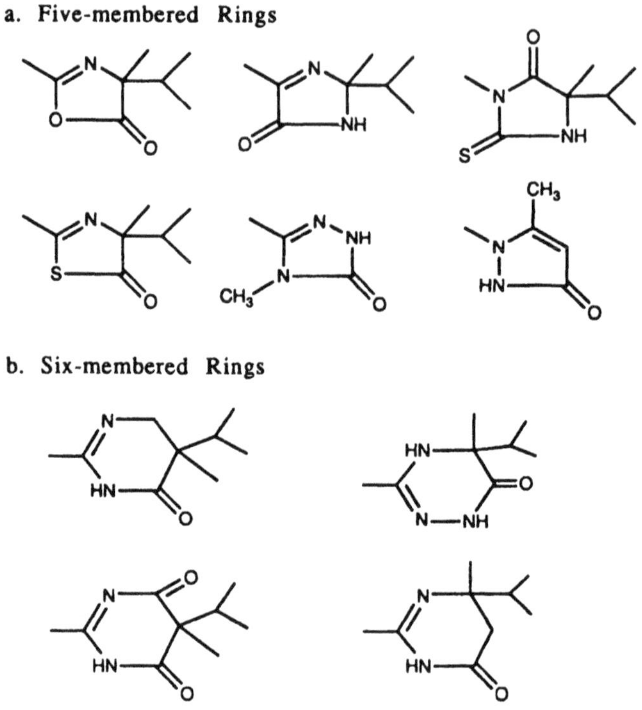

a. Five-membered Rings

b. Six-membered Rings

FIGURE 6.   Imidazolinone replacements.

4. Imidazolidinones

The reduced product of each prepared imidazolinone (e.g., structure **1**) had roughly the same or slightly less activity than the parent. These compounds were also surmised to be pro-herbicides which rely upon *in vivo* oxidation for activation.

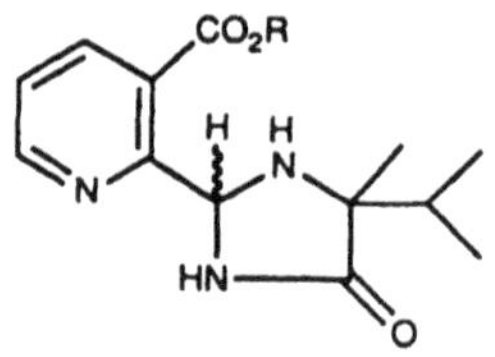

1

5. Ring Replacements for the Imidazolinone Ring

To date no other ring system has given appreciable activity at the 2.0 kg/ha level. This includes the heterocycles shown in Figure 6.

The poor activity shown by these other ring systems implies an intimacy between the imidazolinone ring and the molecular site of action in the plant.

C. THE BACKBONE

Numerous studies have been conducted on optimizing both the ring type and substituent pattern of the backbone portion of this herbicidal series.

1. The Ring System

The original benzene ring remains superior to any five-membered series, as is shown by the list in Figure 7. This is surprising, because thiophene is frequently a successful

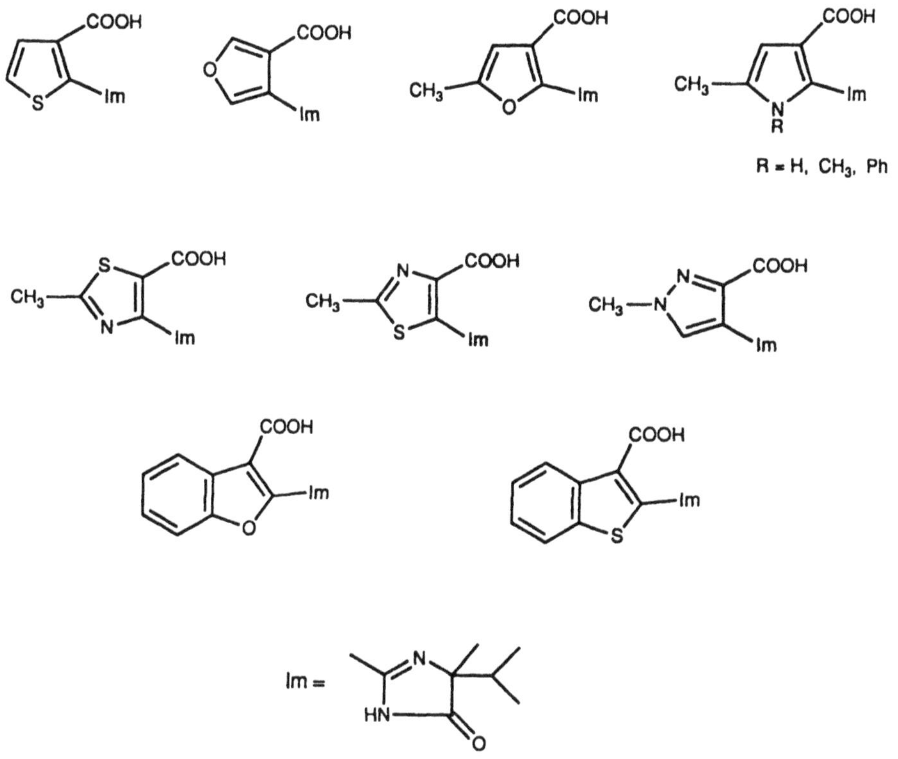

FIGURE 7. Five-membered ring analogs.

isosteric replacement for benzene in analog synthesis. Of all the backbones studied, pyridine has been the most active, with a 10-fold increase over the benzene series. The position of the nitrogen relative to the imidazolinone and carboxylic acid was crucial to high activity.

Several other six-membered heterocyclic rings have been prepared and are shown in Figure 8. Imazapyr was approximately ten times more active than its other isomers. The pyridazine system, **2**, was the best of the two-nitrogen systems. The dithiane had appreciable activity, but because the acid, **4**, decarboxylated readily, the ester, **3**, was more active in this case.

Fusion of rings to the pyridine nucleus, i.e., **5**, provided a large number of heterocyclic analogs of good activity. These backbones can be regarded as *ortho*-disubstituted analogs of the parent and as such will be more thoroughly discussed in the next section.

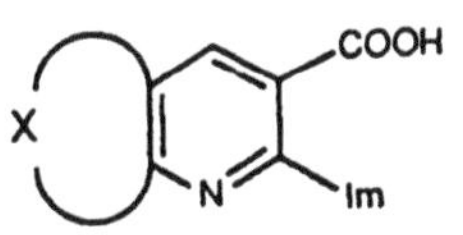

5

2. Substituents on the Backbone

The benzene backbone series, **6**, was the first to be discovered, and the four remaining positions on the ring were examined for the effect of substituents on activity. A general trend was the overall loss in activity for any electron-withdrawing group, regardless of position. On the other hand, an electron-rich group such as methoxy was favorable, especially at position 3. Substitution of any group was unfavorable at position 6, and this negative

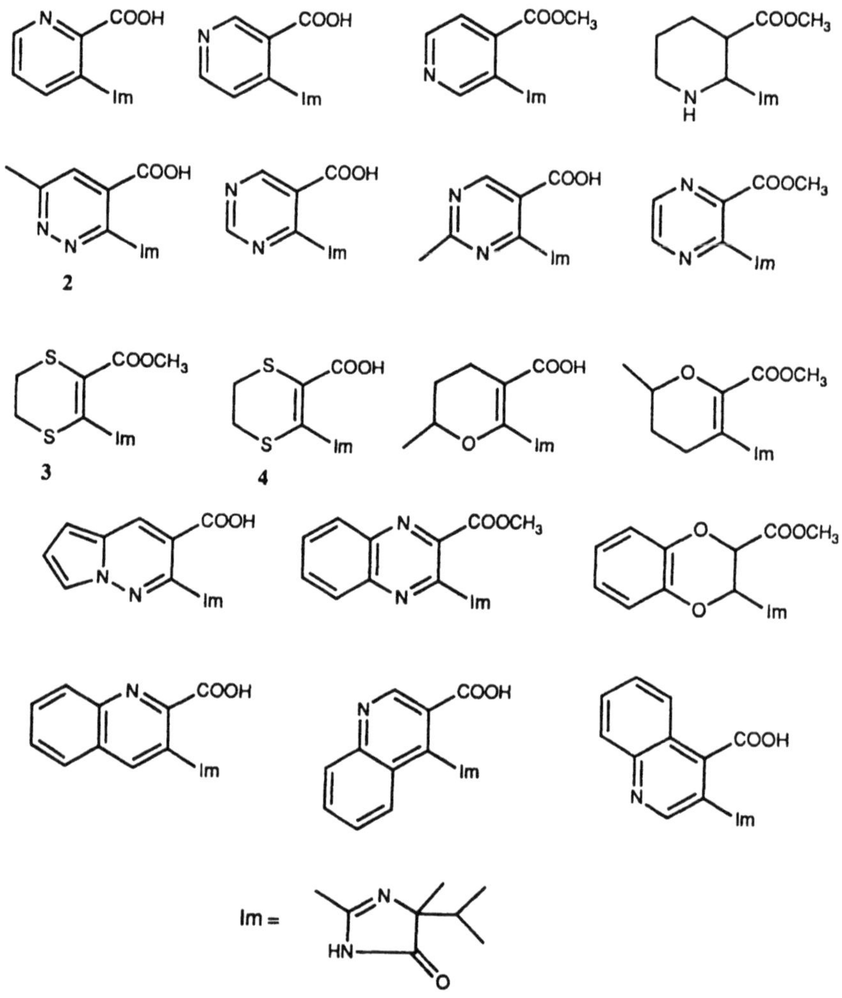

FIGURE 8.   Miscellaneous six-membered ring analogs.

effect *ortho* to the carboxylic acid has now been observed on all backbones. A steric interaction is implicated, since larger substitutions increase the negative effect on activity.

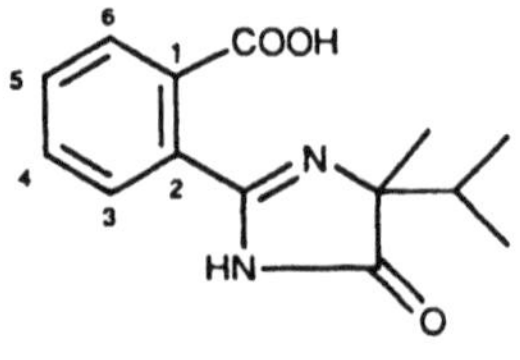

The technical materials in imazamethabenz-methyl, a mixture of two methylated regioisomers, exemplify the differences that occur for slightly different compounds. The 5-methyl (*meta*) isomer has good activity against *Avena fatua* and *Alopecurus myosuroides*, but poor activity against *Sinapis arvensis*. On the other hand, *S. arvensis* is quite sensitive to the 4-methyl (*para*) compound, whereas *A. fatua* and *A. myosuroides* are quite tolerant

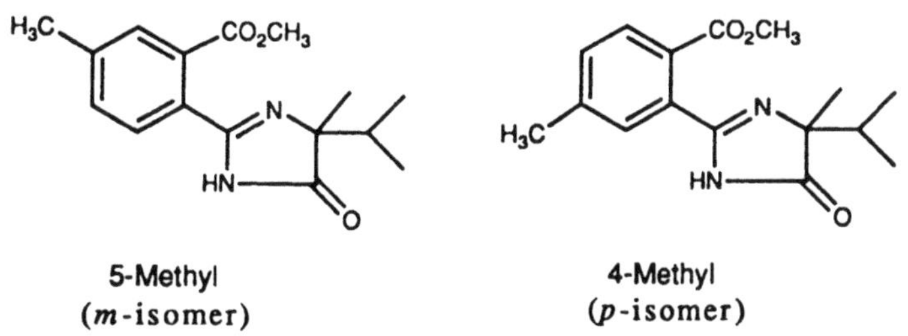

FIGURE 9. The two isomers of imazamethabenz-methyl.

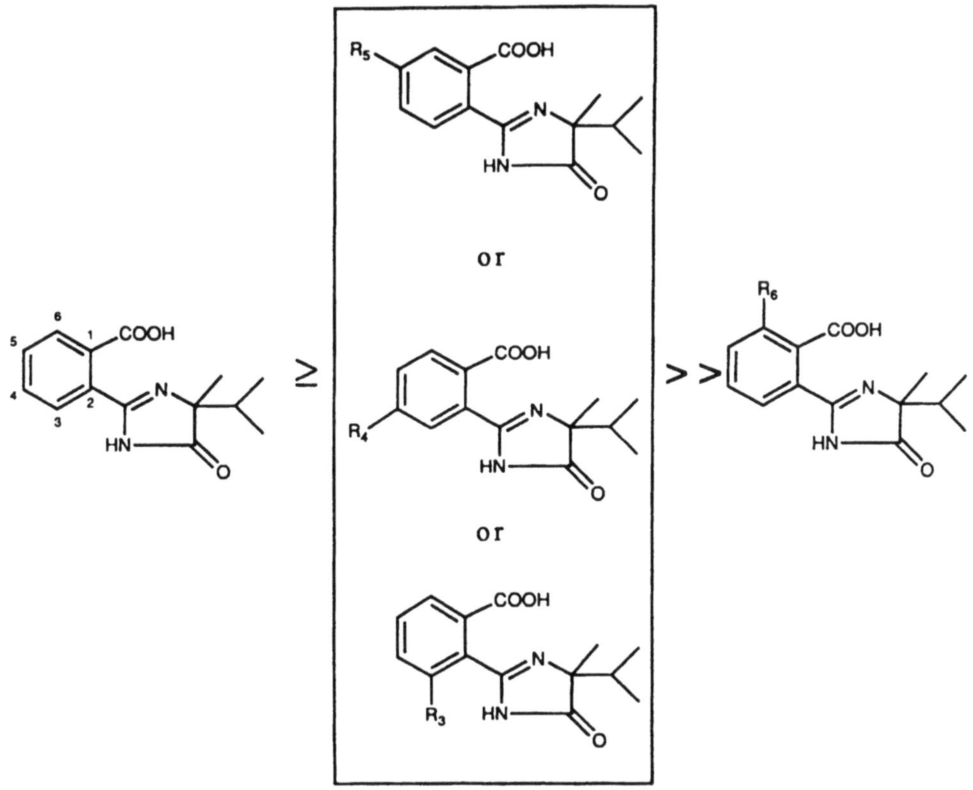

FIGURE 10. Effect of substituents on activity of imidazolinylbenzoic acids.

(Figure 9). Fortunately, this mixture of compounds provides a sufficient spectrum of activity for commercial use because the isomers are more easily manufactured as the mixture.

The general effect of substituent position is shown in Figure 10.

Similar to the benzene backbones, a group at the 4-position (*ortho* to the carboxylic acid) in the pyridine was also detrimental to the activity. On the other hand, groups at positions 5 or 6 could give favorable results, depending on the nature of the group. Alkyl groups at the 5-position, for example, afforded especially active herbicides. Figure 11 is a representative graph of some 5- and 6-substituted compounds.

Substitution at the pyridine nitrogen could also be accomplished in the form of quaternized pyridinium salts, **7**, which were inactive, and N-oxides, **8**, which were relatively more active postemergence than preemergence. The effect was transient, however, and it may be that the N-oxide is eventually reduced back to the parent in the soil.

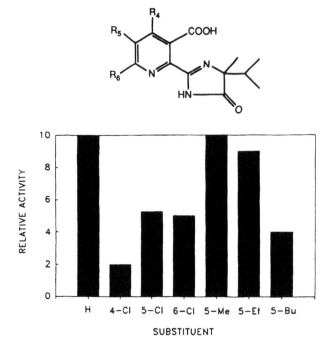

FIGURE 11.   Pyridine substituents.

7   **8**

Some 5,6-disubstituted pyridines were examined and, more importantly, 5,6-ring-fused pyridine compounds were prepared. Despite the fact that the naphthalene analogs of the benzene compounds were of low activity, the quinoline compound imazaquin was prepared. Although imazaquin had less activity than the pyridine parent, it possessed high soybean selectivity and has become the commercial product Scepter herbicide.[3] An extensive study of substituted quinolines was completed and some results are summarized in Figures 12 and 13. A comparison of chloro substituents reveals the 5- or 6-position to be the best. Of a representative group of substituents at position 6, the fluoro analog was nearly equal to the parent.

3. Quantitative Structure-Activity Analysis of Pyridine Imidazolinones

Following the synthesis of the first pyridine compounds, a quantitative structure-activity analysis (QSAR) was performed in an attempt to understand the factors contributing to activity and to guide future synthesis.[4] The use of whole plant data in regression analysis has been successful in other herbicidal compound series.[5] This approach was taken as soon as a sufficient number of imidazolinone analogs was prepared. The limits, goals, and methods of this study were as follows:

1.   The regression was limited to structures in which the carboxylic acid and pyridine backbone were present while the substituents at positions 5 and 6 were varied.

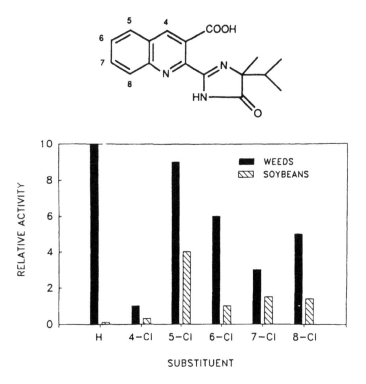

FIGURE 12. Positional effect on activity of substituted quinolines.

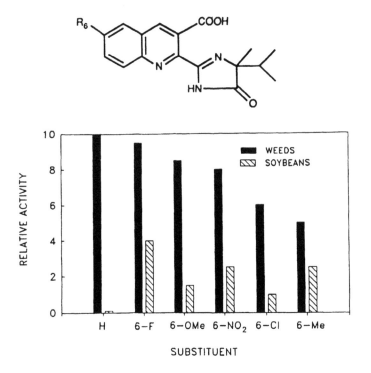

FIGURE 13. 6-substituted quinolines.

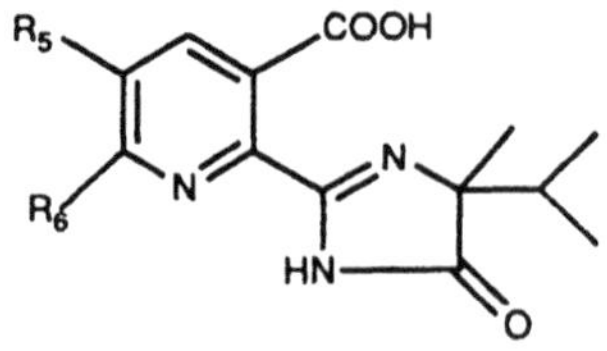

R_5	R_6	R_5, R_6
Me	n-Pr	H, H
Et	i-Pr	Me, Me
n-Pr	Ph	-CH=CH-CH=CH-
i-Pr	CF_3	-(CH_2)n- n= 3,4,5
n-Bu	OMe	
CH_2OH	OEt	
Ph	Oi-Pr	
CH_2CH_2Ph	OPh	
CH=CHPh	SMe	
	$N(Me)_2$	

FIGURE 14.   The QSAR compound dataset.

2.   Two kinds of biological data were used: the ED_{85} postemergence greenhouse results on *Convolvulus arvensis* and the ED_{85} postemergence data averaged for ten different weed species. (The ED_{85} is the lowest rate at which a rating of 8 is obtained, with 9 representing complete kill.)

3.   Standard linear regression techniques were used; well-known physicochemical parameters for each substituent were obtained either from the literature[6] or by using Medchem software* (for Log P and MR), and equations were generated using the Hansch3X program.**

4.   Data sets of compounds were examined in four stages. Thus, a set of 5-mono-substituted pyridines was analyzed first, followed by 6-mono-substituted compounds. These were then combined with 5,6-disubstituted compounds, and finally 5,6-ring-fused compounds were added to this set. This final set is listed in Figure 14.

5.   From the many equations of activity as a function of physicochemical parameters, the best fits were selected on the basis of the correlation coefficient (R^2) for each equation and 95% confidence intervals for each variable used.

Although the compounds in this series had a limited range of parameter values, relationships were sought primarily for suggesting new areas for synthesis. The following parameters were found to be significant in the study:

1.   Pi (π) — the substituent partition coefficient
2.   Sigma p (σ_p) — the Hammett electronic sigma value

*   Available from Daylight Chemical Information Systems, 3951 Claremont St., Irvine, CA 92714.
** A Fortran program obtained from Pomona College MedChem Project.

3. B_4 — Verloop's Sterimol[7] parameter, which describes the maximum width of a substituent

4. L — Verloop's Sterimol[7] parameter, which describes substituent length

Somewhat different results were obtained from the two sets of biological data. Equation 1, which described the *Convolvulus* data, contained sigma-p, L, and B_4 terms, while Equation 2 was quadratic in π, with an additional B_4 term.

$$Log1/C_{Convolvulus} = 2.28 - 0.48\ B_4\ (6) - 2.64\ \sigma_p\ (5) - 0.30\ L\ (5)$$

95% Confidence intervals:

$$\pm 0.60\ \pm\ 0.14\qquad \pm 1.39\qquad\quad \pm 0.10$$

$$R^2 = 0.76\qquad N = 27\qquad S = 0.33 \tag{1}$$

$$Log1/C_{average} = 0.63 + 0.63\ \pi(5 + 6) - 0.13\ B_4\ (6) - 0.49\ \pi^2(5 + 6)$$

95% Confidence intervals:

$$\pm 0.25\ \pm\ 0.27\qquad \pm 0.08\qquad\quad \pm 0.11$$

$$R^2 = 0.84\qquad N = 28\qquad S = 0.28$$

$$ideal\ \pi = 0.65 \tag{2}$$

The number(s) in parentheses refer to the location of the substituent to which this parameter applies.

Structural preference for herbicidal activity can be inferred from these equations. Equation 1 implies that short, electron-donating groups at the 5-position (alkyl, alkoxy) and small (narrow) groups at the 6-position (H, methoxy) are preferred. In addition to the same 6-position preference, Equation 2 suggests that the optimum lipophilicity for a compound occurs for substituents with a combined π value of 0.65 (methyl = 0.58) for the 5 and 6 positions. These structural effects influence binding at the site of action, metabolism, and transport. It is quite usual for uptake and transport to be dependent on π, and an optimum value is to be expected. Although it is tempting to ascribe the electronic and steric terms to the binding-site interactions only, this can be verified solely by studies at the enzyme level. A schematic representation of the findings of the regression is depicted in Figure 15.

Subsequently, a large number of compounds have been prepared which fall into the dataset covered by the equations, with results generally consistent with prediction. The exceptions are typically due to extrapolation into data space not included in the original set. For example, in Equation 1, highly electron-withdrawing groups at position 5 tend to be underpredicted and highly electron-donating groups tend to be overpredicted. In addition, the metabolically reactive group OH is overpredicted. However, when classified by lipophilicity, size, and electronics, the relative position of compounds to each other is in fairly good agreement with these models. Despite limitations, the study has been useful in guiding further synthesis. Figures 16 and 17 summarize results for subsequently prepared compounds.

4. Heteropyridines

An example of the predictive utility of this method occurred as a result of the observation that the B_4 value at position 6 could be minimized by ring fusion and heteroatom substitution.

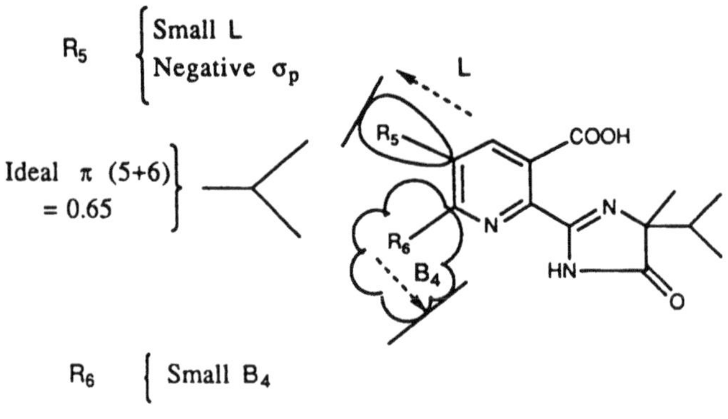

FIGURE 15. Summary of preferred parameter values for imidazolinyl pyridinecarboxylic acids.

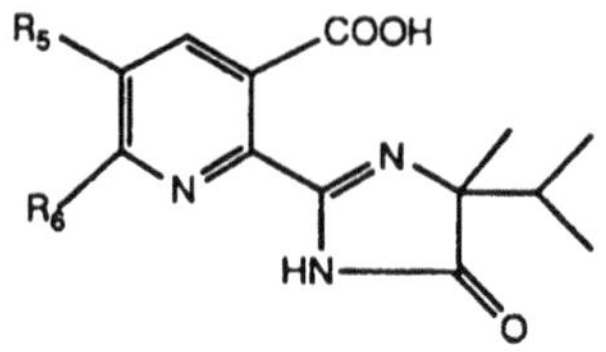

Activity level:

Highly active
Moderately active
Slightly active

R$_5$	R$_6$	R$_5$, R$_6$

a. Well-predicted analogs

MeO	**Me**	**Me, F**
EtO	**F**	
***cyclo*-C$_3$H$_5$**		
Cl	OH (=O)	Et, Cl
HC≡C		Et, MeO
MeSO$_2$	*1-methylpyridinium-4-yl*	

b. Over-predicted analogs

		NH$_2$, Me
(Me)$_2$N		
OH	*MeSO$_2$*	
	CN	

c. Under-predicted analogs

CH$_3$C≡C		**CH$_3$C(O), Me**
NO$_2$		*NO$_2$, Me*
		CN, Me

FIGURE 16.   Predicted bindweed activity determined by using Equation 1.

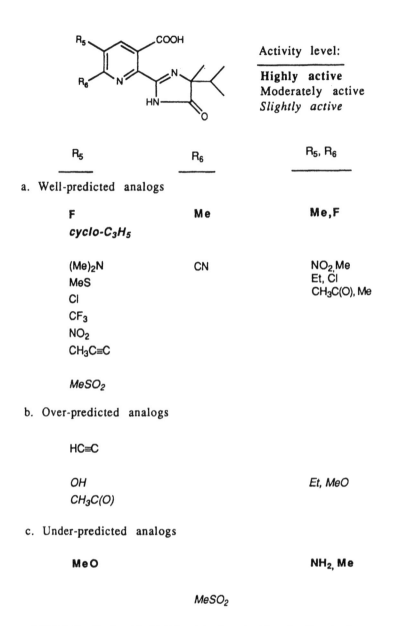

45

Activity level:

Highly active
Moderately active
Slightly active

R_5	R_6	R_5, R_6

a. Well-predicted analogs

F	**Me**	**Me,F**
cyclo-C₃H₅		
(Me)₂N	CN	NO₂, Me
MeS		Et, Cl
Cl		CH₃C(O), Me
CF₃		
NO₂		
CH₃C≡C		
MeSO₂		

b. Over-predicted analogs

HC≡C		
OH		*Et, MeO*
CH₃C(O)		

c. Under-predicted analogs

| **MeO** | | **NH₂, Me** |
| *MeSO₂* | | |

FIGURE 17. Predicted herbicidal activity determined by using Equation 2.

It was hypothesized that the negative steric effect was directional in nature; that is, a bulky group towards the pyridine nitrogen might be interfering with binding and/or the transport properties by a shielding effect. Therefore, this effect might be lessened by tying the 6-substituent back to the 5-position as is the case in the 5,6-ring fused compounds. The substituent with the smallest B_4 value is hydrogen at position 6, but the alkoxy and alkylthio groups also have small B_4 values. Accordingly, the oxygen and sulfur heterocycles were logical synthesis targets. The size of the fused ring was also predicted to affect B_4, in the order $5 < 6 < 7$. The thienopyridine compound was being prepared at the time, and it showed promising herbicidal activity. This gave further impetus to the synthesis of the other isomers and the furopyridines. 1,2-Dihydro analogs were also prepared in both series. Consistent with prediction, the [2,3-*b*] compounds were more active than the [3,2-*b*] isomers, and the furopyridines were generally more active than the thienopyridines (Figure 18).

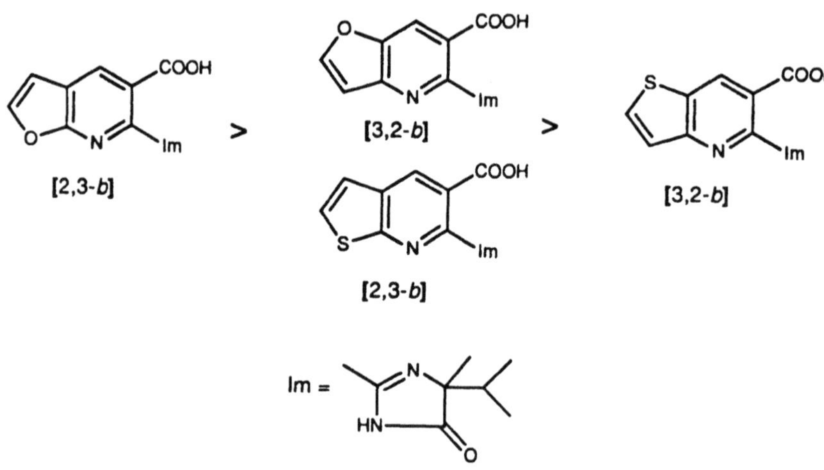

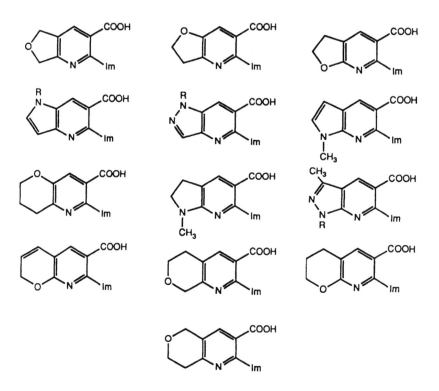

FIGURE 18.   Furo- and thienopyridines: relative order of herbicidal activity.

FIGURE 19.   Highly active imidazolinyl hetero[*b*]pyridine carboxylic acids.

The fact that these compounds were novel, often selective, and had greater activity than imazaquin prompted a large effort devoted to the preparation of other heteropyridine backbone analogs. Examples of these ring systems and their activity appear in Figures 19 and 20. Many of these compounds were selective to soybean, and among the most active compounds were the pyrrolo[3,2-*b*]pyridines. This isomer is more active than the corresponding [2,3-*b*] compound, which is N-substituted and has a larger B_4 value. The six-membered series remained less active than the five, but the pyranopyridines and naphthyridines had the same activity as the quinoline (imazaquin) series. The compounds containing more than one hetero

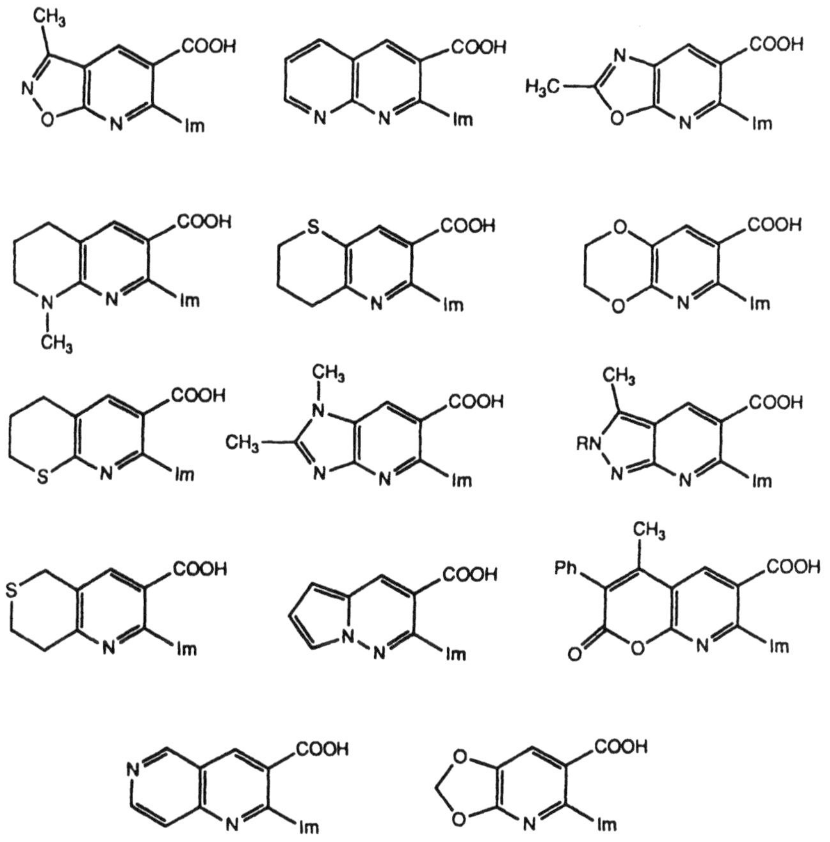

FIGURE 20. Imidazolinyl hetero[*b*]pyridines with moderate to lower herbicidal activity.

atom in the fused ring were not as active as predicted; this may be due to a greater propensity to metabolize in the plant or to hydrolyze in the soil. Patents and patent applications have been published in this area.[8,9]

5. Other Backbones

Recent patent literature disclosures by three companies — Sumitomo,[10] Idemitsu Kosan,[11] and Celamerck GmbH[12] — describe herbicidal activity for maleic and fumaric acid derivatives. The best compounds appear to be the Z-2,3-dimethyl acrylic acid derivatives **9.** The level of activity reported is about 0.25 to 0.5 kg/ha. Replacement of methyl with other groups, e.g., phenyl, lessened the activity considerably.

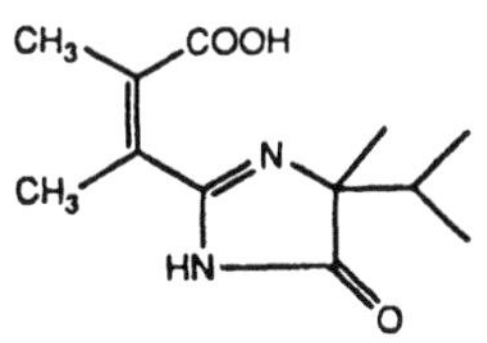

9

D. THE CYCLIC "PRO-HERBICIDES"

Cyclized derivatives (Figure 21), which are the historical predecessors to the imidazolinones, might also be seen as acid derivatives, e.g., cyclic amides. The ease with which

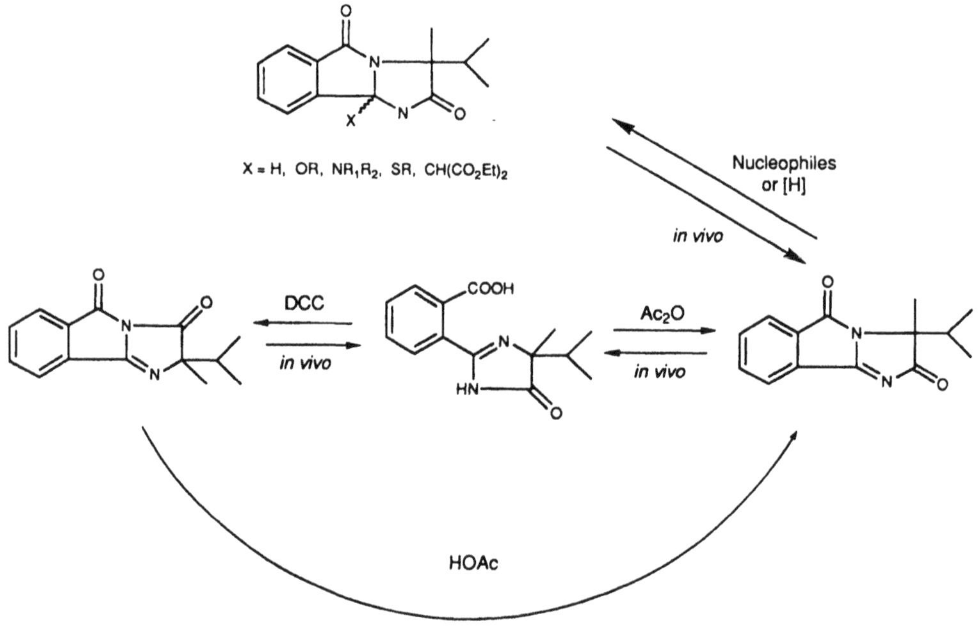

FIGURE 21.   Cyclic "pro-herbicides".

these rings open implies the *in vivo* activation in plants to the corresponding imidazolinone. Note that of the two isomers possible, the 2,5-dione is more stable than the 3,5-dione; the latter can be converted to the former by heating in acetic acid. The 2,5-dione is less susceptible to nucleophilic ring opening but is subject to the side reaction of 1,4 addition at the amidine carbon. Its activity is understandably less than the 3,5 isomer, but the products of the 1,4 addition also are active, and alkoxy, amino, and alkylthio analogs are the subject of patents.[13-17] Even carbon nucleophiles such as malonate have been prepared.[18-21] Hydrogenation gives the dihydro compounds.

III. STRUCTURAL FACTORS INFLUENCING SELECTIVITY

Rapid detoxification of the imidazolinones by metabolism in tolerant species appears to be the single reason for selectivity. At least two modes of detoxification are possible: The destruction of the imidazolinone ring itself and the side chain hydroxylation of alkyl groups. Additional selectivity is obtained as a result of differences in ester hydrolysis.

A difference of commercial importance is observed between wheat and several of its weed pests when treated with imazamethabenz-methyl, shown in Figure 22.[22] Two structural features, the methyl ester and the methyl attached to the benzene ring, work together to provide a commercial level of selectivity in the product. Weeds such as *Avena fatua* quickly hydrolyze the ester group to the phytotoxic acid, while wheat has less capacity to perform this hydrolysis. On the other hand, wheat hydroxylates the methyl side chain more readily than the weeds, which results in detoxification of the compound. The net result of these competing pathways is that the wheat survives, while the weeds die.

Imazethapyr and other 5-alkylated pyridine compounds also undergo hydroxylation of the alkyl group, but in soybeans rather than in wheat. Imazaquin, in contrast, is rapidly metabolized in soybeans to a large number of materials, none containing the imidazolinone ring.[23] Other quinoline backbone compounds behave similarly in soybeans. Weed susceptibility to imazaquin has been shown to be dependent upon the rate of metabolism in each species.[24] As expected, the slower the metabolism, the more susceptible the weed.

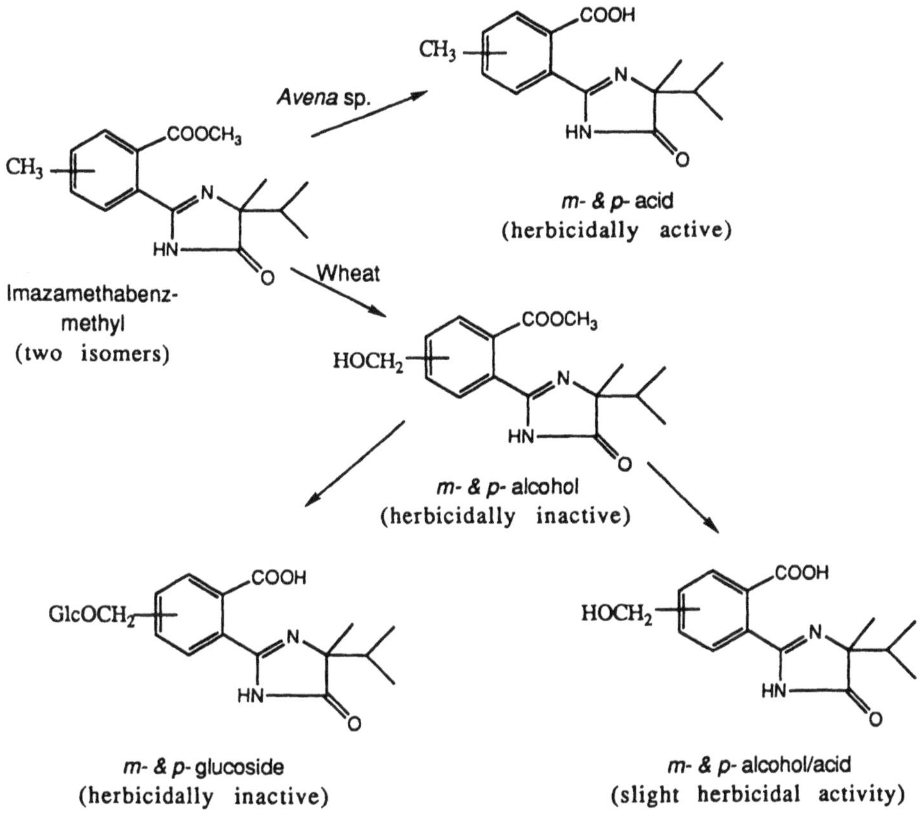

FIGURE 22. Metabolic pathways for imazamethabenz-methyl in susceptible *Avena* vs. tolerant wheat. (From Brown, M. A., Chiu, T. Y., and Miller, P., *Pestic. Biochem. Physiol.*, 27, 24, 1987. With permission.)

Wheat selectivity has been observed in the greenhouse for some of the thienopyridine backbone compounds. This may be attributable to selective oxidation of the sulfur atom to inactive sulfoxides or sulfones.

Significant greenhouse selectivity has been observed in corn with the dihydro "cyclized pro-herbicides" shown in Figure 21. However, this phenomenon was only observed under controlled conditions with sterile soil; reversion to the parent, nonselective imidazolinone apparently occurs rapidly under actual field conditions, and no useful selectivity was found.

IV. TOWARDS THE UNDERSTANDING OF INTRINSIC ACTIVITY

As mentioned at the outset, the activity of a large number of compounds is most easily compared in a standard greenhouse test. It was fortunate that this test correlated well with field activity for many of the compounds tested, especially in light of the discovery of the mode of action of the imidazolinones. It is now known that the imidazolinones disrupt the biosynthesis of valine, leucine, and isoleucine by inhibition of the enzyme acetohydroxyacid synthase (AHAS).[25] While pure enzyme has yet to be prepared, practical level of purity has been achieved which permits an assay of large numbers of compounds for their AHAS inhibitory activity. The data are not yet precise enough to obtain good QSARs, but it has provided significant information about the series.

At the enzyme level, only structural factors actually involved in the inhibition should be important; the uptake, translocation, metabolism, and phloem mobility play little role in enzyme activity. Several observations can now be made from the data:

1.   Good inhibitory activity requires a free carboxylic acid group and an N-unsubstituted imidazolinone ring. Esters, tricycles, and the like have very low intrinsic AHAS activity, as do N-acyl imidazolinones.
2.   The *ortho* disposition of the acid to the imidazolinone ring is strictly required.
3.   In contrast to the situation for herbicidal activity, the position of the nitrogen in the pyridine ring is not particularly important for enzyme activity.
4.   The order of activity for the various backbones is benzenes > quinolines > pyridines; this is in contrast to what is found in whole-plant studies. The trend of more lipophilic compounds being stronger inhibitors does not seem to have a definite optimum or upper limit for obtaining the best inhibitory activity.
5.   Similar to the herbicidal activity, the R ($+$) antipodes are more active than the S ($-$) isomer by a factor of about 8 to 10. Thus, while one isomer is much more active than the other, it does not account for all activity.

Discrepancies between enzyme and whole-plant activity are presumably due to transport and/or metabolism. Parameters such as lipophilicity (π) and pK_a (which correlates to σ) may be more significant at the whole-plant level, whereas steric factors, especially on the imidazolinone ring are probably more important at the enzyme site of action. This rationale explains why modifications of this ring result in much less active compounds, in contrast to modifications of the backbone or its substituents (the "carrier" portion of the molecule).

Finally, the discovery of other herbicidal AHAS inhibitors suggests a comparison of them to imidazolinones for similarities. The sulfonylureas and the sulfonyltriazolopyrimidine compounds are similar to each other, but imidazolinones seem to have very little in common with them. Kinetic data, appearing in later chapters, also show differences among these AHAS inhibitors. For the present, it is reasonable to assume that inhibition of the enzyme by these classes of compounds is a result of interaction at different sites on the enzyme. The use of molecular modeling could benefit the design of better AHAS inhibitors, but until more details of the structure of AHAS and its inhibitor-binding sites are known, this approach is not feasible.

V. SUMMARY

The structure-activity trends in the imidazolinone series have proven to be fairly restrictive; aside from some changes in substituents on the backbone and some interchangeability of six-membered backbones, alterations led to inactivity. Despite the one example of a maleic acid backbone, Figure 1 remains a useful representation of the lowest common denominator for the most active compounds. Selectivity, which principally is a result of metabolism, can be enhanced by the incorporation of specific structural features into active molecules. Differences between enzyme inhibition and whole-plant activity has been ascribed to additional transport and uptake factors.

REFERENCES

1. **Los, M.,** O(5-Oxo-2-imidazolin-2-yl)arylcarboxylates: a new class of herbicides, in *Pesticide Synthesis through Rational Approaches,* Magee, P., Kohn, G. K., and Menn, J. J., Eds., ACS Symp. Ser. No. 255, American Chemical Society, Washington, D.C., 1984, 29.
2. **Guaciaro, M. A., Los, M., Russell, R., Wepplo, P., Lences, B. L., Lauro, P. C., Orwick, P. L., Umeda, K., and Marc, P.,** o-(5-Thiono-2-imidazolin-2-yl)aryl carboxylates: synthesis and herbicidal activity, in *Synthesis and Chemistry of Agrochemicals,* Baker, D. R., Fenyes, J. G., Moberg, W. K., and Cross, B., Eds., ACS Symp. Ser. No. 355, American Chemical Society, Washington, D.C., 1987, 87.
3. **Cross, B., Johnson, J. L., Los, M., and Orwick, P. L.,** o-(5-Oxo-2-imidazolin-2-yl)aryl carboxylates: a new class of herbicides, 185th Natl. Meet., American Chemical Society, Seattle, Washington, D.C., March 20 to 25, 1983.
4. **Ladner, D. and Cross, B.,** Quantitative structure-activity relationships of imidazolinyl-pyridine carboxylic acid herbicides, presented at 6th Int. Congr. Pest. Chem. (IUPAC) Ottawa, Canada, August 10 to 15th, 1986, 1C-07.
5. **Cross, B., Hoffman, P. P., Santora, G. T., Spatz, D. M., and Templeton, A. R.,** The design of postemergence phenylurea herbicides using physicochemical parameters and structure-activity analyses, *J. Agric. Food Chem.,* 31, 260, 1983.
6. **Hansch, C. and Leo, A., Eds.,** *Substituent Constants for Correlation Analysis in Chemistry and Biology,* John Wiley & Sons, New York, 1979.
7. **Verloop, A., Hoogenstraaten, W., and Tipker, J.,** Development and application of new steric substituent parameters in drug design, in *Drug Design,* Vol. 7, Ariens, E., Ed., Academic Press, New York, 1976, 165.
8. **Los, M., Ladner, D., and Cross, B.,** (2-Imidazolin-2-yl)thieno- and -furo[2,3-b] and [3,2-b]pyridines and Intermediates for the Preparation thereof, and Use of Said Compounds as Herbicidal Agents, U.S. Patent 4,650,514, 1987.
9. **Cross, B., Los, M., Doehner, R., Ladner, D., Johnson, J., Jung, M., Kamhi, V., Tseng, S., Finn, J., and Wepplo, P.,** Novel Fused Pyridine Compounds, Intermediates for the Preparation and Use of Said Compounds as Herbicidal Agents, European Patent Application 227,932A, 1987.
10. **Kohsaka, H. and Takase, M.,** Butenoic Acid Derivatives, and their Production and Use, Australian Patent 8661073-A, 1987.
11. **Uemura, M., Sakamoto, M., and Kikkawa, N.,** Herbicidal and Plant-growth Regulating Imidazoline Derivatives, U.S. Patent 4,726,835, 1988.
12. **Buck, W., Garrecht, M., Schneider, G., and Drandarevski, C.,** Herbicidal Imidazolinones, U.S. Patent 4,723,989, 1988.
13. **Los, M.,** Imidazoisoindolediones and the Use thereof as Herbicidal Agents, U.S. Patent 4,017,510, 1977.
14. **Los, M.,** Dihydroimidazoisoindolediones and the Use thereof as Herbicidal Agents, U.S. Patent 4,041,045, 1977.
15. **Los, M.,** Dihydroimidazoisoindolediones and the Use thereof as Herbicidal Agents, U.S. Patent 4,110,103, 1978.
16. **Los, M.,** Dihydroimidazopyrrolopyridine, -quinoline, Thieno- and Furo[2,3-b]pyridine, Dihydrothieno- and -furo[2,3-b]pyridine, Thieno- and Furo[3,2-b]pyridine and Dihydrothieno- and -furo[3,2-b]pyridine Herbicides and Method for Preparation thereof, U.S. Patent 4,701,208, 1987.
17. **Los, M., Ladner, D., and Cross, B.,** (2-Imidazolin-2-yl)thieno- and -furo[2,3-b] and [3,2-b]pyridines and Use of Said Compounds as Herbicidal Agents, U.S. Patent 4,752,323, 1988.
18. **Draber, W., Eue, L., Santel, H.-J., and Schmidt, R.,** Imidazo-pyrrolo-pyridine Derivatives Useful as Herbicidal Agents, U.S. Patent 4,565,566, 1986.
19. **Obrecht, J.-P. and Urech, P.,** 2H-Imidaz[1',2':1,2]pyrrolo[3,4-b]pyridine und deren Verwendung als Unkrautbekämpfungsmittel, European Patent Application 183,993, 1986.
20. **Dürr, D., Brunner, H.-B., and Szczepanski, H.,** Heterocyclisch Kondensierte Pyridin-Verbindungen als Herbizide, European Patent Application 195,745, 1986.
21. **Hunt, D.,** Herbicidally Active Imidazopyrrolo-pyridine (or benzene) Derivatives, U.S. Patent 4,717,414, 1988.
22. **Brown, M. A., Chiu, T. Y., and Miller, P.,** Hydrolytic activation versus oxidative degradation of Assert herbicide, an imidazolinone aryl-carboxylate, in susceptible wild oat versus tolerant corn and wheat, *Pestic. Biochem. Physiol.,* 27, 24, 1987.
23. **Shaner, D. L.,** personal communication, 1990.
24. **Shaner, D. L. and Robson, P.,** Absorption, translocation and metabolism of AC 252,214 in soybean (*Glycine max*), common cocklebur (*Xanthium strumarium*), and velvetleaf (*Abutilon theophrasti*), *Weed Sci.,* 33, 469, 1985.
25. **Shaner, D., Anderson, P., and Stidham, M.,** Imidazolinones. Potent inhibitors of acetohydroxyacid synthase, *Plant Physiol.,* 76, 545, 1984.

Chapter 5

ABSORPTION AND TRANSLOCATION OF THE IMIDAZOLINONE HERBICIDES

Desiree L. Little and Dale L. Shaner

TABLE OF CONTENTS

I. INTRODUCTION

In order for a herbicide to kill plants, the chemical must be absorbed and translocated to the site of action. In the case of the imidazolinone herbicides, the site of action is the acetohydroxyacid synthase (AHAS) enzyme, which is concentrated in plant meristematic tissues. Absorption of the imidazolinones by roots and shoots and the subsequent translocation to the meristems is presented in this chapter following a brief review of the morphology and physiology of the transport system.

II. MORPHOLOGY AND PHYSIOLOGY OF THE TRANSPORT SYSTEM

Imidazolinones enter plants through the root or shoot tissue and are then translocated throughout the plant via the xylem — the water transport system — or the phloem — the sugar transport system. To understand the mechanism for mobility of the imidazolinones in plants, a morphological and physiological foundation of the transport systems first must be established.

A. ROOT ABSORPTION AND TRANSLOCATION OF HERBICIDES

Herbicides present in the soil solution are available for absorption by plant roots. The first stage of absorption is penetration of the epidermis, which leads to the root cortex. Movement from the cortex to the vascular tissues is limited, however, due to the presence of the Casparian strip. The Casparian strip is a single layer of cells whose anticlinal walls are made impermeable to water and solutes by the deposition of suberin and lignin.[1] For any compound to penetrate this layer, the xenobiotic must be able either to diffuse across the Casparian strip or to bypass these fatty deposits by crossing directly through the plasma membranes of this cell layer. In either case, the physical and chemical properties of the penetrating compound influence its rate of diffusion across these lipophilic barriers.

Many herbicides are mobile in both the apoplast, the nonliving portions of the plant that include the extracellular spaces and the xylem, and the symplast, the living portions of the plant that include the cell protoplasts and the phloem. When absorbed by roots, the mobility of these herbicides is usually greater in the apoplast than in the symplast; evapotranspiration at the leaf surfaces pulls water and its solutes from the roots, maintaining a steep concentration gradient of the xenobiotic between the environment and the root apoplast. Therefore, as the plant transpires, the xenobiotic flows with the water into the root xylem, slowed only by the layer of cells comprising the Casparian strip, and then moves to the leaf surfaces.

B. LEAF ABSORPTION AND TRANSLOCATION OF HERBICIDES

Although herbicide reaching the soil may be available for root uptake, postemergence-applied herbicides are primarily absorbed by the foliage. Plant leaves are complex organs composed of several different layers of photosynthetic cells lying beneath a layer of epidermal tissue. The epidermis secretes, on its exterior surface, a cuticle composed of cutin, pectin, and waxes. The exact composition of the cuticle, in particular the types and amount of wax, depends on the species, the plant part, the age of the tissue, and the environmental conditions both during and after cuticle development.[2] Although the cuticular and epicuticular waxes vary widely in their composition, all serve as a hydrophobic barrier to foliarly applied xenobiotics. The hydrophobicity is reduced by breaks in the cuticle[3] and by the presence of pectins and cutins which can be hydrated under high moisture conditions. It is through these areas of reduced hydrophobicity that water-soluble xenobiotics are primarily absorbed.

Once a compound penetrates the cuticle, the herbicide can either remain in the apoplast of the leaf or move into the symplast. For the compound to translocate out of the treated

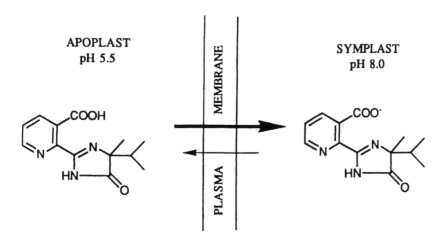

APOPLAST
pH 5.5

MEMBRANE

SYMPLAST
pH 8.0

PLASMA

FIGURE 1. In the acidic environment outside of plant cell membranes, a significant portion of the imidazolinone molecules are nonionized, making them sufficiently lipophilic to diffuse through the membranes. Once inside the cell, the alkaline conditions cause the imidazolinones to dissociate. The ionized species is too polar to diffuse out of the cells and thus becomes trapped within the symplast.

leaf, however, it must enter the phloem and move with the carbohydrate flow. As carbohydrates are actively loaded into the leaf phloem, an osmotic gradient is established that drives water and any dissolved compounds into the phloem. Mass flow then controls movement of the water and solutes to the developing sinks or storage organs.[2] The relative sink strengths, as determined by their size and activity, control the total amount of photosynthates translocated, the distribution of the photosynthates and, therefore, the amount of herbicide translocated and its distribution within the plant.[4]

III. MECHANISM OF CELLULAR UPTAKE OF IMIDAZOLINONES

The imidazolinones kill plants by being retained within the phloem until the compounds reach the meristems. The mechanism by which they are absorbed and retained within plant cells, including the phloem, is called ion trapping.[5] Xenobiotics, such as the imidazolinones, that change in lipophilicity within the physiological pH range of 4 to 8 are subject to ion trapping. The imidazolinones are weak acids with a pK_a between 3.8 and 4.0, which means that a substantial number of these molecules are undissociated at a low pH and, as such, are more lipophilic than the dissociated form which predominates at higher pH levels.

The pH gradient which occurs within plants is typically pH 4.5 to 5.0 outside the cell and 7.0 to 7.5 inside the cell, or as high as 8.0 in the phloem.[2] Although the imidazolinones are water-soluble compounds, under acidic conditions the molecules are relatively more lipophilic and, as such, are able to diffuse across cell membranes. Once inside, the molecule becomes charged in the alkaline environment with a concomitant decrease in lipophilicity. Because the charged form of the imidazolinone is slower to diffuse across the lipophilic plasma membrane, the compound accumulates, or is trapped, within the cell (Figure 1). Should the entrapment occur within nonphloem cells, cytoplasmic streaming may cause the imidazolinone to move through the plasmodesmata, the strands of cytoplasm which connect cells, to the phloem. Once in the phloem and translocated to the site of action, the imidazolinones inhibit AHAS, causing death of the meristematic cells.

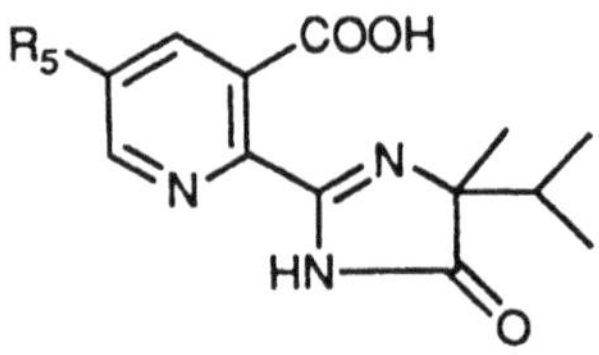

FIGURE 2. Substitution pattern of imidazolinones used in structure-activity studies.

TABLE 1
Physicochemical Parameters Affecting Root Absorption of Imidazolinones Substituted in the 5-Position of the Pyridine Ring

Species	Equation[a]	r^2
Sunflower	$Log\ C = 0.40\pi + 1.34$	0.80
	$Log\ C = -0.10\pi^2 + 0.46\pi + 1.47$	0.87
Corn	$Log\ C = 0.53\pi + 1.18$	0.78
	$Log\ C = -0.17\pi^2 + 0.63\pi + 1.39$	0.87

[a] Hydroponically grown sunflower and corn seedlings were exposed to 10 μM (pH 4) imidazolinone solutions spiked with ^{14}C label for 8 h. The amount of herbicide absorbed by the plants (nmoles absorbed per gram fresh weight, or C) was strongly correlated with π, the parameter representing lipophilicity of the substituents. There was little or no effect on absorption by the electronic and steric parameters. Data set included 12 analogs.

TABLE 2
Root Absorption and Translocation of Three ^{14}C-Imidazolinones in Sunflower

Treatment[a]	Kow[b]	% Absorbed	Distribution of absorbed radioactivity (%)			
			Root	Stem	Old leaves	New leaves
Imazaquin	7.7	65.6	84.4	8.4	3.2	4.1
Imazethapyr	1.4	54.4	87.4	4.7	3.2	4.7
Imazapyr	0.1	19.2	95.8	0.6	1.6	2.0

[a] Herbicides were supplied at a 10 μM concentration (spiked with ^{14}C-imidazolinone) in half-strength nutrient solution (pH 4.5).
[b] Partition coefficient between *n*-octanol and citrate-dibasic sodium phosphate buffer (pH 4.0).

IV. ABSORPTION AND TRANSLOCATION OF IMIDAZOLINONES WITHIN WHOLE PLANTS

A. ROOT ABSORPTION OF THE IMIDAZOLINONES

Absorption of the imidazolinones by plant roots is a function of the lipophilicity of the molecule as shown with a series of analogs substituted in the 5-position (Figure 2). The physical and chemical parameters evaluated in these studies include lipophilicity, electronic nature (field and resonance effects), and steric effects of the substituents on the imidazolinone molecule. Regression analyses of absorption on these parameters show that, in both sunflower and corn, root absorption is highly correlated with lipophilicity (Table 1). In terms of the commercial products, imazaquin is absorbed by roots to a greater extent than is imazethapyr and to an even larger extent than is imazapyr. Such a relationship corresponds with their relative lipophilicity (Table 2).

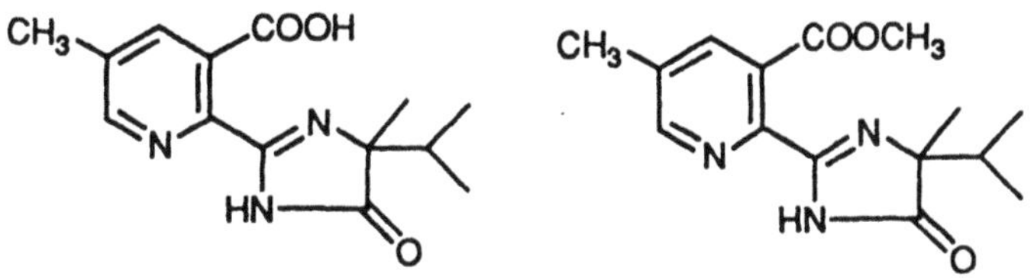

FIGURE 3. Chemical structures of AC 263,222 and AC 263,223.

TABLE 3
Root Absorption of an Acid Imidazolinone and Its
Corresponding Methyl Ester in Soybean Seedlings

Compound[a]	Kow[b]	% absorbed[c]
AC 263,222	1.2	4.5
AC 263,223	8.8	10.1

[a] AC 263,223 is a methyl ester of AC 263,222.
[b] Partition coefficient between *n*-octanol and citrate-dibasic sodium phosphate buffer (pH 4.0).
[c] 10 μM concentrations of the herbicides, spiked with ^{14}C-imidazolinone, in half-strength nutrient solution were supplied to the roots for 4 h. Plants were harvested and radioactivity was counted following combustion.

For some imidazolinone analogs, the dissociable acid group is esterified. An example is AC 263,223, the methyl ester of AC 263,222 (Figure 3). Esterification greatly increases lipophilicity and, with this increase, there is enhanced root absorption (Table 3).

B. TRANSLOCATION OF IMIDAZOLINONES FROM ROOTS

Translocation from the roots to the shoots also appears to be correlated with the lipophilicity of the imidazolinones. For example, in terms of movement of radioactivity from the roots to the rest of the plant, mobility was greatest for imazaquin and least for imazapyr when supplied to the root system of sunflowers (Table 2). Furthermore, nearly half of the translocated imazaquin remained in the stem, probably due to more rapid diffusion from the xylem into the surrounding stem cells than that of imazapyr, which was less likely to diffuse across membranes. The lipophilicity and behavior of imazethapyr were midway between those of imazaquin and imazapyr.

Imazapyr, which is rapidly translocated through the xylem and accumulated in leaf tips, is available for phloem loading and redistribution within the plant. Evidence for such mobility is demonstrated by an experiment in which excised corn shoots were pulsed with a solution of radioactive imazapyr and then chased with distilled water. At the start of the chase, 38.5% of the radioactivity was in the middle leaf and 26.8% was in the stem section containing the leaf sheaths and apical meristem (Figure 4). Four hours later, the percentages were 9.0 and 44.6, respectively, showing that imazapyr was initially translocated through the xylem to the leaf tips and then retranslocated through the phloem to the base of the shoot.

However, labeled imazaquin pulse chased into excised leaves of *Brachiaria platyphylla* and corn was not extensively redistributed. As can be seen in Figure 5, little change occurred from the original distribution of imazaquin for either species after the initial pulse of radiolabeled imazaquin. In a second study, corn and *B. platyphylla* were harvested over a

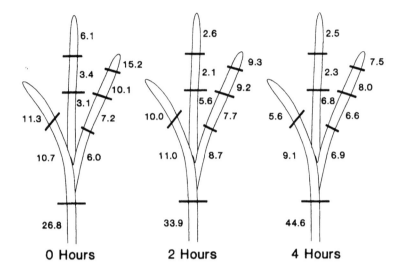

FIGURE 4. Excised corn seedlings were pulsed with ^{14}C-imazapyr for 1.8 h and then harvested at 0, 2, and 4 h after the water chase. At harvest, shoots were divided as shown and extracted. The extracts were counted for radioactivity, and translocation is expressed as percent of absorbed.

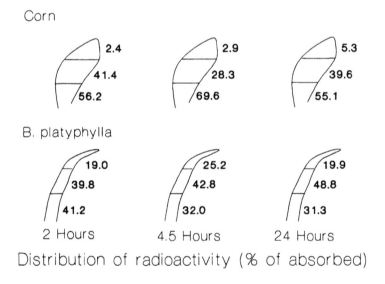

FIGURE 5. Excised leaf tips of corn and *Brachiaria platyphylla* seedlings were pulsed with ^{14}C-imazaquin for 2.5 h and then chased with nutrient solution for 2, 4.5, and 24 h. At harvest, leaves were divided into 2 cm sections and extracted, and the radioactivity in the extracts was determined. (Radioactivity is expressed as percent of absorbed).

longer time period, but again, there was little change in the imazaquin distribution (Figure 6). The lack of redistribution of imazaquin in comparison with that of imazapyr could be due to faster trapping of imazaquin in stem cells, whereas imazapyr may be retained in the xylem for a longer period allowing it to reach the leaves where it may then be phloem loaded and retranslocated. Another explanation may be that imazaquin is more rapidly metabolized than imazapyr to some less mobile compound, thereby preventing any redistribution of radioactivity.

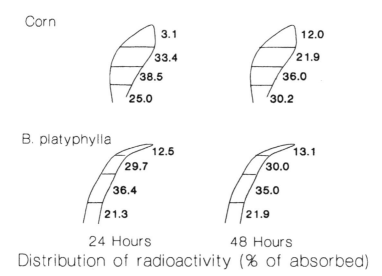

Corn

3.1
33.4
38.5
25.0

12.0
21.9
36.0
30.2

B. platyphylla

12.5
29.7
36.4
21.3

13.1
30.0
35.0
21.9

24 Hours   48 Hours
Distribution of radioactivity (% of absorbed)

FIGURE 6. Excised leaf tips of corn and *Brachiaria platyphylla* seedlings were pulsed with ^{14}C-imazaquin for 2 h, and then chased with nutrient solution for 24 and 48 h. At harvest, leaves were divided into 2 cm sections and extracted, and the radioactivity in the extracts was determined. (Radioactivity is expressed as percent of absorbed).

In summary, the more lipophilic the imidazolinone, the faster it is absorbed into plant roots and the more readily it will translocate to the shoots. Yet, the lipophilic imidazolinones which are acids — for example, imazaquin — will rapidly diffuse from the transpiration stream and become trapped in adjacent tissue, while the more polar acid imidazolinones, such as imazapyr, travel farther along the transpiration stream. Therefore, the extent to which an imidazolinone is absorbed and translocated to the meristems is a function of the lipophilicity of the herbicide.

C. FOLIAR ABSORPTION OF THE IMIDAZOLINONES

In addition to being active through soil applications, the imidazolinones are also herbicidal when applied directly to foliage. One of the factors affecting the ability of the imidazolinones to penetrate the cuticle, which is the primary barrier to foliar absorption, is the environment. Any condition which hastens drying of the spray droplets, such as low humidity, winds, and high temperatures, can potentially decrease the total amount of herbicide absorbed (Chapter 8). The loss in uptake occurs because precipitation of the technical material precludes its penetration. Therefore, to achieve maximal uptake, herbicides should be applied when the weather conditions are optimal for slow drying of the droplets.

The sensitivity of plants to the imidazolinones is species dependent. Selectivity is often a function of metabolism (Chapter 7), but differences in foliar absorption between species have also been observed. For example, Shaner and Robson[6] found that limited foliar absorption of imazaquin by *Abutilon theophrasti* accounted, in part, for its tolerance to this herbicide, and that rapid foliar absorption of imazaquin was associated with *Xanthium strumarium*, a more sensitive species (Table 4). Furthermore, foliar absorption by *A. theophrasti* decreased with increasing plant age (Table 5), which may explain the loss in control of this species when imazaquin is applied late in the growing season.

To maximize weed control and improve foliar penetration of the herbicides, several types of modifications can be made either to the formulation or to the tank mixture. One technique is the use of adjuvants. The addition of adjuvants, which include surfactants, spray modifiers, and utility modifiers,[7] to the formulation or to the spray tank increases the effectiveness of the active ingredient by increasing the amount of technical material contacting

TABLE 4
Absorption of ^{14}C-Imazaquin Applied to the Foliage of Weed Species

Weed species	Days after application	Absorption[a] (% of applied)
Abutilon theophrasti	1	8.5
	3	21.0
Xanthium strumarium	1	46.0
	3	60.5

[a] Radiolabeled (^{14}C) imazaquin was applied to one leaf per plant. Treated leaves were excised at 1 or 3 d after treatment, the leaves washed, and the radioactivity in the leaf wash counted.

From Shaner, D. L. and Robson, P. A., *Weed Sci.*, 33, 469, 1985. With permission.

TABLE 5
Absorption of ^{14}C-Imazaquin Applied to the Foliage of *Abutilon theophrasti* at Various Growth Stages

Growth stage	Absorption[a] (% of applied)
Cotyledon	30.0
Four-leaf	21.0

[a] Radiolabeled (^{14}C) imazaquin was applied to the foliage of *A. theophrasti* at the growth stages indicated. Treated leaves were excised at 1 d after treatment, the leaves washed, and the radioactivity in the leaf wash counted.

From Shaner, D. L. and Robson, P. A., *Weed Sci.*, 33, 469, 1985. With permission.

the foliage. This is accomplished, for example, by surfactants which lower the surface tension of the spray droplet and the interfacial tension between the droplet and leaf tissue. Additionally, surfactants may have humectant properties that lengthen the drying time, thus increasing the amount of herbicide that can penetrate the leaf surface.

Many surfactants, particularly nonionic surfactants, have been shown to increase uptake of the imidazolinones into various plant species. Field tests have confirmed this work as well (Chapters 17 through 20). However, not all surfactants increase herbicidal activity, and some have antagonized the effects of the imidazolinones. For example, cationic surfactants that cause localized phytotoxicity at the site of contact decrease absorption of imazameth-abenz-methyl (Figure 7). These surfactants destroy cell membranes; thus, the herbicide cannot be trapped by these cells or translocated from the site of application.

Another type of modification which increases herbicidal activity is to formulate the imidazolinones as salts. Presumably the salts slow precipitation of the technical material by keeping the herbicide soluble for a longer period of time as the droplets dry. Differences in absorption between salts can also occur. In one study, imazapyr was applied to sugar cane as two different salts (sodium and calcium) in the presence of various concentrations of a nonionic surfactant. The results showed that increasing the surfactant concentration to 0.25% v/v enhanced absorption and that more of the Na$^+$ salt than of the Ca^{2+} salt of imazapyr was absorbed (Figure 8). Therefore, in addition to the use of nonionic surfactants, formulating the imidazolinones as certain salts further enhances herbicidal activity.

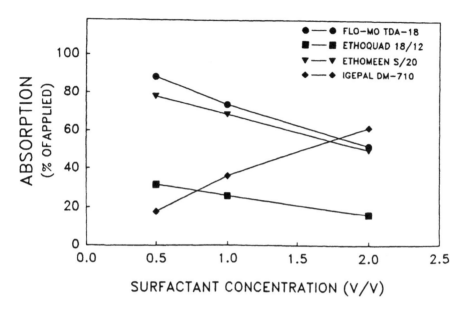

FIGURE 7. Foliage of excised *Avena fatua* seedlings was treated with droplets of ^{14}C-imaza-methabenz-methyl prepared with 0.5, 1.0, and 2.0% v/v of the cationic surfactants Flo-Mo TDA-18, Ethoquad 18/12, and Ethomeen S/20, and the nonionic surfactant Igepal DM-710. At 1 DAT (days after treatment), leaves were washed and extracted, and the radioactivity in the washes and extracts was determined.

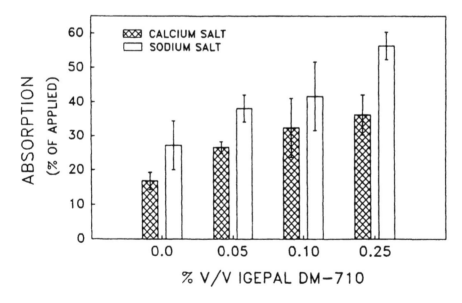

FIGURE 8. The calcium and sodium salts of imazapyr (spiked with ^{3}H-imazapyr) were prepared as 50 g a.e./400 l in 0, 0.05, 0.10, and 0.25% v/v Igepal DM-710. Solutions were applied as droplets to foliage of sugar cane plants grown from seed pieces. At 4 DAT, treated leaves were excised and washed. Radioactivity in the washes was counted.

Although the mechanism of action is not yet understood, liquid fertilizers increase postemergence weed control of imidazolinones on some weed species when the fertilizers are used as spray adjuvants. For example, control of *Ipomoea lacunosa* by imazaquin and imazethapyr was enhanced by the addition of ammonium sulfate ($[NH_4]_2SO_4$) or potassium

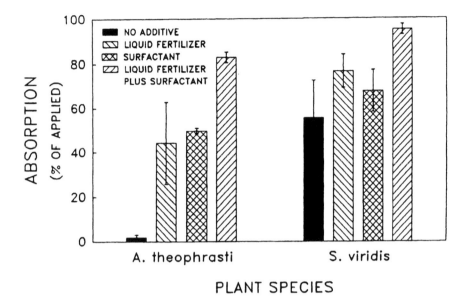

FIGURE 9.   Leaves of *Abutilon theophrasti* and *Setaria viridis* seedlings were treated with droplets of ^{14}C-imazethapyr prepared in 5% 10-34-0, 0.25% v/v Tween 20, or a combination of the adjuvants. Plants were harvested 48 h later, and the treated leaves were excised and washed. Radioactivity in the washes was determined.

TABLE 6

Translocation of Three ^{14}C-Labeled Imidazolinones
Applied to the Foliage of Sunflower

	% of absorbed radioactivity[a]	
% Compound	Treated leaf	Rest of plant
Imazapyr	94.6	5.4
Imazethapyr	88.0	12.0
Imazaquin	84.4	15.6

[a]   Radiolabeled (^{14}C) solutions were applied to leaves of sunflower seedlings. 4 h after treatment, the treated leaves were excised and washed. Radioactivity in the treated leaves and in the remainder of the plant was determined.

phosphate (KH$_2$PO$_4$).[8] Furthermore, foliar absorption of imazethapyr by *A. theophrasti* was enhanced by adding either a surfactant or a liquid fertilizer, but it was increased to an even larger extent by adding both (Figure 9). Although control of *Setaria viridis* by imazethapyr was not significantly enhanced by the single addition of a surfactant or liquid fertilizer in this particular test, control was significantly better when imazethapyr was applied with both additives. Therefore, control of some species, such as *A. theophrasti* and *S. viridis*, may be markedly improved by the addition of liquid fertilizer to the tank mixture.

D. TRANSLOCATION OF IMIDAZOLINONES FROM SHOOTS

In contrast with translocation from the roots, foliar translocation of the imidazolinones does not consistently appear to correlate with lipophilicity. For example, in sunflower, translocation of imazaquin out of the treated leaf is greater than that of imazethapyr or imazapyr at 4 h after treatment (Table 6). In corn, however, more imazapyr than imazaquin

TABLE 7
Foliar Absorption and Translocation of ^{14}C-Labeled
Imazapyr and Imazaquin in Corn

| Compound | DAT | Distribution of absorbed radioactivity[a] | |
		Treated leaf	Rest of plant
Imazapyr	3	18.3	81.7
	5	12.5	87.5
Imazaquin	3	39.6	60.4
	5	37.0	63.0

[a] Corn seedlings growing in nutrient culture were sprayed with 50 g
a.e./ha of either compound at 400 1/ha. Immediately after drying,
^{14}C-labeled imazapyr or imazaquin was applied to a single leaf per
plant. At harvest, the leaves treated with radiolabeled herbicide
were excised and washed, and the radioactivity in the wash and in
the plant parts was determined.

TABLE 8
Translocation of ^{14}C-Labeled Imazapyr in
Imperata cylindrica

DAT	% translocated of the absorbed[a]
1	26.4
2	38.9
4	57.0
9	60.7

[a] Young plants, grown from rhizomes, were sprayed with
50 g a.e./ha of imazapyr at 400 1/ha. Immediately after
drying, ^{14}C-labeled imazapyr was applied to the second
oldest leaf. At harvest, leaves treated with ^{14}C-imazapyr
were excised and then washed. Radioactivity in the leaf
wash, in the treated leaf, and in the rest of the plant was
determined.

From Shaner, D. L., *Trop. Pest Manage.*, 34, 388, 1988.
With permission.

had translocated from the treated leaf at 3 and 5 d after treatment (Table 7). The reason for
these seemingly conflicting results may be that these species metabolize each imidazolinone
differently while in the symplast and that the rates of translocation of the metabolites differ
from those of the parent herbicides. Because of the complexity of the responses, foliar
translocation will be considered separately for each of the imidazolinones.

1. Imazapyr

Once absorbed by annual and perennial species, imazapyr rapidly translocates out of
treated leaves to the rest of the plant, with translocation typically lasting for several days.
Movement of imazapyr in a perennial grass, *Imperata cylindrica,* reached a maximum by
4 d after treatment[9] (Table 8). In another perennial grass, *Sorghum halepense,* imazapyr
translocation out of the treated leaf was measured within 6 h (Table 9) and reached a maximum
by 5 d after treatment. Furthermore, imazapyr translocated both above and below the treated
leaf. When translocated below the treated leaf, some herbicide leaked from the roots (Table
9) and was then available for absorption by other plants (Table 10).

TABLE 9

Translocation of ^{14}C-Imazapyr from Treated Leaves of *Sorghum halepense*

	Distribution of absorbed radioactivity[a]						
DAT	Tips of treated leaves	Treated zone	Base of treated leaves	Shoot above treated leaves	Shoot below treated leaves	Roots	Leakage
0.25	1.7	70.5	5.0	11.1	9.2	2.5	
1.0	3.0	56.3	7.1	18.0	12.2	3.4	
2.0	7.1	37.0	8.0	16.9	19.6	11.7	
5.0	12.7	19.9	12.9	23.1	13.5	11.1	7.0
8.0	12.4	24.4	8.7	27.5	16.7	4.0	6.4

[a] Young plants, grown from rhizomes, were sprayed with 0.5 kg a.e./ha imazapyr at 374 l/ha. After the spray solution had dried, two leaves per plant were treated with ^{14}C-imazapyr. When harvested, sections of the leaves treated with radiolabeled imazapyr were excised and washed. The plants were further divided into sections of the blades above and below the treated zones, the shoot above the treated leaves, the shoot below the treated leaves, and the roots. Radioactivity in these tissues and in the soil was determined.

TABLE 10

Foliar Translocation of ^{14}C-Labeled Imazapyr in Corn

	Distribution of Absorbed Radioactivity[a]			
DAT	Treated leaf	Rest of treated plant	Sand media	Untreated plant
3	19.4	57.2	21.1	2.3
6	11.3	46.2	37.4	5.1

[a] Two of three corn plants growing in a pot received a foliar treatment of ^{14}C-imazapyr. At harvest, the treated leaves and the sand were washed. The amount of radioactivity in the two washes, in the treated leaves, in the rest of the treated plants, and in the untreated plant was determined.

2. Imazaquin

Differences in translocation of imazaquin may be related to differences in susceptibility among species. Imazaquin is very active on corn but only weakly active on *Brachiaria platyphylla* and *Avena fatua;* translocation in the two weed species did not change after 1 day following treatment, while translocation in corn continued to increase (Table 11). The rapid cessation of translocation in the weed species may be due to metabolism of imazaquin to nonmobile products while in the treated leaf.

3. Imazethapyr

Imazethapyr is often more active than imazaquin when applied postemergence to weeds. This difference in activity may be related to absorption and translocation. More than 80% of the applied dose of imazethapyr was absorbed by *Xanthium strumarium* at 1 d after treatment (Table 12), in comparison with 46% of the imazaquin (Table 4). For *A. theophrasti*, the figures are 20.8% and 8.5% for imazethapyr and imazaquin, respectively. In both cases imazethapyr was distributed above and below the treated leaf.

However, a high degree of absorption is not always correlated with herbicidal activity. When imazethapyr was applied to soybean, a tolerant species, more than 90% of the applied dose was absorbed by 1 d after treatment, with approximately 20% translocated from the treated leaf at all time points (Table 12). Most of this radioactivity is likely to be the glucose conjugate of AC 288,511 (Figure 10), a nontoxic metabolite of imazethapyr (Chapter 11).

Absorption and translocation of imazethapyr has also been studied in *A. fatua* and *B. platyphylla*. Absorption of imazethapyr increased over the course of the experiment for both

TABLE 11
Foliar Absorption and Translocation of ^{14}C-Labeled Imazaquin in Corn, *Brachiaria platyphylla*, and *Avena fatua*

Species	DAT	Treated leaf	Distribution of absorbed radioactivity[a]		
			Above treated leaf	Below treated leaf	Leakage from roots
Corn	1	60.1	15.2	19.7	5.0
	3	34.2	21.9	16.3	27.7
	6	27.3	22.7	12.3	37.7
B. platyphylla	1	82.6	7.5	3.0	3.6
	3	86.6	6.4	3.1	4.0
	6	88.3	6.1	1.7	4.0
A. fatua	1	84.8	10.7	2.9	1.6
	3	85.1	8.1	3.3	3.7
	6	82.2	7.4	5.5	5.1

[a] Corn, *A. fatua*, and *B. platyphylla* seedlings growing in nutrient culture were treated with ^{14}C-imazaquin. At harvest, the treated leaves were washed, and the radioactivity in the rinse and in the various plant parts was determined.

TABLE 12
Absorption and Translocation of ^{14}C-Labeled Imazethapyr Applied to the Foliage of Soybean, *Abutilon theophrasti*, and *Xanthium strumarium*

Species	DAT	% absorbed	Distribution of absorbed radioactivity[a]		
			Treated leaf	Above treated leaf	Below treated leaf
Soybean	1	92.1	83.3	6.1	10.7
	4	94.4	80.5	10.2	9.7
	7	97.0	79.0	9.3	11.7
A. theophrasti	1	20.8	92.0	4.1	3.9
	4	43.4	95.4	1.5	3.1
	7	49.5	94.8	3.0	2.2
X. strumarium	1	82.8	85.8	6.7	7.5
	4	88.7	66.8	17.1	16.1
	7	88.8	80.2	10.9	8.9

[a] Radiolabeled (^{14}C) imazethapyr was applied to one unifoliate leaf of soybean and to single leaves of *A. theophrasti* and *X. strumarium* seedlings. At harvest, the treated leaves were washed, and the radioactivity in the wash and in the plant parts was determined. (The leaf opposite the treated unifoliate leaf of soybean was included in the upper portion upon dissection.)

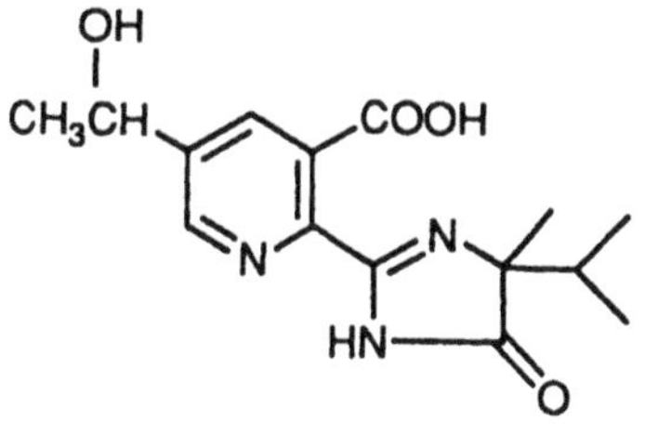

FIGURE 10. Chemical structure of AC 288,511.

TABLE 13
Foliar Absorption and Translocation of ¹⁴C-Labeled Imazethapyr in
Brachiaria platyphylla and *Avena fatua*

			Distribution of absorbed radioactivity[a]		
Species	DAT	% absorbed	Treated leaf	Above treated leaf	Below treated leaf
B. platyphylla	1	60.4	83.1	6.8	10.1
	3	58.3	74.1	16.4	9.5
	7	93.1	47.1	44.9	8.0
A. fatua	1	24.2	82.0	3.8	14.1
	3	42.9	75.6	11.3	13.1
	7	68.1	89.6	4.4	6.0

[a]   Radiolabeled (¹⁴C) imazethapyr was applied to leaves of *A. fatua* and *B. platyphylla* seedlings. At harvest, the treated leaves were washed, and the radioactivity in the wash and in the various plant parts was determined.

species but translocation and metabolism differed between them (Table 13). In *B. platyphylla*, translocation increased with time, unlike that for imazaquin (Table 11), with the radioactivity (shown to be the parent compound) increasing in tissues above the treated leaf. Similar to imazaquin, translocation of radioactivity in *A. fatua* did not change substantially over time, indicating that perhaps a nonmobile metabolite, possibly AC 288,511, was being formed. Therefore, not only are there differences in translocation and metabolism between imidazolinones but also differences in these processes between plant species.

4. Imazamethabenz-methyl

The high degree of lipophilicity of imazamethabenz-methyl enhances its foliar uptake while the lack of a dissociable acid group limits its phloem mobility. For example, when applied to the second leaf of *A. fatua* seedlings, 51.6% of the absorbed dose translocated to the tip of the treated leaf, 43.3% remained in the treated area, 2.6% moved to the lower portion of the treated leaf, and 2.5% was found in the remainder of the plant. Similarly, work by Hinshalwood[10] showed that even after 7 d only 0.55% to 1.47% of the applied imazamethabenz-methyl was translocated out of the treated leaf of *A. fatua*. These data suggest that this highly lipophilic compound flows with the transpirational water with little trapping of the xenobiotic in the surrounding cells. When metabolized to its acid form, imazamethabenz-methyl can be trapped in the phloem and then translocated from the treated leaves (Chapter 7).

E. FACTORS AFFECTING PHLOEM TRANSLOCATION OF THE IMIDAZOLINONES

Translocation of the imidazolinones can be affected by the activity of other herbicides. In a study of the combination of imazamethabenz-methyl and propanil (especially formulated to overcome compatibility problems), no translocation of the imidazolinone from the treated leaf occurred for the combination treatment in comparison to imazamethabenz-methyl applied alone (Table 14). The reason for the lack of translocation of the ·herbicide combination is related to the mode of action of propanil. Propanil is a photosynthesis inhibitor which causes localized phytotoxicity of sprayed leaves. This phytotoxicity prevents metabolism of the ester imidazolinone to its corresponding acid and interferes with sugar transport by stopping photosynthesis, thereby preventing phloem translocation of the herbicide. A similar type of antagonism occurs when imazapyr is mixed with paraquat, another photosynthesis inhibitor.

However, even in the absence of another herbicide, phloem translocation of the imidazolinones usually ceases between 3 to 5 d after treatment. For example, when *Sorghum halepense* was treated with imazapyr, phloem translocation of the herbicide from the treated

TABLE 14
Translocation and Metabolism of ^{14}C-Imazamethabenz-methyl in *Avena fatua* When Applied Alone and in Combination with Propanil

Treatment[a]	DAT	Translocation out of leaf	Metabolism of imazamethabenz-methyl		
			%imazamethabenz-methyl	%acid metabolite	% other compounds
Assert	2	0.67	86.3	8.0	5.7
	4	0.40	58.8	7.6	33.6
Assert + Propanil	2	0.00	97.8	1.7	0.6
	4	0.00	96.7	1.0	2.4
Assert Solution	—	—	97.4	0.6	2.1

[a] The Assert-propanil combination was a product of the Formulation Group. A sample of this product and the Assert treatment were spiked with ^{14}C-imazamethabenz-methyl. Foliar application was made to *Avena fatua* seedlings at an Assert rate of 400 g a.i./ha delivered at 100 l/ha. At harvest, treated leaves were washed, and the radioactivity in the plant extracts was counted. Metabolism was monitored by TLC.

leaf stopped within 5 days (Table 9). This same pattern occurs with other imidazolinones on many other species. In all cases, the treated leaves are not visibly damaged and appear capable of photosynthesis. If they are still capable of photosynthesis, why does translocation stop?

An explanation for this phenomenon is that the mode of action of the imidazolinones limits their own translocation. Devine et al.[11] found that translocation of sulfonylureas (another class of AHAS inhibitors) from the treated leaves of *Fagopyrum tataricum* is limited by the availability of the branched-chain amino acids. A deficiency of these amino acids induced by either class of inhibitors will cause the meristems to stop dividing, resulting in a loss of sink strength. Because the meristems are no longer sinks, phloem translocation of photosynthates, as well as translocation of the inhibitor, to the meristems ceases.

In order to study this limitation, the herbicidal effects of the imidazolinones must be eliminated. To do this, plants were supplemented with the branched-chain amino acids to prevent the effects of the imidazolinones on AHAS. When isoleucine, leucine, and valine were supplied to the roots of sunflowers, both foliar absorption of imazaquin (Figure 11A) and translocation (Figure 11B), particularly translocation above the treated leaf (Figure 11C), were increased as compared with nonsupplemented plants. These results indicate that a deficiency of these amino acids inhibits translocation and support the theory that the imidazolinones inhibit their own mobility.

Whereas the total amount of imidazolinone translocated within plants is a function of their mechanism of action, the pattern of distribution is determined by the relative sink strengths. An important ramification of these two factors is the herbicidal effect on tillers following treatment of the main stem of grasses. Application of imidazolinones will kill the main culm of susceptible grasses but may release the tillers for growth because the apical meristem of the main culm is the primary sink and, as discussed above, no further translocation occurs subsequent to its death. Another consequence of the death of the apical meristem is that tiller meristems are no longer under apical dominance and are able to grow. These tillers are not dependent on the main culm for photosynthates; therefore, they do not import the imidazolinones and, unless the initial dose reaching the tillers before the apical meristem senesces is sufficient to induce their death, the tillers will develop.

An example of the lack of tiller dependence on the main stem following treatment with the imidazolinones is provided by work from Hinshalwood.[10] The addition of a surfactant to imazamethabenz-methyl applied to leaves of *A. fatua* increased the total radioactivity absorbed and translocated out of the treated leaves, but did not increase the amount trans-

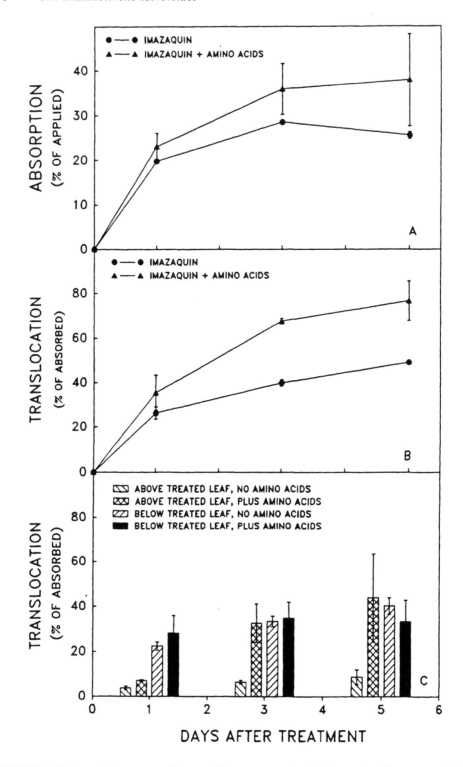

FIGURE 11.   Hydroponically grown sunflower seedlings were treated with 50 g a.e./ha of imazaquin in 0.25% v/v Igepal DM-710 delivered at 400 l/ha using a laboratory belt sprayer. Immediately after spraying, ^{14}C-imazaquin solution was applied as droplets to one leaf per plant and the hydroponic solution was replaced with a nutrient solution supplemented with 1 mM valine, leucine, and isoleucine. Plants were harvested at 1, 3, and 5 DAT, the treated leaves washed, and all plant parts extracted. Radioactivity in the extracts was then counted.

TABLE 15
Absorption and Translocation of Foliar-Applied ^{14}C-Imazamethabenz-methyl in *A. fatua*

Surfactant added to imazamethabenz-methyl[a]	% absorbed	Mobility based on % of applied dose		
		Total translocated[b]	Translocation from treated leaf	Translocation to the tillers
None	22.99	12.74	0.55	0.05
Assist	85.21	62.71	1.15	0.11
Igepal DM-710	96.93	64.79	1.04	0.11
Tween 20	85.83	57.84	1.47	0.13

[a] Formulated Assert, which contained no added surfactant or to which 0.5% v/v Assist Oil Concentrate, Igepal DM-710, or Tween 20 had been added, was spiked with ^{14}C-imazamethabenz-methyl and applied to the foliage of *A. fatua* seedlings.

[b] Total translocation includes the amount of compound moving to the leaf tips from the treated zone, the amount moving below the treated zone within the same leaf, and the amount moving to other plant parts.

located to the tillers (Table 15). These results suggest that improved control of this weed species will be achieved only by increasing the amount of imidazolinone reaching the tiller meristems, either by changing the relative sink strengths or by better placement of the herbicide treatments.

REFERENCES

1. **Raven, P. H., Evert, R. F., and Curtis, H., Eds.,** *Biology of Plants,* 2nd ed., Worth Publishers, New York, 1976, 425.
2. **Hess, F. D.,** Herbicide absorption and translocation and their relationship to plant tolerances and susceptibility, in *Weed Physiology,* Vol. 2, *Herbicide Physiology,* Duke, S. O., Ed., CRC Press, Boca Raton, 1985, 191.
3. **Price, C. E.,** Movement of xenobiotics in plants — perspectives, in *Advances in Pesticide Science,* Vol. 3, Geissbuhler, H., Ed., Pergamon Press, Oxford, 1978, 401.
4. **Wolswinkel, P.,** Phloem unloading and "sink strength": the parallel between the site of attachment of *Cuscuta* and developing legume seeds, *Plant Growth Regul.,* 2, 309, 1984.
5. **Van Ellis, M. R. and Shaner, D. L.,** Mechanism of cellular absorption of imidazolinones in soybean (*Glycine max*) leaf discs, *Pestic. Sci.,* 23, 25, 1988.
6. **Shaner, D. L. and Robson, P. A.,** Absorption, translocation, and metabolism of AC 252,214 in soybean (*Glycine max*), common cocklebur (*Xanthium strumarium*), and velvetleaf *(Abutilon theophrasti),* *Weed Sci.,* 33, 469, 1985.
7. **McWhorter, C. G.,** The physiological effects of adjuvants on plants, in *Weed Physiology,* Vol. 2, *Herbicide Physiology,* Duke, S. O., Ed., CRC Press, Boca Raton, 1985, 141.
8. **Wills, G. W. and McWhorter, C. G.,** Influence of inorganic salts and imazapyr on control of pitted morningglory (*Ipomoea lacunosa*) with imazaquin and imazethapyr, *Weed Technol.,* 1, 328, 1987.
9. **Shaner, D. L.,** Absorption and translocation of imazapyr in *Imperata cylindrica* (L.) Raeuschel and effects on growth and water usage, *Trop. Pest Manage.,* 34, 388, 1988.
10. **Hinshalwood, A.,** personal communication, 1988.
11. **Devine, M. D., Bestman, H. D., and Vanden Born, W. H.,** Differential phloem mobility of chlorsulfuron and clopyralid: effect of chlorsulfuron on translocation *per se,* *Weed Sci. Soc. Am.,* 27 (Abstr), 70, 1987.

Chapter 6

IMIDAZOLINONE-ACETOHYDROXYACID SYNTHASE INTERACTIONS

Mark A. Stidham and Bijay K. Singh

TABLE OF CONTENTS

I. INTRODUCTION

Research on the mode of action of the imidazolinone herbicides led to the conclusion that the sole mechanism was the inhibition of the enzyme acetohydroxyacid synthase (AHAS, acetolactate synthase, ALS, E.C. 4.1.3.18). Five key experiments led to this conclusion. First, the treatment of susceptible maize cell cultures with imidazolinones resulted in reducing the levels of only three amino acids: valine, leucine, and isoleucine.[1] These amino acids share four enzymes in their biosynthetic pathway; AHAS is the first in the sequence (Figure 1). Second, supplementation of herbicide-treated maize with the three amino acids resulted in the alleviation of herbicidal symptoms.[1,2] Third, AHAS extracted from susceptible maize was inhibited by the imidazolinones *in vitro*.[3] Fourth, imidazolinone-treated plants were found to have reduced AHAS activity.[4] Fifth, AHAS from imidazolinone-tolerant maize derived from cell culture selection was not inhibited by the imidazolinones, and tolerance cosegregated with insensitive AHAS.[5,6] Many other observations support this hypothesis of the mode of action of the imidazolinone herbicides. Thus, a discussion of AHAS and the interactions of the imidazolinones with this enzyme is essential for a complete description of the imidazolinone herbicides.

II. AHAS BACKGROUND

AHAS is required for the biosynthesis of two acetohydroxyacids: acetolactate and acetohydroxybutyrate (Figure 2). AHAS catalyzes the condensation of pyruvate either with a second pyruvate to yield acetolactate or with 2-oxobutyrate to yield acetohydroxybutyrate. Figure 3 shows the structures of the cofactors of AHAS and the amino acid end products of the pathway. Both thiamine pyrophosphate and divalent magnesium activate the enzyme, while flavin adenine dinucleotide (FAD) acts as a stabilizing cofactor with no role in catalysis.[4,7] AHAS is the first enzyme common to valine, leucine, and isoleucine biosynthesis and is feedback inhibited by the pathway end products.[8-10] AHAS has been found in bacterial extracts,[11,13-15] yeast and other fungi,[12,16-20] archaebacteria,[21] algae,[22,23] and plants.[3,5,6,8-10,24-33] The majority of studies on AHAS have been performed on the enzyme extracted from microbial sources.

A. MICROBIAL AHAS

In enterobacteria, as many as six AHAS isozymes are encoded by different genes. Isozyme I of *E. coli*[13] and isozyme II of *Salmonella typhimurium* have been purified to homogeneity.[14] AHAS II was carefully studied by DuPont researchers after discovering that AHAS is the site of action of the sulfonylurea herbicides.[34] Although the sulfonylureas are structurally unrelated to the imidazolinones, the studies on AHAS inhibition by sulfonylureas reveal much about the mechanism of AHAS. Inhibition of AHAS II by sulfonylureas is characterized as slow, tight binding. Stopped-flow kinetic studies showed that the first step in catalysis, the binding and decarboxylation of pyruvate, occurs unimpeded in the presence of sulfometuron methyl (SM). It is the second step, the binding and condensation with the second pyruvate, that is inhibited by SM.[35] AHAS II activity was not influenced by the redox state or the reduction potential of the FAD bound to the enzyme;[7] however, the absorption spectrum of the FAD bound to the enzyme changes during catalysis, and these changes were diminished by SM.[36] Also, substitution of reduced FAD (FADH$_2$) for FAD resulted in an increase in the binding constant for SM. Thus, the binding of SM, which results in the inhibition of AHAS, must occur proximal to the FAD binding site and near the binding site of the second pyruvate (or the binding site of the 2-oxobutyrate). The binding of SM first occurs with low affinity followed by formation of the final high-affinity complex.[34]

Certain conclusions about the structure of the catalytic site of AHAS can be inferred from these studies. First, FAD must serve a structural role for the enzyme while being

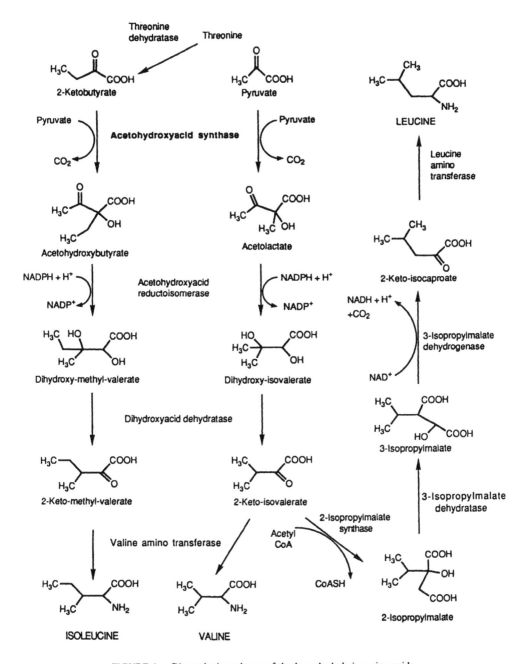

FIGURE 1. Biosynthetic pathway of the branched-chain amino acids.

associated with conformational changes in the enzyme occurring during catalysis. Second, the FAD binding site must be proximal to the second pyruvate binding site. Third, the sulfonylurea binding site must be proximal to both the FAD and the second pyruvate binding site. Fourth, all of these sites must be somewhat removed from the first pyruvate binding site. The implications of these studies on the mechanism of plant AHAS inhibition by the imidazolinones remain to be investigated.

Inhibition of bacterial AHAS by the imidazolinones has been documented in only one report. Schloss et al.[7] determined the binding constants of imazaquin with the three bacterial isozymes (3.6 mM, 0.32 mM, and 3.2 mM for isozymes I, II, and III, respectively).

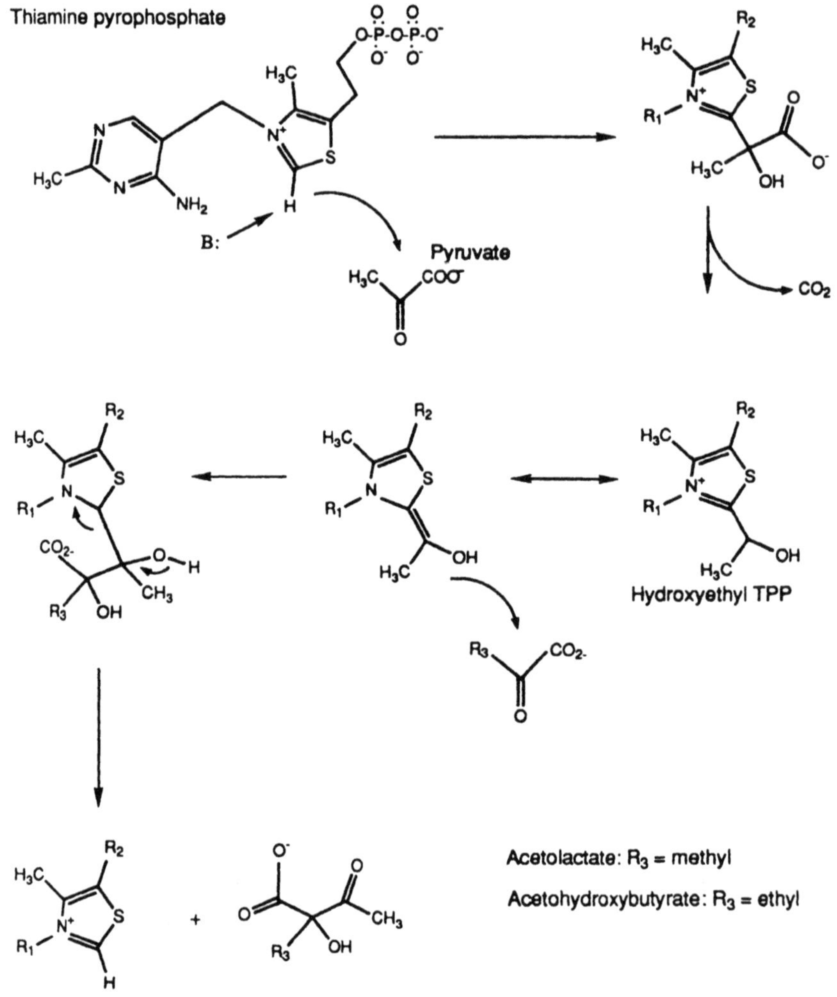

FIGURE 2.   Proposed reaction mechanism for AHAS.

Equilibrium binding studies using AHAS II and radiolabeled SM proved that SM could be displaced by imazaquin or ubiquinone-0. The amino acid sequences of all of the AHAS isozymes are similar to the sequence of pyruvate oxidase,[37,38] a mechanistically related enzyme, which catalyzes the oxidative decarboxylation of pyruvate. FAD undergoes cyclical oxidation and reduction during the cycle, and the reoxidant for $FADH_2$ is ubiquinone. Schloss speculated that the herbicide binding sites of AHAS are derived from the vestigial ubiquinone binding site of the ancestral pyruvate oxidase.[7]

B. GENERAL PROPERTIES OF PLANT AHAS

AHAS occupies the same central role in valine, leucine, and isoleucine biosynthesis in plants as it does in microbes. Feedback of all three amino acids inhibits the enzyme, but a cooperative inhibition is observed when both valine and leucine are present.[9] AHAS occurs in small quantities in plant tissues and, like the bacterial isozymes, the plant enzyme requires FAD for stability.[4,6,32]

Like a majority of the amino acid biosynthetic enzymes, AHAS is nuclear encoded and plastid localized.[8,28] AHAS gene sequences from *Arabidopsis* and tobacco have been published[39] and show a high degree of homology at the amino acid level both between plant

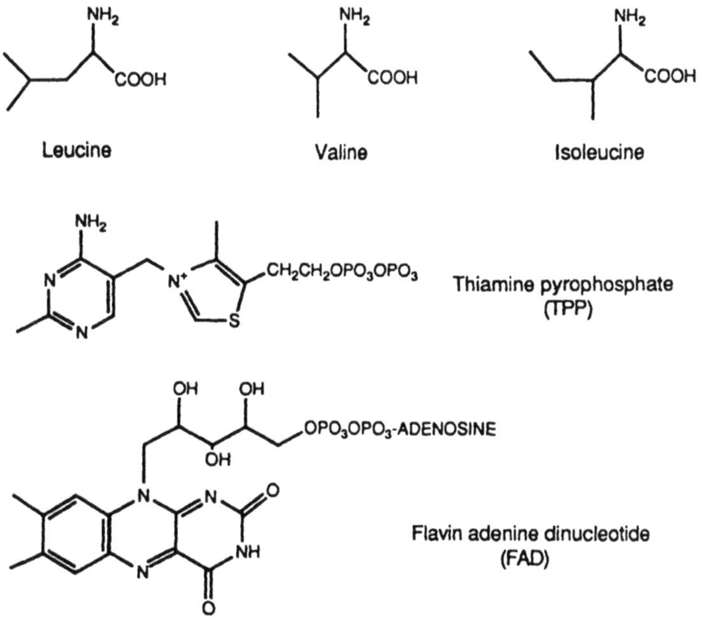

FIGURE 3. Structures of end products of AHAS pathway and cofactors of AHAS.

species and with microbial enzymes. A European patent application on the AHAS gene has been published, showing sites of mutation in the gene that confer resistance to sulfonylurea and imidazolinone herbicides.[40]

III. AHAS INHIBITION BY THE IMIDAZOLINONES

A. *IN VITRO* EXPERIMENTS WITH AHAS

Initial experiments on the inhibition of AHAS by the imidazolinones were conducted on AHAS from corn roots. Substrate-inhibitor studies suggest that the inhibition of AHAS by imazapyr is uncompetitive with respect to pyruvate (Figure 4). Uncompetitive inhibition implies that imazapyr binds to AHAS only after the formation of the ternary enzyme-pyruvate-TPP complex.

Muhitch et al.[4] reported on a number of experiments that demonstrated a more complex interaction of imazapyr with AHAS than was assumed in the earlier studies. When AHAS activity was measured over an extended assay period (4 h) in the presence of various imazapyr concentrations, inhibition was found to increase with time (Figure 5). These results suggest that the equilibrium between imazapyr and the enzyme is reached slowly, a feature typical of many tight-binding inhibitors. The initial and final K_i values for the inhibition of AHAS from black Mexican sweet corn suspension cell cultures by imazapyr were 15 and 0.9 μM, respectively.

The structural requirements for inhibition of AHAS by the imidazolinones correlate, to a certain degree, with the structural requirements for herbicidal activity. Figure 6 shows a summary of some of these structural requirements. Individually, neither the nicotinic acid ring nor the imidazolinone ring is inhibitory. Neither the des-carboxy imidazolinone nor the esters of the nicotinic acid are inhibitory. In contrast, among those imidazolinones containing nitrogen in the aromatic ring, only the nicotinic imidazolinone is an effective herbicide.

Among the herbicidal imidazolinones, the benzene imidazolinone is the best inhibitor, followed by the quinoline and the pyridine (Figure 7). In the pyridine series, all of the

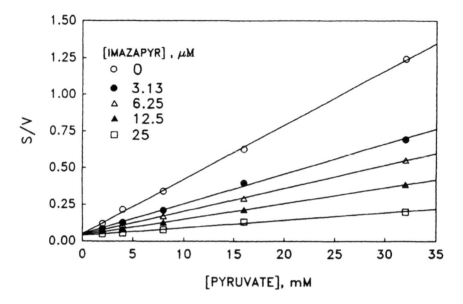

FIGURE 4.   Effects of imazapyr on AHAS dependence on pyruvate (Hanes-Woolf plot).

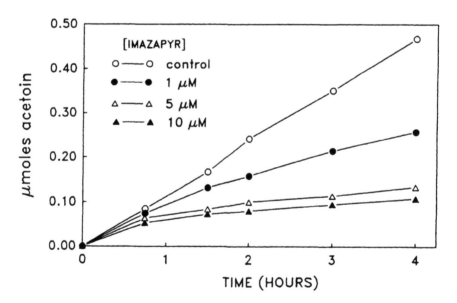

FIGURE 5.   Time course for inhibition of AHAS by imazapyr. (Adapted from Muhitch, M. J., Shaner, D. L., and Stidham, M. A., *Plant Physiol.*, 83, 451, 1987. With permission.)

positional isomers that have adjacent imidazolinone and carboxylic acid substitutions on the ring are active inhibitors of AHAS. In the imidazolinone ring, the chiral carbon at the attachment of the isopropylmethyl group is important in determining enzyme inhibition: the *R* isomer is about 10 times more inhibitory than the *S* isomer.

B. EFFECTS OF IMIDAZOLINONE TREATMENT ON THE LEVEL OF EXTRACTABLE AHAS

When AHAS is extracted from corn tissue treated with imazapyr, the amount of extractable AHAS is dractically reduced compared to that in untreated tissue (Table 1). This

Active

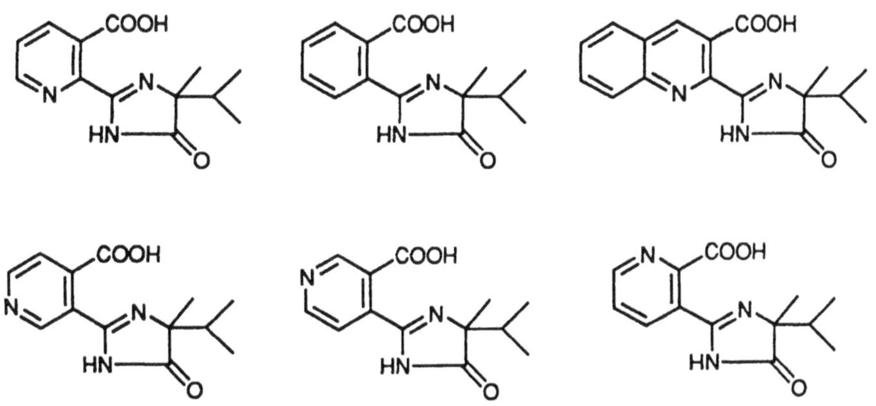

Inactive

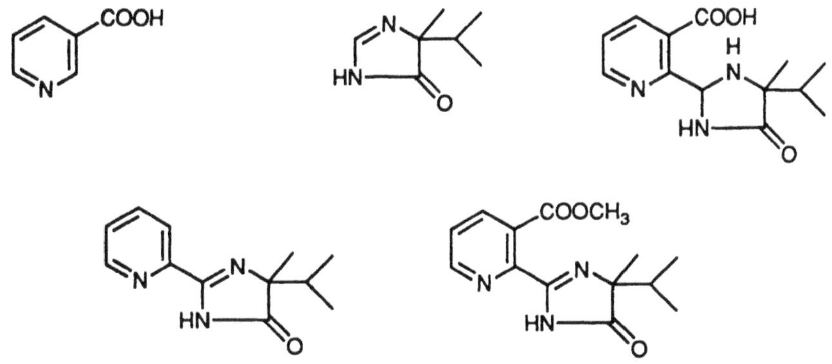

FIGURE 6. Structural requirements for AHAS inhibition.

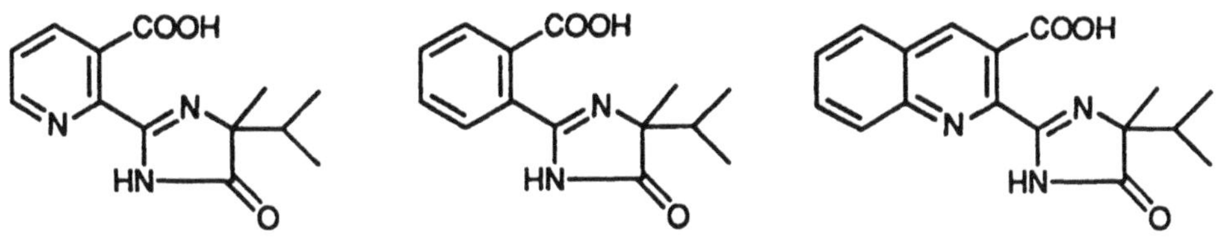

$I_{50} = 4$ µM $I_{50} = 0.1$ µM $I_{50} = 1$ µM

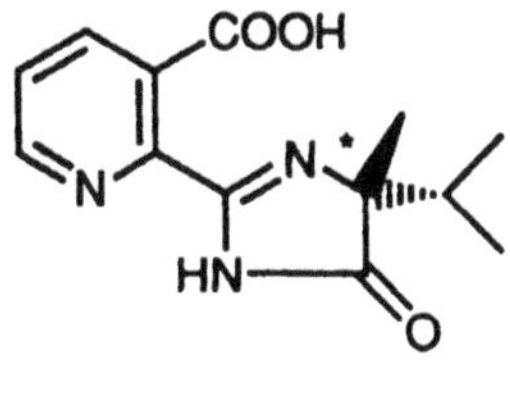

R, S : $I_{50} = 4$ µM

R : $I_{50} = 2.4$ µM

S : $I_{50} = 18$ µM

FIGURE 7. Relative inhibition of AHAS by imidazolinones.

TABLE 1
Effect of Imazapyr on Extractable AHAS in Excised Maize Shoots

| | Extractable AHAS activity (4 h after treatment) | |
Treatment	μg Acetoin/mg protein/h	% of control
Control	3.73	—
Control + Imazapyr spike[a]	3.80	101
Imazapyr fed[b]	0.37	10

[a]   100 nM imazapyr/ml in extraction buffer
[b]   57 nM imazapyr/ml internal concentration of maize leaves

From Stidham, M. A. and Shaner, D. L., *Pestic. Sci.*, 29, 335, 1990. With permission.

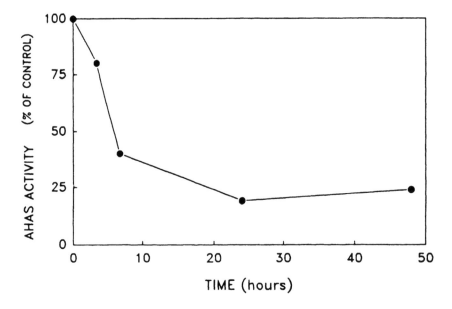

FIGURE 8.   *In vivo* inhibition of AHAS by imazaquin. (From Shaner, D., Singh, B., and Stidham, M., *J. Agric. Food Chem.*, 38, 1279, 1990. With permission.)

decrease is unexpected because preparation of AHAS from the plant tissue involves dilution steps that should result in a dissociation of reversibly bound imazapyr. Decreases in extractable AHAS are not observed when untreated tissue is extracted in a buffer containing high concentrations of imazapyr. Thus, it appears that the imidazolinone effect on extractable AHAS is not due to the mixing of AHAS with imazapyr during extraction. Radiolabeled imazapyr was used in an excised shoot experiment to quantify the amount of imazapyr carried over in the AHAS extraction procedure. The imazapyr associated with the protein was 0.05 nm/mg protein; this level translated to an imazapyr concentration of 50 nM in the assay. Since this concentration of imazapyr is roughly 100 times less than the concentration required for 50% inhibition of the enzyme, the imazapyr carried over could not account for the *in vitro* inhibition of the extracted AHAS.

The reduction in extractable AHAS caused by imazapyr occurs when either excised leaves or intact plants are used; this effect is observed with other imidazolinones as well. Figure 8 shows the amount of AHAS activity extracted from corn treated with imazaquin in a soil drench. Eight hours after application, the extractable AHAS in plants treated with

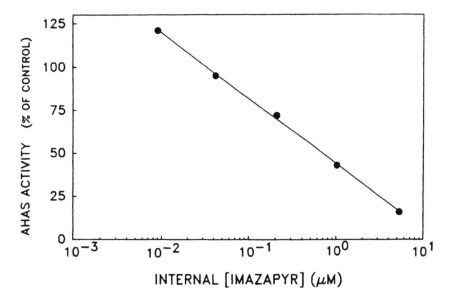

FIGURE 9. Effect of imazapyr on extractable AHAS from excised corn. (From Shaner, D., Singh, B., and Stidham, M., *J. Agric. Food Chem.*, 38, 1279, 1990. With permission.)

imazaquin is reduced by 60% relative to that in the untreated plants, and one day after treatment, the quantity is 20% of that in the untreated plant. The effect is specific for AHAS, since the levels of other enzymes are unaffected after imidazolinone treatment.[4]

The dependence of AHAS inactivation on imidazolinone concentration was determined using an excised leaf system, wherein corn leaves were placed in solutions containing different levels of imazapyr and then extracted for AHAS activity 4 h after complete uptake of the solution. The extractable AHAS activity decreased as the semilogarithm of the calculated internal imazapyr concentration (Figure 9). The I_{50} value calculated from the plot was one tenth the I_{50} value calculated from *in vitro* experiments on the enzyme. Since the enzyme was extracted 4 h after exposure to the herbicide, the drop in the I_{50} value could reflect the slow tight-binding inhibition observed *in vitro*. Alternatively, the herbicide could concentrate in the chloroplast via weak-acid trapping.[41] Since the calculation of the internal imazapyr concentration is averaged over the entire cell volume, accumulation in the chloroplast would give higher concentrations at the site of action.

A number of interpretations could explain the effect of the imidazolinones on the amount of extractable AHAS. One interpretation could be that upon prolonged exposure *in vivo* the imidazolinones form an irreversible complex with AHAS. Another interpretation is that AHAS-binding of imidazolinones results in the destabilization of AHAS, which may or may not be reflected in degradation of the enzyme *in vivo*. Antibodies raised against AHAS will be valuable in determining which of the above interpretations is correct.

IV. AHAS FROM VARIOUS SPECIES

AHAS activity in the intact plants ranged widely among different species (Table 2). The greatest AHAS activity was found in pea. However, compared to AHAS activity from intact plants, the BMS (Black Mexican Sweet) corn cells grown in suspension culture had even higher AHAS activity. On a fresh weight basis, BMS cells had 2.8-fold higher AHAS activity than the next highest source (pea). Similarly, on a protein basis, BMS cells had over 10-fold higher activity than any intact plant species used in this survey. It is important to note that actively dividing BMS cells were used in this study. Therefore, the higher

TABLE 2
Acetohydroxyacid Synthase from Different Species

Plant	Protein (mg)	Total AHAS (μmol/h)	Specific activity of AHAS	
			(μmol/mg protein/h)	(μmol/g fresh wt/h)
Matricaria	50	2.0	0.04	0.2
Flax	59	3.3	0.06	0.3
Sunflower	81	5.7	0.07	0.6
Arabidopsis	56	6.5	0.12	0.7
Ambrosia	90	6.7	0.07	0.7
Barley	97	8.9	0.09	0.9
Wheat	120	9.4	0.08	0.9
Rape	59	12.4	0.21	1.2
Amaranthus	68	12.8	0.19	1.3
Spinach	53	16.6	0.31	1.7
Tobacco	51	17.4	0.34	1.7
Sorghum	66	18.1	0.28	1.8
Soybean	97	21.0	0.22	2.1
Mustard	51	21.4	0.42	2.1
Corn	76	28.4	0.37	2.8
Lima bean	110	28.9	0.26	2.9
Pea	117	58.6	0.50	5.9
BMS cells	61	164.1	5.36	16.4

From Singh, B. K., Stidham, M. A., and Shaner, D. L., *J. Chromatogr.*, 444, 251, 1988. With permission.

specific activities of exponentially growing cell cultures are consistent with studies on lima bean in which the younger leaves (undergoing rapid cell division) have higher specific activities of AHAS than do the older leaves. This aspect is discussed later in this chapter.

AHAS activity from all species showed feedback inhibition by valine, leucine, and isoleucine (Table 3) as was reported earlier in other studies.[8-10] A combination of valine and leucine was most inhibitory. There were significant differences in the kinetics of inhibition between species (Figure 10); however, maximum inhibition was similar to the values presented in Table 3.

AHAS from all species was highly sensitive to inhibition by imazapyr and imazethapyr, but in general, imazethapyr was a stronger inhibitor than imazapyr (Table 4). There were large differences in the I_{50} for herbicides among species. These differences may be attributed to the inherent differences in the AHAS protein itself and/or differences in the stability of the enzyme. Interestingly, AHAS activity from pea, lima bean, and soybean appeared to be the most tolerant to the herbicides. These same species are also tolerant of both herbicides at the whole-plant level because of their ability to metabolize these compounds.

Besides their morphological and physiological characteristics, plant species used in this survey were chosen based on their sensitivity to imidazolinones at the whole-plant level. For example, *Amaranthus* is one of the most sensitive species, whereas *Ambrosia* is one of the less sensitive ones. Previous work has demonstrated that detoxification of an imidazolinone is the basis for tolerance of some of these species (e.g., soybean). However, tolerance to an imidazolinone herbicide can also result from a higher level of AHAS activity or from the presence of a form of AHAS that is insensitive to inhibition by imidazolinone. Results from this survey showed large differences between species in the specific activity of AHAS (Table 2) and in the sensitivities of these activities to herbicides (Table 4). However, no correlation was found between glasshouse control rates (the minimum amount of herbicide required for complete or nearly complete kill of plants) of imazapyr and imazethapyr and the I_{50} of AHAS activities for the respective herbicides (Figure 11; $r^2 = 0.03$ for imazapyr and 0.0001 for imazethapyr). Similarly, no correlation was found between these control

TABLE 3
Effects of Valine, Leucine, and Isoleucine on AHAS Activity

Plant	Inhibition caused by various amino acids at 1 mM (%)						
	L[a]	V[b]	I[c]	L+I	V+I	L+V	L+V+I
Flax	26	15	32	29	37	42	37
Amaranthus	39	35	10	45	25	40	45
Soybean	33	44	33	51	36	50	46
Barley	55	26	20	45	31	49	48
Pea	41	9	21	46	27	41	49
Lima bean	24	17	6	33	30	55	53
BMS cells	26	5	12	44	19	57	59
Wheat	60	49	33	62	45	61	59
Rape	54	30	28	61	34	62	62
Mustard	46	29	19	61	26	65	62
Sunflower	52	40	34	60	44	63	63
Arabidopsis	52	47	33	64	50	67	67
Spinach	48	43	32	64	47	71	71
Tobacco	50	43	31	65	45	70	71
Corn	49	34	16	57	43	76	76
Sorghum	69	61	41	69	66	72	77
Ambrosia	66	54	40	70	58	74	77
Matricaria	72	54	50	73	54	75	79

[a] L = Leucine.
[b] V = Valine.
[c] I = Isoleucine.

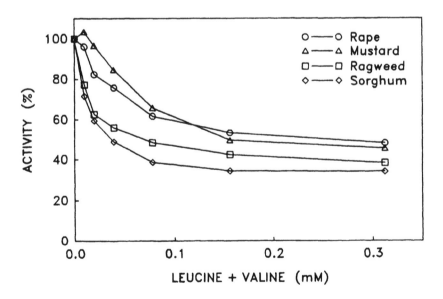

FIGURE 10.   Inhibition of AHAS with leucine + valine.

rates and the specific activities of AHAS in different species (Figure 12; $r^2 = 0.08$ for imazapyr and 0.15 for imazethapyr). The absence of any relationship between the control rates and other characteristics of AHAS in these species (Figures 11 and 12) implies that AHAS does not significantly contribute to the natural tolerance of different species to imidazolinones at the whole-plant level.

TABLE 4
Inhibition of AHAS from Various Species by Different Herbicides

Plant	Imazapyr (μM)	I_{50} Imazethapyr (μM)	Sulfometuron methyl (nM)
Matricaria	9.5	7.9	74.1
Flax	2.8	1.7	50.1
Sunflower	4.7	2.9	31.3
Arabidopsis	5.1	1.8	7.3
Ambrosia	14.1	5.4	21.9
Barley	7.6	4.0	38.0
Wheat	17.8	5.7	32.4
Rape	2.9	1.3	20.5
Amaranthus	3.5	2.0	21.0
Spinach	1.4	0.4	4.0
Tobacco	3.0	0.8	7.9
Sorghum	5.9	2.1	27.6
Soybean	19.4	51.3	95.5
Mustard	2.6	1.5	17.8
Corn	4.0	1.2	23.4
Lima bean	7.5	18.4	160.0
Pea	31.7	83.2	53.7

From Singh, B. K., Newhouse, K. E., Stidham, M. A., and Shaner, D. L., in *Biosynthesis of Branched-Chain Amino Acids*, Barak, A., Schloss, J. V., and Chipman, D. M., Eds., VCH Publishers, New York, 1990, 357. With permission.

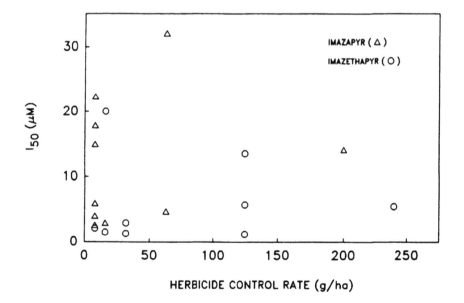

FIGURE 11.   Correlation between glasshouse control rates and I_{50} for AHAS from different species. (From Singh, B. K., Newhouse, K. E., Stidham, M. A., and Shaner, D. L., in *Biosynthesis of Branched-Chain Amino Acids*, Barak, A., Schloss, J. V., and Chipman, D. M., Eds., VCH Publishers, New York, 1990, 357. With permission.)

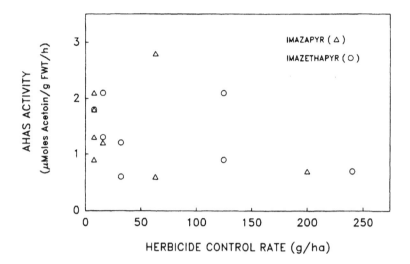

FIGURE 12. Correlation between glasshouse control rates and total AHAS activity from different species. (From Singh, B. K., Newhouse, K. E., Stidham, M. A., and Shaner, D. L., in *Biosynthesis of Branched-Chain Amino Acids*, Barak, A., Schloss, J. V., and Chipman, D. M., Eds., VCH Publishers, New York, 1990, 357. With permission.)

V. DEVELOPMENTAL REGULATION OF AHAS ACTIVITY

There are two principal mechanisms of metabolic regulation in microbes: variation of enzyme concentration and alteration of enzyme activity.[42,43] In plants, the contribution of these mechanisms to the coordination of metabolism is not yet clear. The only known mechanism of regulation of AHAS in plants is feedback inhibition of the enzyme by valine, leucine, and isoleucine, which are the end products of the pathway. In order to understand growth-dependent regulation of AHAS in plants, AHAS activity was examined in lima beans at different stages of growth.

The first pair of leaves from lima bean plants at different ages were used for this study. On the first sampling day, 5 d after planting, these unifoliate leaves had emerged from the cotyledons. The fresh weight of leaves increased with advancing leaf age and was maximal on the last sampling day, 12 d after planting, when this leaf was fully expanded. The leaf growth measured in this way showed a typical sigmoidal pattern. Soluble protein content per leaf showed a similar increase in response to leaf growth (Figure 13).

AHAS activity per leaf increased about threefold from days 5 to 7 after planting, then remained relatively stable throughout the sampling period. However, the specific activity of AHAS declined sharply from days 5 to 10 after planting, then remained steady (Figure 13). The lack of further increase in AHAS activity per leaf after several days of growth suggests that no apparent net increase in AHAS activity occurs after about 2 d of leaf growth. Since soluble protein content per leaf continues to increase with advancing leaf age (Figure 13), lack of further increase in AHAS activity after day 7 is not due to a cessation of protein synthesis. The first few days of leaf growth involve rapid cell division, whereas most of the subsequent growth can be attributed to cell expansion. Therefore, it appears that net synthesis of AHAS occurs only during cell division. Consistent with this theory were the results of experiments using a tissue culture system as discussed earlier. The results of these experiments, however, do not rule out the possibility that increases in the enzyme activity observed here were due to enzyme activation. Quantification of AHAS protein using an antibody is required to resolve this issue.

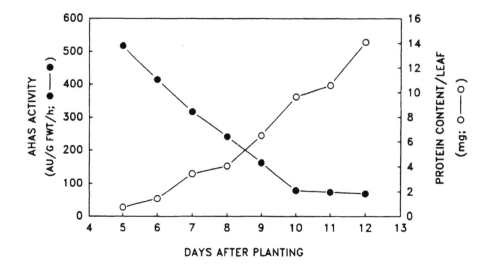

FIGURE 13.   Total soluble protein and AHAS activity in lima bean leaves.

In a previous study, the properties of AHAS isolated from shoots of peas appeared to differ from the properties of the enzyme isolated from seeds: AHAS activity from seeds was less sensitive to inhibition by the pathway end products, either alone or in combination.[10] This observation may reflect a type of growth-dependent regulation similar to that observed with homoserine dehydrogenase: its sensitivity to inhibition by threonine may vary significantly during plant growth.[44] The present study did not reveal differences in the sensitivity of AHAS to inhibition by the leucine-valine combination at any stage of growth. This observation could suggest that there is no growth-dependent difference in the regulatory properties of AHAS in lima bean plants. However, it is possible that the observation in the previous case is tissue specific[10] and not a growth-dependent response. Further studies are needed to evaluate the possibility that the differences in the sensitivity of AHAS to feedback inhibition by amino acids appear at a later stage of leaf growth.

VI. DIFFERENT FORMS OF AHAS

Microorganisms may possess as many as six isozymes of AHAS with different sensitivities to feedback inhibition by amino acids.[43] Similarly, bacterial AHAS isozymes differ in their sensitivities to sulfonylureas.[7] Even though genetic evidence for the presence of AHAS isozymes in tobacco was present,[45] these isozymes could not be resolved by chromatography. Using BMS corn cells from suspension cultures as the enzyme source, we were successful in separating different forms of AHAS with different physical and kinetic characteristics.

Two peaks of AHAS activity were seen after anion exchange chromatography on a Mono Q* column (Figure 14). The major peak, designated AHAS I, was eluted during gradient elution and contained about 90% of the total recovered enzyme activity. A minor peak of AHAS activity, designated AHAS II, which contained nearly 10% of the total AHAS activity, was recovered in the unbound fraction. The proportions and recoveries of enzyme activity of the two peaks of AHAS activity were similar to those observed during chromatofocusing.[33] Rechromatography of each of these two peaks separately on the Mono Q column yielded single peaks of AHAS activity that eluted at their original retention times.

*   Trademark of Pharmacia, Piscataway, NJ.

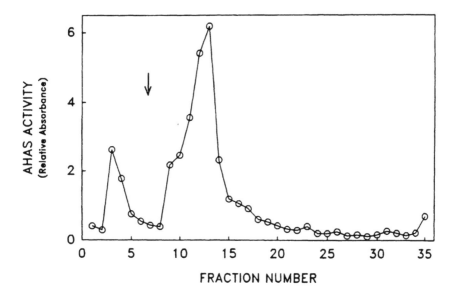

FIGURE 14. Anion exchange chromatography of AHAS from BMS cells on a Mono Q column. The arrow indicates the beginning of salt gradient. (From Singh, B. K., Stidham, M. A., and Shaner, D. L., *J. Chromatogr.*, 444, 251, 1988. With permission.)

TABLE 5
Physical and Kinetic Properties of AHAS I and AHAS II

Property	AHAS I	AHAS II
Molecular weight	193,000	55,000
pH optimum	6—7	7
K_m for pyruvate (mM)	5	8
I$_{50}$		
Leucine + valine (mM)	0.1	a
Imazapyr (μM)	2.0	1.5
Sulfometuron methyl (nM)	10.0	10.0

a   Less than 10% inhibition.

From Singh, B. K., Stidham, M. A., and Shaner, D. L., *J. Chromatogr.*, 444, 251, 1988. With permission.

The properties of the two peaks of AHAS activity were significantly different. AHAS I was sensitive to inhibition by both leucine and valine as well as to inhibition by imazapyr. In contrast, AHAS II was insensitive to inhibition by leucine and valine but was more sensitive to inhibition by imazapyr than was AHAS I. Further characterization revealed differences in the molecular weights and in other kinetic properties of the two activities.

The native molecular weights of AHAS I and AHAS II were estimated to be 193,000 and 55,000, respectively, by chromatography on an HPLC gel filtration column (Table 5). AHAS I had a broad pH optimum between pH 6 and 7. The pH response was not affected by the type of buffer system used. In contrast, AHAS II had a very distinct pH optimum at pH 7 in phosphate buffer (Table 5) but showed very little activity in MES (2-[*N*-morpholino]ethanesulfonic acid) or Tris (*tris*[hydroxymethyl]aminomethane). The pyruvate saturation curves of AHAS I and AHAS II are hyperbolic, and both forms of AHAS were

TABLE 6
Effects of Valine, Leucine, and Isoleucine on the Activity of
AHAS I and AHAS II

Form of	Inhibition caused by various amino acids at 1 mM (%)						
AHAS	L[a]	V[b]	I[c]	L + I	V + I	L + V	L + V + I
AHAS I	34	34	16	59	50	66	63
AHAS II	4	0	0	0	10	8	0

[a] L = Leucine.
[b] V = Valine.
[c] I = Isoleucine.

From Singh, B. K., Stidham, M. A., and Shaner, D. L., *J. Chromatogr.*, 444, 251, 1988. With permission.

saturated at about 100 mM pyruvate.[33] The Hanes-Woolf plot of the data gave a K_m value of 5 mM pyruvate for AHAS I and 8 mM for AHAS II (Table 5).

Feedback inhibitors — valine, leucine, and isoleucine — were used to test the sensitivity of the two forms of AHAS to these inhibitors. The degree of inhibition of the enzyme activity was compared at 1 mM for each amino acid. AHAS I was inhibited by all of the amino acids, singly and in combination (Table 6), and to the greatest degree (66%) by a combination of leucine and valine. In contrast, AHAS II is relatively unaffected by these amino acids (Tables 5 and 6).

The kinetics of inhibition of AHAS I and AHAS II by imazapyr differed significantly.[33] AHAS II was much more sensitive to the herbicide than was AHAS I and was almost completely inhibited (>90%) by imazapyr. In contrast, the maximum inhibition of AHAS I caused by imazapyr (100 μM) was approximately 70%. Despite these differences in the degree of inhibition, imazapyr concentrations required for 50% inhibition of the enzyme (I_{50}) were the same for both isozymes (Table 5).

The origin and relationship of the two enzyme forms are unclear at this time. Recent results suggest that the two types of AHAS observed represent different aggregation states of the same enzyme with AHAS I being the multimeric form and AHAS II the monomeric form. Additional experiments are in progress to determine the origin and relationship of different forms of AHAS in plants.

VII. PROPERTIES OF AHAS FROM IMIDAZOLINONE-TOLERANT CORN

The development of imidazolinone-tolerant corn is extensively reviewed in Chapter 10. AHAS activity from tolerant corn lines is detailed in this section.

Tissue culture selection methods were used to obtain imidazolinone tolerance in corn,[5] and a number of mutant lines were selected. Three lines, XA17, XI12, and QJ22, were regenerated into plants, and conventional breeding techniques were used to obtain the homozygous tolerant plants that were used in the study. An inbred line, B73, was used as the sensitive control. Two imidazolinones — imazaquin and imazethapyr — and one sulfonylurea — sulfometuron methyl — were used to examine the sensitivity at the whole-plant level as well as at the enzyme level.

The inbred corn line B73 was killed at low rates of each of the herbicides tested (Figure 15; only data for imazaquin presented). Similar to this whole-plant response, all of these herbicides inhibited AHAS activity from B73 (Figure 16; only data for imazaquin presented). Sulfometuron methyl was the most active herbicide as well as the most potent enzyme inhibitor.

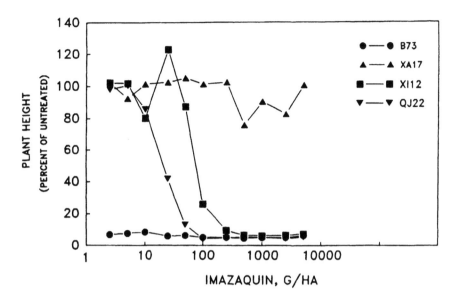

FIGURE 15. Effects of imazaquin on plant growth. (From Singh, B. K., Newhouse, K. E., Stidham, M. A., and Shaner, D. L., in *Prospects for Amino Acid Biosynthesis Inhibitors in Crop Protection and Pharmaceutical Chemistry*, Monogr. No. 42, Copping, L. G., Dalziel, J., and Dodge, A. D., Eds., British Crop Protection Council, Farnham, England, 1989, 87. With permission.)

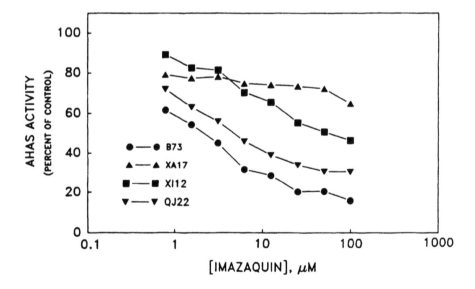

FIGURE 16. Effects of imazaquin on the AHAS activity from wild-type corn (B73) and various herbicide-tolerant mutants. (From Singh, B. K., Newhouse, K. E., Stidham, M. A., and Shaner, D. L., in *Prospects for Amino Acid Biosynthesis Inhibitors in Crop Protection and Pharmaceutical Chemistry*, Monogr. No. 42, Copping, L. G., Dalziel, J., and Dodge, A. D., Eds., British Crop Protection Council, Farnham, England, 1989, 87. With permission.)

Homozygous XA17 was the most tolerant line and could not be killed even at the highest rate of herbicides used (Figure 15; only data for imazaquin presented). AHAS extracted from XA17 plants was insensitive to inhibition by these herbicides (Figure 16; only data for imazaquin presented). In contrast to XA17, the XI12 and QJ22 genotypes were selectively tolerant to the imidazolinone herbicides. Compared to XA17 and XI12, QJ22 has the lowest

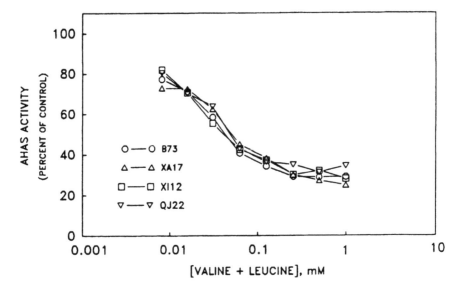

FIGURE 17.   Effects of valine + leucine on the AHAS activity from the wild-type corn (B73) and various herbicide-tolerant mutants. (From Singh, B. K., Newhouse, K. E., Stidham, M. A., and Shaner, D. L., in *Prospects for Amino Acid Biosynthesis Inhibitors in Crop Protection and Pharmaceutical Chemistry,* Monogr. No. 42, Copping, L. G., Dalziel, J., and Dodge, A. D., Eds., British Crop Protection Council, Farnham, England, 1989, 87. With permission.)

degree of tolerance to imidazolinones (see Figures 15 and 16; only data for imazaquin presented). The herbicide tolerance of XI12 and QJ22 shown at the whole-plant level was also displayed by AHAS from these lines.

The herbicide tolerance in the three corn mutants described above differs in both the spectrum of herbicide tolerance and the level of tolerance expressed at the whole-plant level. Herbicide tolerance at the whole-plant level is explained by the presence of a form of AHAS that shows a similar degree of tolerance by the respective herbicides. Insensitivity of AHAS from XA17 to the imidazolinones and sulfonylureas and the sensitivity of AHAS from XI12 and QJ22 to sulfonylureas suggests that the binding sites of the imidazolinones and sulfonylureas are not the same, but that they may share a common binding domain. However, the possibility that AHAS extracted from XA17 contains two separate mutations, one in each herbicide-binding domain, cannot be ruled out. Characterization of these mutants at the gene level will explain the differential response of these genotypes to various herbicides.

An interesting property of AHAS from each of these genotypes is the identical response to inhibition by the combination of valine and leucine, feedback inhibitors of the enzyme (Figure 17). It has been speculated that the imidazolinones are analogs of the feedback inhibitors and therefore they bind at the feedback inhibitor bind site.[46,47] Experimental results presented here clearly demonstrate that the feedback inhibitor-binding site is separate from the herbicide-binding site.

VIII. CONCLUSIONS

A complete understanding of the mechanism of action of the imidazolinone herbicides begins at the level of AHAS. Sensitivity of a plant to the imidazolinones is a result of the sensitivity of AHAS from that plant to the imidazolinones. Although natural tolerance to these herbicides cannot be attributed to AHAS, it is clear that plants obtained through tissue culture selection methods are tolerant to imidazolinones because of the presence of an imidazolinone-insensitive form of AHAS.

Many questions remain concerning AHAS, the biosynthesis of leucine, isoleucine, and valine, and the inhibition of AHAS by the imidazolinones. These questions include the molecular interactions between the imidazolinones and AHAS, the rate of turnover of AHAS, tissue localization of AHAS, and AHAS isozymes. An increased understanding of the role of AHAS in amino acid biosynthesis will help answer the important remaining question of why plants die upon inhibition of AHAS.

REFERENCES

1. **Anderson, P. C. and Hibberd, K. A.,** Evidence for the interaction of an imidazolinone herbicide with leucine, valine, and isoleucine metabolism, *Weed Sci.,* 33, 479, 1985.
2. **Shaner, D. L. and Reider, M. L.,** Physiological responses of corn (*Zea mays*) to AC 243,997 in combination with valine, leucine, and isoleucine, *Pestic. Biochem. Physiol.,* 25, 248, 1986.
3. **Shaner, D. L., Anderson, P. C., and Stidham, M. A.,** Imidazolinones: potent inhibitors of acetohydroxyacid synthase, *Plant Physiol.,* 76, 545, 1984.
4. **Muhitch, M. J., Shaner, D. L., and Stidham, M. A.,** Imidazolinones and acetohydroxyacid synthase from higher plants, *Plant Physiol.,* 83, 451, 1987.
5. **Shaner, D. L. and Anderson, P. C.,** Mechanism of action of the imidazolinones and cell culture selection of tolerant maize, in *Biotechnology in Plant Science — Relevance to Agriculture in the Eighties,* Zaitlin, M., Day, P., and Hollaender, A., Eds., Academic Press, New York, 1985, 287.
6. **Singh, B. K., Newhouse, K. E., Stidham, M. A., and Shaner, D. L.,** Acetohydroxyacid synthase-imidazolinone interaction, in *Prospects for Amino Acid Biosynthesis Inhibitors in Crop Protection and Pharmaceutical Chemistry,* Monogr. No. 42, Copping, L. G., Dalziel, J., and Dodge, A. D., Eds., British Crop Protection Council, Farnham, England, 1989, 87.
7. **Schloss, J. V., Ciskanik, L. M., and Van Dyke, D. E.,** Origin of the herbicide binding site of acetolactate synthase, *Nature,* 331, 360, 1988.
8. **Miflin, B. J.,** Cooperative feedback control of barley acetohydroxyacid synthetase by leucine, isoleucine, and valine, *Arch. Biochem. Biophys.,* 146, 542, 1971.
9. **Miflin, B. J.,** The location of nitrite reductase and other enzymes related to amino acid biosynthesis in the plastids of root and leaves, *Plant Physiol.,* 54, 550, 1974.
10. **Miflin, B. J. and Cave, P. R.,** The control of leucine, isoleucine, and valine biosynthesis in a range of higher plants, *J. Exp. Bot.,* 23, 511, 1972.
11. **Bauerle, R. H., Freundlich, M., Stormer, F. C., and Umbarger, H. E.,** Control of isoleucine, valine and leucine biosynthesis. II. Endproduct inhibition by valine of acetohydroxyacid synthase in *Salmonella typhimurium, Biochim. Biophys. Acta,* 92, 142, 1964.
12. **Bussey, H. and Umbarger, H. E.,** Biosynthesis of branched-chain amino acids in yeast. Regulation of synthesis of the enzymes of isoleucine and valine biosynthesis, *J. Bacteriol.,* 98, 623, 1969.
13. **Eoyang, L. and Silverman, P. M.,** Purification and subunit composition of acetohydroxy acid synthase I from *Escherichia coli* K-12, *J. Bacteriol.,* 157, 184, 1984.
14. **Schloss, J. V., Van Dyke, D. E., Vasta, J. F., and Kutny, R. M.,** Purification and properties of *Salmonella typhimurium* acetolactate synthase isozyme II from *Escherichia coli* HB101/pDU9, *Biochemistry,* 24, 4952, 1985.
15. **Sutton, A., Newman, T., Francis, M., and Freundlich, M.,** Valine-resistant *Escherichia coli* K-12 strains with mutations in the *ilv*B operon, *J. Bacteriol.,* 148, 998, 1981.
16. **Glatzer, L., Eakin, E., and Wagner, R. P.,** Acetohydroxyacid synthetase with a pH optimum of 7.5 from *Neurospora crassa* mitochondria: characterization and partial purification, *J. Bacteriol.,* 112, 453, 1972.
17. **Kuwana, H. and Date, M.,** Solubilization of valine-sensitive acetohydroxyacid synthetase from *Neurospora* mitochondria, *J. Biochem.,* 77, 257, 1975.
18. **Magee, P. T. and Hereford, L. M.,** Multivalent repression of isoleucine-valine biosynthesis in *Saccharomyces cerevisiae, J. Bacteriol.,* 98, 857, 1969.
19. **McDonald, R. A., Satyanarayana, T., and Kaplan, J. G.,** Biosynthesis of branched-chain amino acids in *Schizosaccharomyces pombe:* properties of acetohydroxyacid synthetase, *J. Bacteriol.,* 114, 332, 1973.
20. **McDonald, R. A., Satyanarayana, T., and Kaplan, J. G.,** Biosynthesis of branched-chain amino acids in *Schizosaccharomyces pombe:* regulation of the enzymes involved in isoleucine, valine, and leucine synthesis, *Can. J. Biochem.,* 52, 51, 1974.

21. **Xing, R. Y. and Whitman, W. B.,** Sulfometuron methyl-sensitive and -resistant acetolactate synthases of the archaebacterium *Methanococcus* spp., *J. Bacteriol.,* 169, 4486, 1987.

22. **Borstslap, A. C.,** Interactions between the branched-chain amino acids in the growth of *Spirodela polyrhiza, Planta,* 151, 314, 1981.

23. **Oda, Y., Nakano, Y., and Kitaoka, S.,** Properties and regulation of valine-sensitive acetolactate synthase from mitochondria of *Euglena gracilis, J. Gen. Microbiol.,* 128, 1211, 1982.

24. **Chaleff, R. S. and Mauvais, C. J.,** Acetolactate synthase is the site of action of two sulfonylurea herbicides in higher plants, *Science,* 224, 1443, 1984.

25. **Davies, M. E.,** Acetolactate and acetoin synthesis in ripening peas, *Plant Physiol.,* 39, 53, 1964.

26. **Durner, J. and Boger, P.,** Acetolactate synthase from barley (*Hordeum vulgare* L.): purification and partial characterization, *Z. Naturforsch.,* 43c, 850, 1988.

27. **Durner, J. and Boger, P.,** Inhibition of purified acetolactate synthase from barley *Hordeum vulgare* L. by chlorsulfuron and imazaquin, in *Prospects for Amino Acid Biosynthesis Inhibitors in Crop Protection and Pharmaceutical Chemistry,* Monog. No. 42, Copping, L. G., Dalziel, J., and Dodge, A. D., Eds., British Crop Protection Council, Farnham, England, 1989, 85.

28. **Jones, A. V., Young, R. M., and Leto, K. J.,** Subcellular localization and properties of acetolactate synthase, target site of the sulfonylurea herbicides, *Plant Physiol.,* 77, S-293, 1985.

29. **Ray, T. B.,** Site of action of chlorsulfuron, *Plant Physiol.,* 75, 827, 1984.

30. **Relton, J. M., Wallsgrove, R. M., Bourgin, J-P., and Bright, S. W. J.,** Altered feedback sensitivity of acetohydroxyacid synthase from valine-resistant mutants of tobacco (*Nicotiana tabacum* L.), *Planta,* 169, 46, 1986.

31. **Singh, B. K., Newhouse, K. E., Stidham, M. A., and Shaner, D. L.,** Acetohydroxyacid synthase-imidazolinone interaction, in *The Biosynthesis of Branched-Chain Amino Acids,* Barak, A., Schloss, J. V., and Chipman, D. M., Eds., VCH Publishers, New York, 1990, 357.

32. **Singh, B. K. and Schmitt, G. K.,** Flavin adenine dinucleotide causes oligomerization of acetohydroxyacid synthase from Black Mexican Sweet corn cells, *FEBS Lett.,* 258, 113, 1989.

33. **Singh, B. K., Stidham, M. A., and Shaner, D. L.,** Separation and characterization of two forms of acetohydroxyacid synthase from Black Mexican Sweet corn cells, *J. Chromatogr.,* 444, 251, 1988.

34. **La Rossa, R. A. and Schloss, J. V.,** The herbicide sulfometuron methyl is bacteriostatic due to inhibition of acetolactate synthase, *J. Biol. Chem.,* 259, 8753, 1984.

35. **Ciskanik, L. M. and Schloss, J. V.,** Reaction intermediates of the acetolactate synthase reaction: effect of sulfometuron methyl, *Biochemistry,* 24, 3357, 1985.

36. **Schloss, J. V.,** Interaction of the herbicide sulfometuron methyl with acetolactate synthase: a slow binding inhibitor, in *Flavins and Flavoproteins,* Bray, R. C., Engel, P. C., and Mayhew, S. C., Eds., Walter de Gruyter & Co., Berlin, 1984, 737.

37. **Chang, Y-Y. and Cronan, J.E., Jr.,** Common ancestory of *Escherichia coli* pyruvate oxidase and the acetohydroxy acid synthases of the branched-chain amino acid biosynthetic pathway, *J. Bacteriol.,* 170, 3937, 1988.

38. **Grabau, C. and Cronan, J. E., Jr.,** Nucleotide sequence and deduced amino acid sequence of *Escherichia coli* pyruvate oxidase, a lipid activated flavoprotein, *Nucleic Acids Res.,* 14, 5449, 1986.

39. **Mazur, B. J., Chui, C-F., and Smith, J. K.,** Isolation and characterization of plant genes coding for acetolactate synthase, the target enzyme for two classes of herbicides, *Plant Physiol.,* 85, 1110, 1987.

40. **Bedbrook, J., Chaleff, R. S., Falco, S. C., Mazur, B. J., and Yadav, N.,** Nucleic Acid Fragment Encoding Herbicide-resistant Plant Acetolactate Synthase, Eur. Patent Appl. 0257993, 1988.

41. **Van Ellis, M. R. and Shaner, D. L.,** Mechanism of cellular absorption of imidazolinones in soybean (*Glycine max*) leaf discs, *Pestic. Sci.,* 23, 25, 1988.

42. **Umbarger, H. E.,** Regulation of amino acid metabolism, *Annu. Rev. Biochem.,* 38, 323, 1969.

43. **Umbarger, H. E.,** Biosynthesis of branched-chain amino acids, in *Escherichia coli and Salmonella typhimurium: Cellular and Molecular Biology,* Vol. 1, Neidhardt, F. C., Ingraham, J. L., Low, K. B., Magasanik, B., Schaechter, M., and Umbarger, H. E., Eds., American Society for Microbiology, Washington, D.C., 1987, 368.

44. **Matthews, B. F., Gurman, A. W., and Bryan, J. K.,** Changes in enzyme regulation during growth of maize. I. Progressive desensitization of homoserine dehydrogenase during seedling growth, *Plant Physiol.,* 55, 991, 1975.

45. **Chaleff, R. S. and Bascomb, N. F.,** Genetic and biochemical evidence for multiple forms of acetolactate synthase in *Nicotiana tabacum, Mol. Gen. Genet.,* 210, 33, 1987.

46. **Brown, M. A., Chiu, T. Y., and Miller, P.,** Hydrolytic activation versus oxidative degradation of ASSERT herbicide, an imidazolinone aryl-carboxylate, in susceptible wild oat versus tolerant corn and wheat, *Pestic. Biochem. Physiol.,* 27, 24, 1987.

47. **Rathinasabapathi, B., Williams, D., and King, J.,** Altered feedback sensitivity to valine, leucine, and isoleucine of acetolactate synthase from herbicide-resistant variants of *Datura innoxia, Plant Sci.,* 67, 1, 1990.

Chapter 7

MECHANISMS OF SELECTIVITY OF THE IMIDAZOLINONES

Dale L. Shaner and N. Moorthy Mallipudi

TABLE OF CONTENTS

I. INTRODUCTION

Plant species have at least four different mechanisms which affect herbicide selectivity. These result in differences in the absorption, translocation, rate of metabolism, and sensitivity of the site of action to the herbicide. The primary mechanisms of natural selectivity to the imidazolinones include the first three, with differential metabolism playing the predominant role. Because of a modification in the sensitivity of the active site to the herbicides, there are plant lines which are tolerant to the imidazolinones, but these have been selected for via cell culture, seed mutagenesis, or other methods. This facet of imidazolinone selectivity is covered in Chapter 10 and will not be considered any further in this chapter.

The selectivity of the registered imidazolinones is shown in Table 1. This chapter summarizes present knowledge about the mechanisms of selectivity for imazamethabenz-methyl, imazapyr, imazaquin, and imazethapyr.

II. IMAZAMETHABENZ-METHYL

Imazamethabenz-methyl is selective on cereals and sunflower. This herbicide is composed of two positional isomers and differs from the other imidazolinones in that it is applied as an ester rather than as a salt of the acid. These differences play a major role in the mechanism of selectivity of imazamethabenz-methyl.

The mechanism of action of the imidazolinones is the inhibition of AHAS (Chapter 6). Pillmoor and Caseley[1] clearly showed that imazamethabenz-methyl kills plants via this mechanism. However, Shaner et al.[2] and Pillmoor and Caseley[1] also found that the active herbicidal form of imazamethabenz-methyl is not the methyl ester that is applied but the acid form of the herbicide that accumulates in the plant due to action of plant esterases. Shaner et al.[2] also discovered that the translocatable form of imazamethabenz-methyl is the acid and not the parent compound. A similar phenomenon occurs with other *Avena fatua* herbicides such as diclofop methyl[3] and flamprop-isopropyl.[4]

The rate of formation of the active acid of imazamethabenz-methyl appears to determine the susceptibility of a particular plant species. Shaner et al.[2] found that wheat sprayed with the acid form of imazamethabenz-methyl was no more tolerant to the herbicide than susceptible species. The selectivity of imazamethabenz-methyl on cereals and sunflower is due to their ability to rapidly metabolize this herbicide to nontoxic forms and the lack of accumulation of the acid form.[5]

Additional information that supports this mechanism of selectivity of imazamethabenz-methyl is the differential activity of the two isomers on *Brassica* spp. The *para* isomer is very active on *Brassica* spp. while the *meta* isomer is almost inactive (Table 2). Metabolism studies show that the active acid *para* isomer accumulates to very high levels in *Brassica* spp., while very little of the acid forms in plants treated with the meta isomer.[2]

The route of metabolism of imazamethabenz-methyl in susceptible species differs significantly from that in tolerant species (Figure 1). In tolerant species there is rapid hydroxylation of the methyl substituent on the benzene ring followed by conjugation to glucose.[5] Hydroxylation of these methyl substituents occurs with both the methyl ester and the acid form of the herbicide. This hydroxylation greatly reduces the herbicidal activity of the compound (Table 2) and probably accounts for the major detoxification step.

In conclusion, the mechanism of selectivity of imazamethabenz-methyl is differential metabolism of the compound by different species. In susceptible species the parent compound is de-esterified to the active acid form which translocates to the meristematic tissue of the plant. In tolerant species very little of the herbicidally active acid is formed. Instead, these species rapidly hydroxylate the methyl substituent on the benzene ring which is then conjugated to a glucose moiety. Both of these steps result in loss of herbicidal activity.

TABLE 1
Selectivity of the Imidazolinone Herbicides

Herbicide	Selective crop
Imazapyr	Rubber trees
	Sugar cane
	Oil palm
	Conifers
Imazaquin	Soybean
Imazethapyr	Leguminous crops
Imazamethabenz-methyl	Wheat
	Barley
	Rye
	Sunflower

TABLE 2
Comparison of Herbicidal Efficacy of Imazamethabenz-methyl and Its Metabolites

Compound	Control rate (kg/ha)[a]			Safe rate (kg/ha)[b]	
	A. fatua	A. myosuroides	B. kaber	Wheat	Barley
Imazamethabenz-methyl	0.6	0.6	0.4	4.0	4.0
para isomer	2.0	2.0	0.3	4.0	4.0
meta isomer	0.4	0.4	4.0	4.0	4.0
Imazamethabenz acid	0.3	0.3	0.1	<0.1	<0.1
Hydroxyimazamethabenz-methyl	>8	>8	>8	>8	>8

[a] Control equals the lowest rate that kills 85% of the plant population.
[b] Safe rate equals the highest rate that gives less than 15% crop injury.

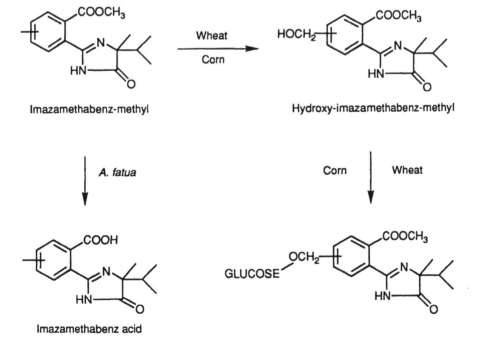

FIGURE 1. Proposed metabolism of imazamethabenz-methyl by wheat, *Avena fatua*, and corn.

TABLE 3
Uptake and Distribution of Imazapyr in Soybean and Corn[a]

Species	Uptake (μg per plant)	Distribution in plant		Control rate[b] (g/ha)
		Shoot (ppm)	Root (ppm)	
Soybean	1.90	0.69	7.80	40.00
Corn	0.51	1.14	0.83	8.00

[a] Plants were grown in sand culture and treated with ^{14}C-imazapyr as a root drench at a rate of 0.05 ppm of herbicide per pot.
[b] Control equals the lowest rate that kills 85% of the plant population.

TABLE 4
Metabolism of Radiolabeled Imazapyr in Excised Leaves of Corn and Soybeans

Metabolites[a]

(% of methanol-acetone extractable radioactivity)

Species	Imazapyr		AC 247,087	
	Day 2[b]	Day 4	Day 2	Day 4
Soybean	87.0	61.1	6.5	19.9
Corn	90.5	87.5	1.6	2.5

[a] Approximately 95% of the total radioactivity was extractable for leaf samples.
[b] Days after exposure to the herbicide.

III. IMAZAPYR

Imazapyr controls a broad spectrum of annual and perennial dicot and monocot weeds and is particularly effective on woody perennials. However, certain species in the conifer, composite, legume, and euphorb families have very high tolerances to imazapyr. The mechanism of selectivity of imazapyr in these species appears to be their ability to metabolize the herbicide to a relatively immobile metabolite.

Corn is approximately five times more sensitive to imazapyr than soybeans (Table 3). When radiolabeled imazapyr was applied to the roots of these two species, more of the radioactive compound translocated to the upper parts of corn than of soybeans, although the soybeans accumulated more radioactivity from the sand (Table 3). Metabolism studies in these two species showed there was very little breakdown of imazapyr in corn. However, soybeans metabolized approximately 20% of the absorbed imazapyr to one major metabolite, AC 247,087 (Table 4 and Figure 2).

In whole-plant studies, AC 247,087 is less phytotoxic than imazapyr (Table 5). This decrease in herbicidal activity appears to be due to the lack of phloem mobility of AC 247,087. When soybean leaves were treated with either imazapyr or AC 247,087, imazapyr rapidly translocated out of the leaves, but AC 247,087 remained within the treated leaf. However, when AC 247,087 was applied directly to the apical meristem of the plant, it was capable of killing the meristem, indicating that this compound does have some herbicidal activity.

The lack of mobility of AC 247,087 is not surprising in light of the mechanism of phloem transport of the imidazolinones (Chapter 5). Phloem trapping of the imidazolinones

FIGURE 2. Proposed metabolism of imazapyr by soybeans.

TABLE 5
Comparison of the Herbicidal Activity of Imazapyr and Its Major Metabolite

Compound	Control rate (g/ha)[a]	
	Corn	Soybean
Imazapyr	8	40
AC 247,087	63	1000

[a] Control equals the lowest rate that kills 85% of the plant population.

depends on a change in the lipophilicity of the compounds with changes in pH. This change requires the presence of a free carboxylic acid group with a pK_a of approximately 4. Since AC 247,087 does not have a free carboxylic acid group, it cannot be trapped in the phloem and, hence, is immobilized within the treated leaf.

The tolerance of other species, such as conifers[5a] and *Euphorbia* spp., also appears to be related to their ability to metabolize imazapyr to AC 247,087. Thus, the selectivity of imazapyr is due to the ability of tolerant species to rapidly metabolize the herbicide to an immobile form which prevents translocation of the herbicide to the growing points of the plant.

IV. IMAZAQUIN

Imazaquin is selective on soybeans and controls a broad spectrum of dicot weeds. The mechanism of selectivity of imazaquin appears to be metabolism of the herbicide to nontoxic metabolites. Work with excised leaves of soybean, *X. strumarium,* and *Abutilon theophrasti* showed a correlation between the half-life of imazaquin in these three species and their sensitivity to the herbicide (Table 6).[6] Wilcut et al.[7] found a similar correlation between the tolerance of peanut, soybean, *Cassia obtusifolia, Desmodium tortuosum,* and *X. strumarium* and their ability to metabolize imazaquin.

Soybean plants rapidly metabolize imazaquin to a wide range of compounds. The first metabolites appear to be AC 271,157 and AC 263,460 (Figure 3), which are subsequently metabolized compounds which are either degraded or incorporated into naturally occurring plant constituents. (A more detailed description of the metabolic pathway of imazaquin is given in Chapter 11.) However, the residues found in soybean seed are insignificant. Soybean crops treated with imazaquin at exaggerated rates ranging from 280 g a.e./ha to 560 g a.e./ha (two to four times the recommended use rate) showed the total radioactive residue levels in seeds to be between 0.01 and 0.05 ppm. Since the total residues are so low in seed, these metabolites will be in negligible amounts at normal application rates. The first

TABLE 6
Herbicidal Activity and Half-Life of Imazaquin in Various
Species

Species	Control rate (g/ha)[a]	Half-life (days)
Soybean	>4000	3.0
Abutilon theophrasti	125	14.5
Xanthium strumarium	40	30.0
Corn	8	2.0
Setaria faberi	125	1.5
Sorghum halepense	125	1.8
Brachiaria platyphylla	>125	1.5

[a] Control equals the lowest rate that kills 85% of the plant population.

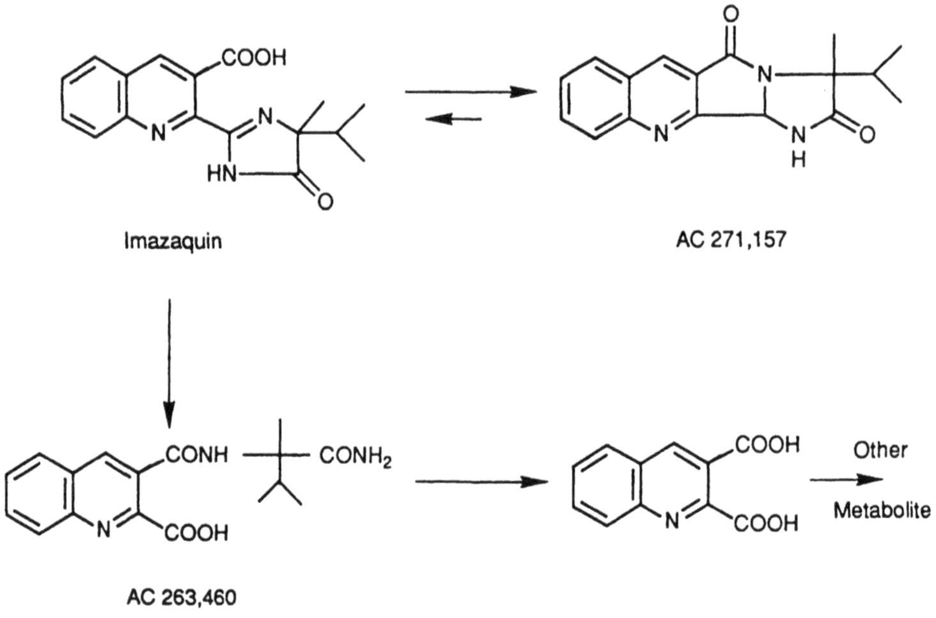

FIGURE 3.   Proposed metabolism of imazaquin by soybeans.

two metabolites of imazaquin are not as herbicidally active as the parent compound (Table 7). These same metabolites are also formed in *A. theophrasti* and *X. strumarium*, but to a much lesser extent. These two species do not metabolize imazaquin rapidly enough to prevent its herbicidal activity.

Another mechanism of selectivity may also be operating in *A. theophrasti*. Greenhouse and field studies have shown that the herbicidal activity of imazaquin on *A. theophrasti* rapidly diminishes once the plant reaches the four-leaf stage (Table 8). Studies with radiolabeled imazaquin showed that two factors change as *A. theophrasti* develops. First, as the plant ages, the amount of herbicide absorbed by the leaves decreases over two fold (Table 8). Second, the metabolic rate of imazaquin increases in older plants, so that the half-life decreases from 14.5 d in plants at the cotyledonary stage to 4.5 d at the four-leaf stage. The combination of decreased uptake and increased metabolism probably accounts for the increased tolerance in older *A. theophrasti* plants.

The tolerance of monocots to imazaquin also varies. Corn is very sensitive to the herbicide

TABLE 7
Comparison of The Herbicidal Activity of Imazaquin and Two Soybean Metabolites

	Control rate (g/ha)[a]			
Compound	Corn	Abutilon theophrasti	Amaranthus spp.	Xanthium strumarium
Imazaquin	8	80	5	40
AC 271,157	>1000	>1000	>1000	>1000
AC 263,460	>1000	>1000	>1000	>1000

[a] Control equals the lowest rate that kills 85% of the plant population.

TABLE 8
Effect of Growth Stage on Susceptibility of Abutilon theophrasti to Imazaquin[a]

Growth stage	Control rate (g/ha)[a]	Absorption (% of applied)	Half-life (days)
Cotyledon	32	30.0	14.5
Four-leaf	1000	8.5	4.5

[a] Control equals the lowest rate that kills 85% of the plant population.

From Shaner, D. L. and Robson, P. A., *Weed Sci.*, 33, 469, 1985. With permission.

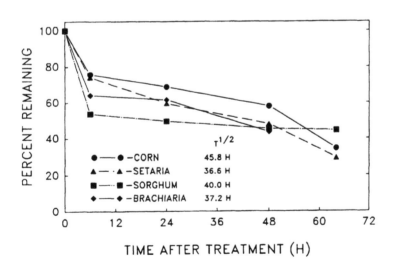

FIGURE 4. Metabolism of ^{14}C-imazaquin by excised leaves of corn, *Setaria faberi*, *Sorghum halepense*, and *Brachiaria platyphylla*.

while *Brachiaria platyphylla* is relatively tolerant (Table 6). The mechanism of selectivity of imazaquin among the monocots, however, is not clear. Metabolism studies with excised leaves have shown very little correlation between the apparent half-life of imazaquin in various grasses and their sensitivity to the herbicide (Table 6). However, closer examination of the data suggests that tolerant grasses initially metabolize imazaquin very rapidly followed by either a cessation or a significant reduction in the rate of metabolism (Figure 4). These

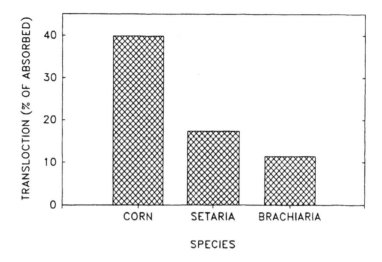

FIGURE 5.   Translocation of ^{14}C-imazaquin in corn, *Setaria faberi,* and *Brachiaria platyphylla* 24 h after foliar application.

results suggest that the herbicide is being sequestered in some way in the tolerant grasses. Supporting evidence for this sequestration is that the amount of radiolabeled material translocated in susceptible grasses is much higher than in tolerant species (Figure 5). However, additional work needs to be done in this area to determine what role, if any, these different factors play in the selectivity of imazaquin.

In conclusion, the mechanism of selectivity of imazaquin appears to be predominantly the ability of tolerant species to metabolize the herbicide rapidly to nontoxic forms. Differential absorption and possibly compartmentalization of the herbicide may also play some role in certain dicot and monocot species.

V. IMAZETHAPYR

Imazethapyr is selective in leguminous crops and controls many important dicot and monocot weeds. The predominant mechanism of selectivity of imazethapyr is differential rates of metabolism. Unlike imazaquin and imazapyr, imazethapyr is metabolized to nontoxic forms via hydroxylation of the ethyl substituent on the pyridine ring followed by conjugation to glucose (Figure 6). The first step of this metabolic pathway occurs in a broad range of species. However, the subsequent metabolite still retains considerable herbicidal activity, although it is less active than the parent compound (Table 9).

The real detoxification step of imazethapyr is the conjugation of the hydroxylated metabolite to glucose. Corn is relatively sensitive to imazethapyr compared to soybeans (Table 10). However, the half-life of the parent compound in both species is approximately equal (Table 10). The difference in the metabolism of imazethapyr between the two species is that soybean plants rapidly conjugate the hydroxylated imazethapyr with glucose while corn plants do not, and only the hydroxylated metabolite accumulates (Figure 6).

However, if hydroxylation of imazethapyr occurs rapidly enough, it can significantly reduce phytotoxicity. When corn is pretreated with naphthalic acid, the toxicity of imazethapyr is greatly reduced (Figure 7). The mechanism of the safening action of naphthalic acid appears to be the induction of a mixed-function oxidase activity which accelerates the rate of hydroxylation of imazethapyr. The half-life of imazethapyr in unsafened corn is approximately 20 h (Figure 8), but is reduced to approximately 2 h if the plants are treated

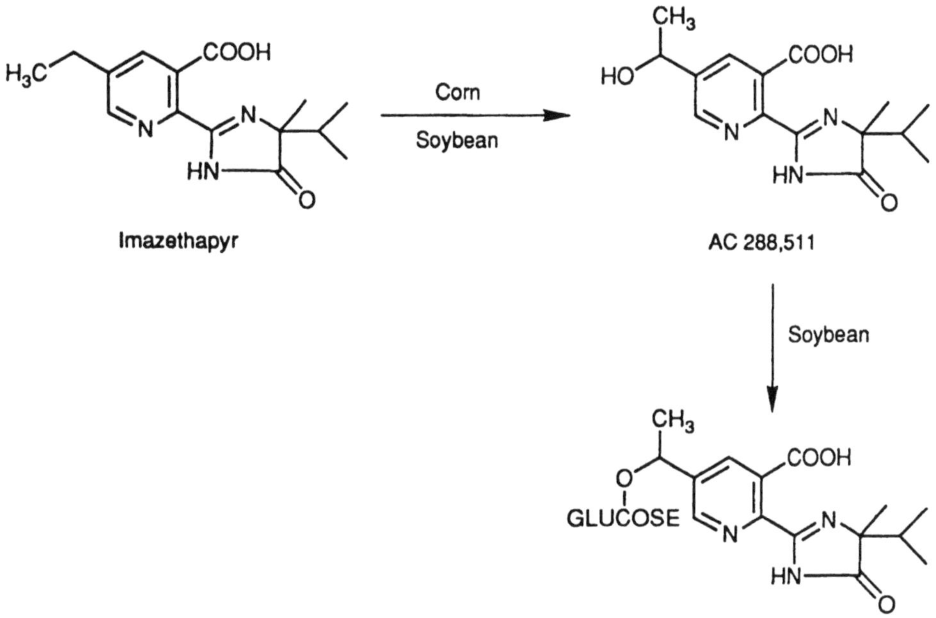

FIGURE 6. Proposed metabolism of imazethapyr in corn and soybeans.

TABLE 9
Comparison of the Herbicidal Activity of Imazethapyr and Its Major Plant Metabolite

Compound	Control rate (g/ha)			AHAS inhibition (I_{50}) (μM)
	Setaria virescens	*Convolvulus* spp.	*Abutilon theophrasti*	
Imazethapyr	13	13	63	1
AC 288,511	1000	125	500	2.5

TABLE 10
Comparison of Herbicidal Activity and Metabolism Rate of Imazethapyr in Soybeans and Corn

Species	Safe rate (g/ha)[a]	Half-life (hours)
Soybean	>500	31
Corn	24	21

[a] Safe rate is the highest rate which results in less than 15% crop injury.

with naphthalic acid (Figure 8). Hydroxylated imazethapyr appears to be relatively immobile in the plant and remains within the root system of safened plants, thus decreasing the concentration of herbicide that reaches the growing points (Figure 9).

The safening effect of naphthalic acid is antagonized by aminobenzotriazole (ABT), a mixed-function oxidase inhibitor. This provides further proof that the safening effect of naphthalic acid is due to an induction of an increased rate of imazethapyr hydroxylation

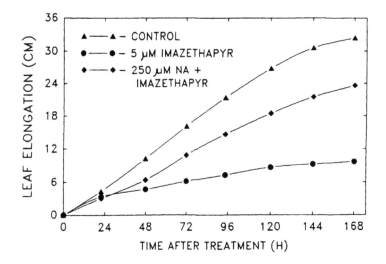

FIGURE 7.   Interaction of imazethapyr and naphthalic acid (NA) on fourth leaf elongation of hydroponically grown corn. NA-treated plants received the 250 μM NA 24 h prior to exposure to 5 μM imazethapyr.

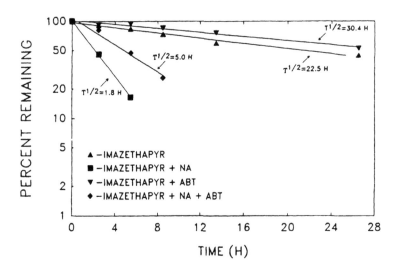

FIGURE 8.   Rate of metabolism of imazethapyr in hydroponically grown corn with and without naphthalic acid (NA) or aminobenzotriazole (ABT). NA = 250 μM naphthalic acid; ABT = 75 μM aminobenzotriazole. Plants were pretreated with either NA, ABT, or both 24 h before exposure to imazethapyr.

(Figure 10). ABT treatment increases the half-life of imazethapyr in plants treated with naphthalic acid by over twofold (Figure 8). This decreased rate of metabolism is accompanied

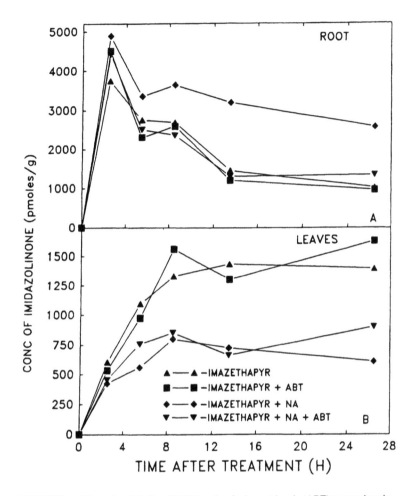

FIGURE 9. Effect of naphthalic acid (NA) and aminobenzotriazole (ABT) on translocation of imazethapyr in hydroponically grown corn. All treatments were applied through the root system. NA = 250 μM naphthalic acid; ABT = 75 μM aminobenzotriazole. Plants were pretreated with either NA, ABT, or both 24 h before exposure to imazethapyr.

by an increase in the amount of herbicide that accumulates in the meristematic regions of the plant (Figure 9).

In conclusion, the primary mechanism of selectivity of imazethapyr appears to be differential metabolism of the herbicide. In susceptible species such as *X. strumarium*, imazethapyr is metabolized very slowly, while in tolerant species it is rapidly hydroxylated and then conjugated with glucose. Species with intermediate tolerance to imazethapyr, such as corn, appear to be able to hydroxylate the herbicide but do not conjugate it. In order for hydroxylation alone to act as the primary detoxification mechanism it has to be very rapid. If it occurs rapidly enough, hydroxylation of imazethapyr immobilizes the herbicide at the site of application. This reduces the toxicity of imazethapyr by limiting the amount of compound that reaches the meristematic regions of the plant.

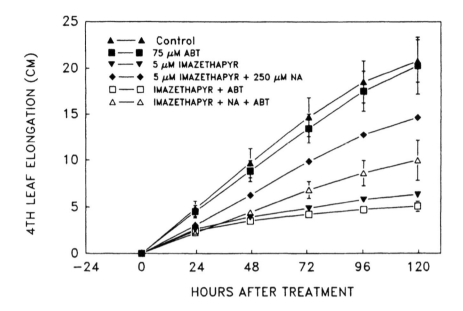

FIGURE 10.   Effect of combinations of naphthalic acid (NA) and aminobenzotriazole (ABT) on the elongation of the fourth leaf of hydroponically grown corn exposed for 24 h to 5 μ*M* imazethapyr. NA = 250 μ*M* naphthalic acid; ABT = 75 μ*M* aminobenzotriazole. Plants were pretreated with either NA, ABT, or both 24 h before exposure to imazethapyr.

REFERENCES

1. **Pillmoor, J. B. and Caseley, J. C.,** The biochemical and physiological effects and mode of action of AC 222,293 against *Alopecurus myosuroides* Huds. and *Avena fatua* L., *Pestic. Biochem. Physiol.,* 27, 340, 1987.

2. **Shaner, D. L., Simcox, P. D., Robson, P., Mangels, G., Reichert, B., Ciarlante, D., and Los, M.,** AC 222,293 — Translocation and metabolic selectivity, in *Br. Crop Prot. Conf. — Weeds,* Vol. 1, British Crop Protection Conference Publications, Croydon, England, 1982, 333.

3. **Shimabukuro, R. H., Walsh, W. C., and Hoerauf, R. A.,** Metabolism and selectivity of diclofop-methyl in wild oat and wheat, *J. Agric. Food Chem.,* 27, 615, 1979.

4. **Jeffcoat, B. and Harries, W. N.,** Selectivity and mode of action of flamprop-isopropyl, isopropyl (±)-2-[*N*-(3-chloro-4-fluorophenyl)-benzoamido]propionate, in the control of *Avena fatua* in barley, *Pestic. Sci.,* 6, 283, 1975.

5. **Brown, M. A., Chiu, T. Y., and Miller, P.,** Hydrolytic activation versus oxidative degradation of Assert herbicide, an imidazolinone aryl-carboxylate, in susceptible wild oat versus tolerant corn and wheat, *Pestic. Biochem. Physiol.,* 27, 24, 1987.

5a. **Minoge, P.,** personal communication.

6. **Shaner, D. L. and Robson, P. A.,** Absorption, translocation and metabolism of AC 252,214 in soybean (*Glycine max*), common cocklebur (*Xanthium strumarium*), and velvetleaf *(Abutilon theophrasti), Weed Sci.,* 33, 469, 1985.

7. **Wilcut, J. H., Wehtje, G. R., Patterson, M. G., and Cole, T. A.,** Absorption, translocation, and metabolism of foliar-applied imazaquin in soybeans (*Glycine max*), peanuts (*Arachis hypogaea*), and associated weeds, *Weed Sci.,* 36, 5, 1988.

Chapter 8

INFLUENCE OF ENVIRONMENTAL FACTORS ON THE BIOLOGICAL ACTIVITY OF THE IMIDAZOLINONE HERBICIDES

Timothy Malefyt and Laura Quakenbush

TABLE OF CONTENTS

I. INTRODUCTION

For herbicides to be a commercial success they must provide acceptable weed control and crop selectivity in a complex environment. Numerous factors can influence the behavior and performance of herbicides whether they are soil applied or used postemergence. These factors can be categorized as constant and variable.

Constant factors include physical or chemical properties of the soil, such as texture, pH, organic-matter content, and water-holding capacity, and physical or chemical properties of the compound, such as water solubility, soil-binding properties, volatility, and soil mobility. Other elements, such as the crop planted, variety used, depth of planting, tillage, and other cultural practices can also be considered constant factors, since they usually do not change once the herbicide is applied and begins working.

Environmental elements, considered variable factors, are subject to change over long periods or from hour to hour. Such changes can influence the biological activity of herbicides depending on whether the compound is applied to the soil or used postemergence. Rainfall, for example, activates many soil-applied herbicides, but can be detrimental if it occurs soon after a postemergence application. The major variable factors known to influence herbicide performance are temperature, light, rainfall, soil moisture, and humidity.

The intent of this chapter is to review and summarize what is known of these variable environmental factors and how they impinge upon the activity of the imidazolinone herbicides.

II. EFFECTS OF TEMPERATURE ON IMIDAZOLINONE ACTIVITY

Several studies indicate that temperature plays an important role in the activity of the imidazolinone herbicides. Differences in crop tolerance and weed control have been observed with imazamethabenz-methyl, imazaquin, imazethapyr, and imazapyr. It is important to note that experiments that examine the effects of temperature on the behavior of a herbicide frequently examine the compound at other-than-normal use rates in order to find differences. For example, a good commercial herbicide should cause little or no crop injury at field-use rates when applied under a wide variety of environmental conditions. Therefore, in order to produce enough injury, the researcher may be forced to test the compound at high rates. The opposite could be true when looking for differences in weed control. A reliable commercial herbicide will provide commercial weed control under a variety of temperatures, but to determine if differences exist, the researcher may have to use suboptimum rates.

A. TEMPERATURE EFFECTS ON IMAZAMETHABENZ-METHYL

Extensive studies on the effects of temperature on the activity of imazamethabenz-methyl have been conducted by Pillmoor and Caseley and by Allen and Caseley at the Weed Research Organization.[1] Most of this work was done postemergence on two species, *Alopecurus myosuroides* and *Avena fatua*, under three different temperature regimens for 14-h day/10-h night cycles: 11/7°C (cool), 16/10°C (medium), and 26/16°C (warm).

Initial experiments on the effect of temperature indicate that *A. fatua* showed less growth inhibition at the warm temperature than at the cool and medium temperatures (Table 1). *A. myosuroides* showed the opposite effect, with greater growth inhibition at the warm than at the medium or cool temperatures (Table 2). The effect of temperature on the activity of imazamethabenz-methyl applied postemergence either to the soil or the foliage was subsequently examined (Table 3). This study, conducted with *A. fatua*, showed that growth inhibition through foliar uptake was more important than through soil uptake. In this experiment, imazamethabenz-methyl had the greatest effect at the medium temperature, re-

TABLE 1
The Influence of Postspray Temperature on the Activity of Foliar-Applied Imazamethabenz-methyl against *Avena fatua*

Dose (g a.i./ha)	Temperature regimens (% of control foliage fresh weight)		
	Low (11/7°C)	Medium (16/10°C)	High (26/16°C)
0	(11.25 g)	(11.65 g)	(10.10 g)
12.5	72	73	111
25	60	56	83
50	38	33	67
100	29	19	41

TABLE 2
The Influence of Postspray Temperature on the Activity of Foliar-Applied Imazamethabenz-methyl against *Alopecurus myosuroides*

Dose (g a.i./ha)	Temperature regimens (% of control foliage fresh weight)		
	Low(11/7°C)	Medium(16/10°C)	High(26/16°C)
0	(6.06 g)	(4.75 g)	(4.19 g)
12.5	93	108	63
25	77	75	46
50	53	49	53
100	82	39	28

gardless of application site. With *A. myosuroides*, growth inhibition through soil uptake was more important than through foliar uptake. Temperature had little effect on the activity of soil applications of imazamethabenz-methyl against *A. myosuroides*, except at the lowest dose of 50 g/ha, where inhibition increased at the warm temperature. Temperature had a significant effect on the foliar activity of imazamethabenz-methyl in this species, with greater activity at the warm temperature for all doses tested (Table 3).

Further experiments were conducted to examine the effects of temperature variation on the activity of imazamethabenz-methyl either pretreatment or posttreatment. Pretreatment temperature had little effect on activity, whereas posttreatment temperature did play an important role. *A. fatua* plants were grown at 16/10°C. Some plants were transferred after treatment to 11/7°C, while others were maintained at 16/10°C. Plants transferred to the cool temperature following treatment showed an increase in activity over those kept at 16/10°C. Exposure to warm temperatures (26/16°C) following treatment showed that imazamethabenz-methyl activity on *A. fatua* gradually decreased with increasing time of exposure. With *A. myosuroides*, a 24 h exposure to warm temperatures was sufficient to achieve the maximum foliar activity. Exposure for longer periods did not significantly increase this activity (Table 4).

The differences in the activity of imazamethabenz-methyl at different temperatures could be the result of spray retention, uptake, translocation, or metabolism. Pillmoor[2] found that *A. fatua* had the greatest spray retention when grown at the medium temperature and the least retention at the cool temperature. *A. myosuroides* had the greatest spray retention at the warm temperature (Table 5). There was some correlation between spray retention and activity at the various temperatures, since foliar activity was greatest in *A. fatua* at the medium temperature and in *A. myosuroides* at the high temperature (Table 3). Spray retention,

TABLE 3
Influence of Temperature on the Activity of Imazamethabenz-methyl
Applied either to the Foliage or the Soil against *A. fatua* and *A. myosuroides*

Dose (g a.i./ha)	Foliar applied			Soil applied		
	26/16°C	16/10°C	11/7°C	26/16°C	16/10°C	11/7°C
			Percent of control foliage fresh weight			
Avena fatua						
Control	(6.37 g)	(9.4 g)	(9.53 g)	(5.74 g)	(9.2 g)	(8.00 g)
12.5	75	45	74	105	80	103
25	69	34	49	87	69	89
50	43	27	44	96	37	70
100	16	24	27	41	30	34
			S.E. = ±6.1 (63 d.f.)			
Average	51	33	48	82	54	74
			S.E. = ±3.3 (9 d.f.)			
			Percent of control foliage fresh weight			
Alopecurus myosuroides						
Control	(5.15 g)	(4.10 g)	(4.61 g)	(4.73 g)	(4.10 g)	(4.86 g)
50	54	79	77	37	58	74
100	19	55	67	41	17	51
200	10	49	56	11	16	20
400	9	38	41	8	16	14
800	24	51	53	6	13	10
			S.E. = ±8.5 (81 d.f.)			
Average	23	54	59	21	24	34
			S.E. = ±5.1 (9 d.f.)			

Note: Application (either spray to foliage or drench to soil) was made at growth stage 13, 22. Destructive assessments were made when the untreated plants reached 6.5-leaf stage (2.5, 3, and 4 weeks after treatment at the high, medium, and low temperatures, respectively).

From Pillmoor, J. B., *Weed Res.*, 25, 433, 1985. With permission.

however, does not fully explain these differences. For example, on *A. myosuroides* spray retention between the medium and warm temperatures was not significantly different, but large differences were noted in activity between these temperatures.

Uptake of imazamethabenz-methyl was influenced by temperature, with more rapid uptake into the lamina of both *A. fatua* and *A. myosuroides* at the warm than at the medium temperature and with the greatest differences observed at 24 h after treatment (Figures 1 and 2). Metabolism of imazamethabenz-methyl was also significantly affected by temperature. Metabolism in both species occurred much more rapidly at the warm than at the medium temperature (Table 6). This is especially evident in the level of metabolites, other than parent and free acid, observed at the high temperature versus the medium temperature. Temperature also influenced the level of parent herbicide converted to the free acid in the plants. The free acid is the compound that actually inhibits acetohydroxyacid synthase and is responsible for killing the plant.[3]

Levels of free acid in the meristematic region of *A. fatua* decreased with time at the warm temperature, as an increased rate of metabolism caused a more rapid production of other metabolites (Table 7). After 7 days, the level of free acid at the warm temperature was five times less than at the medium temperature. *A. fatua*, therefore, appears to detoxify imazamethabenz-methyl more rapidly at warmer temperatures, and this observation could partially explain why imazamethabenz-methyl has less activity at higher temperatures on

TABLE 4
Influence of High Temperatures during the Postspray Period on the Foliar Activity of Imazamethabenz-methyl against A. *fatua* and A. *myosuroides*

Dose (g a.i./ha)	Time at high (26/16°C) temperature postspraying				
	Zero	24 h	48 h	96 h	3 weeks
	Percent of control foliage fresh weight				
Avena fatua					
Control	(9.59 g)	(9.82 g)	(10.37 g)	(10.01 g)	(9.41 g)
12.5	77	73	80	94	91
25	65	58	84	69	97
50	38	30	47	66	74
100	28	22	34	32	62
			S.E. = ±5.6 (99 d.f.)		
Average	52	46	61	65	81
			S.E. = ±2.2 (33 d.f.)		
	Percent of control foliage fresh weight				
Alopecurus myosuroides					
Control	(2.83 g)	(3.68 g)	(3.37 g)	(3.85 g)	(3.74 g)
50	74	50	56	32	41
100	61	43	23	49	31
200	35	29	15	18	27
400	34	15	27	44	16
800	83	28	32	39	24
			S.E. = ±10.9 (131 d.f.)		
Average	57	33	31	37	27
			S.E. = ±5.5 (33 d.f.)		

Note: Plants were grown at the medium (16/10°C) temperatures apart from exposure to high temperature for the periods indicated in the table. Spray application to the foliage was made at growth stage 13, 22 and destructive assessment made at the 6.5-leaf stage (3 weeks for all treatments except when maintained at the high temperatures throughout the postspray period when assessment was made at 3.5 weeks).

From Pillmoor, J. B., *Weed Res.*, 25, 433, 1985. With permission.

TABLE 5
Foliar Spray Retention of A. *fatua* and A. *myosuroides* Grown at Three Temperature Regimens

Dose	Temperature regimen			S.E. (18 d.f.)
	High	Medium	Low	
Avena fatua				
ml per plant	28	40	30	±1.6
ml/g foliage (dry wt)	193	237	131	±11
Alopecurus myosuroides				
ml per plant	11.1	9	3.4	±0.9
ml/g foliage (dry wt)	213	213	110	±15

From Pillmoor, J. B., *Weed Res.*, 25, 433, 1985. With permission.

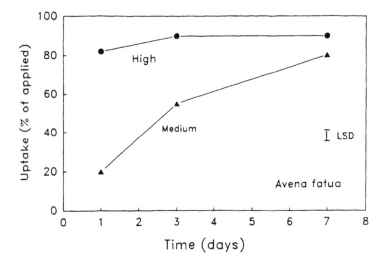

FIGURE 1.   The effect of temperature on the uptake of imazamethabenz-methyl in *A. fatua*. Medium temperature = 11/7°C, high temperature = 16/10°C.

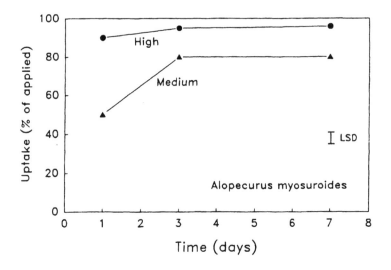

FIGURE 2.   The effect of temperature on the uptake of imazamethabenz-methyl in *A. myosuroides*. Medium temperature = 11/7°C, high temperature = 16/10°C.

this species. At the warm temperature, *A. myosuroides* rapidly converts imazamethabenz-methyl to the free acid (Table 7), but this level of free acid decreases over time, so that by 3 d after treatment there was no difference between the two temperatures, and by 7 d, there was actually less free acid in the meristematic region at the warm temperature. Since more activity was observed with imazamethabenz-methyl on *A. myosuroides* at the warm temperature, it could be inferred that the relatively high level of free acid accumulating within the first 24 h at the meristematic region was enough to kill the plant.

Pillmoor[2] points out that the first 24 h following application of imazamethabenz-methyl could be very important for *A. myosuroides*, since a 24-h exposure to warm temperatures was all the time required to reach the maximum effect following foliar application. Exposure to warm temperatures for longer periods of time did not improve activity. The difference

TABLE 6
Influence of Temperature on the Metabolism of ^{14}C-Imazamethabenz-methyl in the Treated Second Lamina of *A. fatua* and *A. myosuroides*

| | Time (d) | Radioactivity recovered in lamina as[a] (%) | | |
		Parent herbicide	Acid	Other metabolites
Avena fatua				
High	3	26	15	59
	7	6	3	91
Medium	3	51	27	22
	7	29	24	47
S.E. (12 d.f.)		± 1.0	± 1.0	± 1.3
Alopecurus myosuroides				
High	3	26	5	67
	7	11	4	86
Medium	3	54	14	33
	7	18	14	68
S.E. (12 d.f.)		± 4.3	± 2.6	± 4.8

[a] Unabsorbed radioactivity removed by washing with 0.25% Agral.

From Pillmoor, J. B., *Weed Res.*, 25, 433, 1985. With permission.

TABLE 7
Influence of Temperature on the Levels of the Metabolites of ^{14}C-Imazamethabenz-methyl in the Meristem Region of *A. fatua* and *A. myosuroides*

| | Time (d) | Square root of dpm recovered as: | | |
		Parent herbicide	Acid	Other metabolites
Avena fatua				
High (26/16°C)	1	2.9	19.4	37.6
	3	2.9	18.5	35.6
	7	2.9	7.9	33.8
Medium (16/10°C)	1	2.3	13.5	10.7
	3	2.0	17.5	19.5
	7	2.4	18.7	24.0
S.E. (18 d.f.)		± 0.6	± 1.1	± 2.7
Alopecurus myosuroides				
High (26/16°C)	1	10.0	41.6	30.4
	3	4.5	26.6	48.3
	7	5.1	14.9	59.4
Medium (16/10°C)	1	7.9	22.6	17.8
	3	4.8	21.8	28.7
	7	4.8	21.9	33.7
S.E. (18 d.f.)		± 1.4	± 2.6	± 4.0

Note: High specific activity solutions of ^{14}C-imazamethabenz-methyl were applied equally to the first two laminae of *A. fatua* and *A. myosuroides* (1708 × 10^3 dpm per plant, respectively) at growth stage 12, 20.

From Pillmoor, J. B., *Weed Res.*, 25, 433, 1985. With permission.

between the two species at the different temperatures suggests that with *A. fatua*, a continuous supply of free acid at the meristematic region is important for good control, while with *A. myosuroides* the initial level seems to be more important. Pillmore suggests that "the two species must differ in the details of their response to the herbicide at the molecular level."[2]

With imazamethabenz-methyl, temperature may affect the rate of metabolism. The way a plant responds to this change in rate depends on which aspect of metabolism is most affected and how the plant responds to the accumulation of various metabolites. If a plant is highly sensitive to the parent acid, a rapid production of the acid under warm temperature conditions may be enough to kill the plant. Conversely, if the plant has good short-term tolerance to the free acid and also has the ability to further metabolize the free acid, then warm temperatures may actually help the plant survive after application of imazamethabenz-methyl.

B. EFFECTS OF FROST ON IMAZAMETHABENZ-METHYL ACTIVITY

Experiments by Allen and Caseley[1] showed that frost did not affect pretreated frosted *A. fatua* or *A. fatua* plants exposed to 4/2°C day/night temperatures for 2 weeks prior to treatment when subjected to postspray temperatures of 11/7, 16/−2, or 4/−2°C. Plants exposed to −2/−8°C for 24 h pre- and posttreatment, followed by 11/7°C for 72 h, showed greatly reduced activity. Imazamethabenz-methyl showed less herbicidal activity in plants which had frost damage on the foliage than in nonfrost-damaged plants. Experiments with radiolabeled imazamethabenz-methyl showed that frost-damaged plants had greatly reduced translocation out of the first two leaves, which may have caused this reduced activity.

C. TEMPERATURE EFFECTS ON IMAZAQUIN

Experiments by Malefyt, Robson, and Shaner[4] examined the effects of temperature on the herbicidal activity of imazaquin. All experiments were conducted on three species (soybean, *Abutilon theophrasti*, and *Digitaria sanguinalis*) at three temperatures: 18/14°C (cool), 24/20°C (medium), and 30/26°C (warm). All plants were treated postemergence and harvested at the same physiological stage of growth.

In soybeans, there was a slight trend toward increased injury at the coolest temperature, although this injury was only statistically significant at the 750 g/ha rate (Figure 3). Soybeans treated with 750 g/ha showed no difference in fresh weight from the untreated and no herbicide injury symptoms at the warm temperature.

In *A. theophrasti*, a postemergence dosage of 500 g/ha showed no difference in fresh weight between temperatures (Figure 4) and resulted in complete kill of all the plants. At 250 g/ha, a slight reduction in herbicidal activity was noted at the warm temperature, but this was not significant. At 125 g/ha, there was a definite trend toward greater growth reduction at lower temperatures.

In *D. sanguinalis*, no significant differences were observed between temperatures (Figure 5), although there was a trend at the lower rates toward a decrease in activity at higher temperatures.

There was little differences in the absorption and translocation of imazaquin between the various temperatures tested. No significant differences were found between temperatures with soybeans and *A. theophrasti*. *D. sanguinalis* showed significantly less foliar absorption at the high temperature than at the medium and cool temperatures. This difference in absorption did not, however, appear to affect herbicidal activity (Table 8).[4]

Edmund and York[5] examined the effect of temperature on the control of *Cassia obtusifolia* with imazaquin by exposing plants to cool temperature conditions for various periods both before and after treatment. *C. obtusifolia* plants were grown at 32/24°C as the normal temperature and then exposed to 24/18°C for 3-d segments either before or after application of imazaquin. Results of this study showed the greatest differences occurred at 70 g/ha,

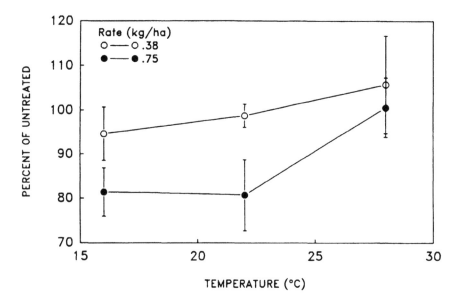

FIGURE 3.   The effect of temperature on the activity of imazaquin in soybean. Points represent the mean of fresh weights of six replications, I = ±S.E. (From Malefyt, T., Robson, P., and Shaner, D. L., in *Aspects of Applied Biology*, No. 4, Association for Applied Biology, Warwick, England, 1983, 265. With permission.)

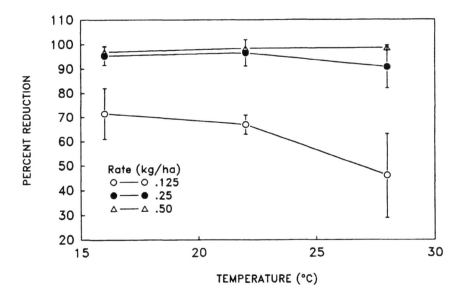

FIGURE 4.   The effect of temperature on the activity of imazaquin in *A. theophrasti*. Points represent the mean of fresh weight of six replications, I = ±S.E. (From Malefyt, T., Robson, P., and Shaner, D. L., in *Aspects of Applied Biology*, No. 4, Association for Applied Biology, Warwick, England, 1983, 265. With permission.)

with only minor differences observed at higher rates of 140 and 280 g/ha. Exposure to low temperatures increased the activity of imazaquin over that in plants which had not been exposed to low temperatures (Table 9).[5] Low temperatures prior to treatment had a greater effect than exposure following treatment. The greatest effect was noted in plants exposed to the low temperature for the longest time.

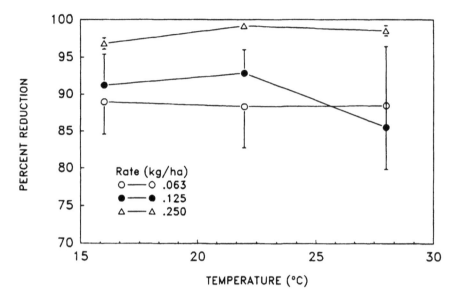

FIGURE 5. The effect of temperature on the activity of imazaquin in *D. sanguinalis*. Points represent the mean of fresh weights of six replications, I = ± S.E. (From Malefyt, T., Robson, P., and Shaner, D. L., in *Aspects of Applied Biology*, No. 4, Association for Applied Biology, Warwick, England, 1983, 265. With permission.)

TABLE 8
Effect of Temperature on Absorption and Translocation of Imazaquin in Soybeans, *A. theophrasti*, and *D. sanguinalis* 3 Days after Treatment

Temperature	Absorption (%)	Distribution in plant (% of absorbed)		
		Above	Treated leaf	Below
Soybeans				
High	56.0 ± 7.8[a]	8.2 ± 2.4	82.0 ± 3.2	9.8 ± 0.8
Medium	62.0 ± 17.9	6.3 ± 1.1	82.2 ± 2.7	10.5 ± 2.6
Low	66.3 ± 14.7	7.2 ± 2.1	83.2 ± 2.7	10.6 ± 2.3
A. theophrasti				
High	8.2 ± 2.4[a]	4.5 ± 2.7	82.7 ± 11.1	12.8 ± 8.8
Medium	13.1 ± 2.4	3.1 ± 1.8	80.4 ± 12.2	18.5 ± 8.1
Low	13.8 ± 0.9	9.4 ± 4.1	74.0 ± 5.6	16.6 ± 10.6
D. sanguinalis				
High	67.9 ± 1.5[a]	8.9 ± 0.1	87.6 ± 0.3	3.6 ± 0.3
Medium	93.3 ± 2.1	7.5 ± 1.6	88.9 ± 2.4	3.7 ± 0.3
Low	81.2 ± 2.0	8.2 ± 1.5	83.3 ± 1.6	8.6 ± 0.3

[a]   ± Standard error.

Wills[6] also examined the effects of temperature, humidity, and surfactant on *Ipomoea lacunosa* treated with imazaquin. After postemergence application of imazaquin to *I. lacunosa* in the four-leaf stage, plants were placed in growth chambers at 18, 27, or 35°C for 48 h and then returned to the greenhouse. After 6 weeks, the control obtained at the cool temperature was superior to that obtained at the medium and warm temperatures (Table 10). Wills also examined the same factors and their influence on absorption and translocation. Results showed significantly greater absorption at the higher temperature, which does not correlate with the control obtained at the whole-plant level.

TABLE 9
Effect of Temperature Regimen on Sicklepod Control with Postemergence Application of Imazaquin

Temperature regimen[a]			Visually estimated control imazaquin rate (g/ha)			Shoot dry weight reduction imazaquin rate (g/ha)		
3 DBT[b]	3 DAT[c]	4-6 DAT[c]	70	140	280	70	140	280
					(%)			
L	H	H	48	73	85	64	77	80
H	L	H	27	74	92	57	80	89
H	H	L	37	73	92	54	78	88
L	L	H	65	84	95	71	82	87
H	L	L	42	89	96	57	88	90
L	L	L	72	87	97	76	83	88
H	H	H	28	65	83	53	76	83

[a] L = Exposure to low temperature, 24/18°C; H = exposure to high temperature, 32/24°C.
[b] DBT = days before herbicide treatment.
[c] DAT = days after herbicide treatment.

From Edmund, R. M. and York, A. C., *Weed Sci.*, 35, 231, 1987. With permission.

TABLE 10
The Effect of Temperature and Humidity on Imazaquin Applied Postemergence to *I. lacunosa*[a]

Temperature	Relative humidity	Control[b] (%)		Absorption (%)	
		No adjuvant	Tween 20	No adjuvant	Tween 20
18	40	23 c	41 c	9 cd	15 c
	100	53 a	79 a	28 b	50 b
27	40	16 d	25 d	3 d	15 c
	100	23 c	49 b	20 bc	61 b
35	40	16 d	27 d	3 d	24 c
	100	31 b	41 c	63 a	91 a

[a] Within columns, numbers followed by the same letter are not different at the 5% level according to Duncan's multiple range test.
[b] Data taken 6 weeks after imazaquin treatments of 280 g/ha.

The above data strongly suggest that the decrease in the activity of imazaquin at warm temperatures in many species is not due to differences in uptake or translocation at these temperatures. It is possible that warm temperatures increase the rate of detoxification, thereby reducing herbicidal activity.

There is a difference in the effect of temperature on phytotoxicity when imazaquin is absorbed via the roots as opposed to the foliage. When corn was grown under a warm temperature regimen of 37/22°C, a solution of 0.1 μM of imazaquin inhibited leaf elongation by 61%. However under cool conditions (22/9°C), 0.5 μM of imazaquin caused the same amount of inhibition. The greater toxicity of imazaquin under warm as compared to cool conditions correlates with the effect of temperature on the translocation of imazaquin to the shoot meristem region. Under warm conditions, approximately five times more herbicide accumulates in the leaf meristem region than in plants grown under cool conditions which correlates very well with the effects on growth. This decrease in the translocation of imazaquin from the roots to the leaves is apparently related to the rate of transpiration of the

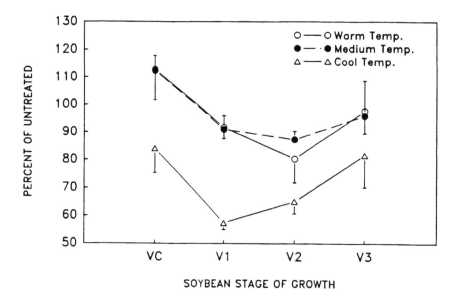

FIGURE 6.   The effect of imazethapyr applied postemergence on shoot fresh weight of soybeans.

plants under the two conditions. Under these cool conditions corn transpired at approximately one fifth the rate of plants grown under warm conditions.

So, unlike the effects of temperature on foliarly applied imazaquin, the herbicidal activity of root-applied imazaquin appears to be more effective under warm than cool conditions because the herbicide is translocated from the roots to the leaves by transpiration.

D. TEMPERATURE EFFECTS ON IMAZETHAPYR

Imazethapyr was tested under various temperatures to determine the effect of temperature on phytotoxicity in soybeans and alfalfa.[7] Soybeans were grown at three temperatures: 15/21°C (cool), 21/27°C (medium), and 27/33°C (warm). Plants were treated with postemergence applications of imazethapyr at four different growth stages; cracking (VC), unifoliate (V1), first trifoliate (V2), and second trifoliate (V3). These experiments showed that the greatest soybean phytotoxicity was at the cool temperature as seen in both shoot growth (Figure 6) and root growth (Figure 7). Within a temperature regimen, phytotoxicity dramatically increased from the VC stage to the V1 stage, and then gradually decreased as the size of the treated plant increased (Figures 6 and 7). Experiments were also conducted on the absorption, translocation, and metabolism of soybeans treated at the V1 growth stage. Temperature did not significantly affect the absorption or translocation of imazethapyr, although translocation was slower at the cooler temperature than at the medium and warm temperatures. The greatest difference between temperatures was the rate of imazethapyr metabolism, with almost a twofold increase in the rate of metabolism for every 5°C increase (Table 11).

Alfalfa was also treated postemergence with imazethapyr at the cotyledon, first, second, fourth, sixth, and eighth trifoliate stages and then grown at 14/21°C (cool) or 23/30°C (warm). As in soybeans, greater phytotoxicity was observed at the cool temperature than at the warm temperature (Figure 8). Phytotoxicity also decreased as the size of the treated plant increased. As was observed in soybeans, the greater phytotoxicity at lower temperatures and in smaller plants may be due to a slower rate of detoxification of the imazethapyr through metabolism.

The effect of temperature on the control of *Anoda cristata*, a relatively sensitive species, was investigated by Herrick.[8] Plants treated with 8 g/ha of imazethapyr and then grown at various temperatures were not significantly affected by temperature.

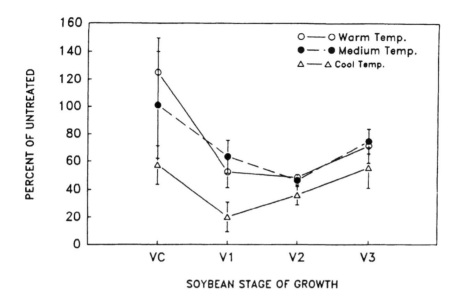

FIGURE 7.   The effect of imazethapyr applied postemergence on root fresh weight of soybeans.

TABLE 11
Effect of Temperature on Half-Life of Imazethapyr in Soybean

Temperature (°C)	Half-life (days)
15/21	1.75
21/27	1.00
27/33	0.51

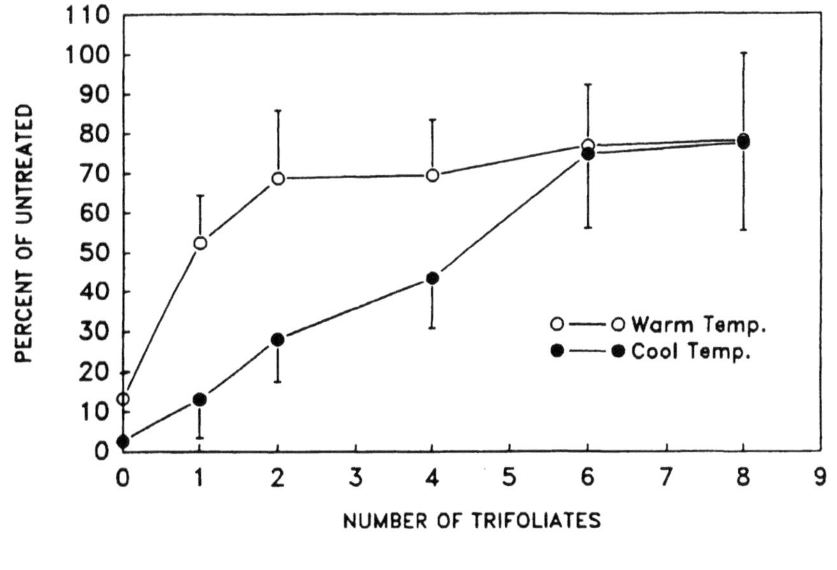

FIGURE 8.   The effect of temperature on imazethapyr activity on alfalfa treated postemergence at five growth stages. Rate tested was 375 g/ha, "cool" = 14/21°C, "warm" = 23/30°C. Values plotted are shoot fresh weights.

As with imazaquin and imazamethabenz-methyl, temperature plays a role in the herbicidal activity of imazethapyr. Temperature has a greater effect on plants, such as soybeans and alfalfa, which are able to effectively metabolize the herbicide than it does on species which do not readily metabolize imazethapyr.

E. TEMPERATURE EFFECTS ON IMAZAPYR

Malefyt (unpublished) has investigated the effect of temperature on imazapyr activity on *Agropyron repens* and *Cyperus rotundus*. *C. rotundus* was grown either at 21 or at 32°C for 7 d prior to and after treatment for the remainder of the experiment. Imazapyr was sprayed postemergence at rates of 32, 63, 125, and 250 g/ha. At 32 g/ha and 21°C, imazapyr gave 90% control; 125 g/ha were required to achieve the same level of control at 32°C. A similar experiment, conducted with *A. repens,* showed that better control was achieved at 21°C than at 32°C. Therefore, it appears that cooler temperatures improve the herbicidal activity of imazapyr as is noted in experiments with imazethapyr and imazaquin.

III. EFFECTS OF RAINFALL ON IMIDAZOLINONE ACTIVITY

Rainfall can significantly affect the activity of all imidazolinone herbicides. Compounds that have both soil and foliar activity, such as the imidazolinones, can be influenced by rainfall with both preemergence and postemergence applications. Most of the research on rainfall, however, has been done with postemergence applications of the imidazolinone herbicides.

A. THE EFFECT OF RAINFALL ON POSTEMERGENCE APPLICATIONS

Pillmoor and Caseley[1] examined the effects of rainfall intensity on the activity of imazamethabenz-methyl on *A. fatua* and *A. myosuroides*. The foliage of the plants was treated with several different dosages of imazamethabenz-methyl and transferred 30 min later to a rainfall simulator which produced 0.25, 1, or 5 mm of rain in 30 min. Results from this experiment indicated that the lowest rainfall intensity did not alter the activity of imazamethabenz-methyl on *A. fatua*. However, an increase in the intensity of the rainfall to 1 and 5 mm/30 min led to a progressive loss in activity, with 30% less activity between no rain and the 5 mm rainfall. This difference was expected, since imazamethabenz-methyl has the greatest effect on *A. fatua* when applied postemergence to the foliage, and the higher-intensity rainfall probably washed the imazamethabenz-methyl off the leaf surface.

Instead of reducing the activity of imazamethabenz-methyl, rainfall increased the foliar activity on *A. myosuroides*. This increase was only significant at the lowest rainfall intensity. Rainfall may have caused a redistribution from the leaf surface into the leaf sheath, which would account for this increase in activity. The rainfall could have also washed some of the herbicide off the leaf surface and into the soil, where it would be available to the plants through root uptake, which has been shown to cause greater phytotoxicity than foliar uptake in *A. myosuroides* (Table 3).

Edmund and York[5] have examined the effect of rainfall on the postemergence activity of imazaquin on *Cassia obtusifolia* and found that imazaquin has its greatest activity on *C. obtusifolia* when absorbed through the roots. The researchers applied imazaquin to both the foliage and the soil prior to rainfall and determined that the timing of rainfall (0.6 cm), from 0.05 h to 24 h, had no detrimental effect on herbicidal activity when compared to plants receiving no rainfall. Since this work showed that imazaquin was primarily effective through root absorption, washing the imazaquin off the foliage is not likely to have a significant effect.

Malefyt (unpublished) examined the effects of rainfall on the postemergence activity of imazaquin on *C. obtusifolia, Sida spinosa,* and *Xanthium strumarium*. In this experiment, some plants received a foliar treatment, while others received both soil and foliar treatments.

TABLE 12
The Effect of Rainfall Timing on the Activity of Imazaquin Applied
Postemergence to *Cassia obtusifolia*, *Sida spinosa*, and *Xanthium strumarium*

Time of rainfall (hours after treatment)	Visual control (%)					
	C. obtusifolia[a]		*S. spinosa*[b]		*X. strumarium*[c]	
	Foliar only	Soil + foliar	Foliar only	Soil + foliar	Foliar only	Soil + foliar
0.08	25	85	25	85	35	95
0.5	25	85	25	85	55	95
1.0	25	85	25	85	70	100
3.0	25	85	25	85	95	100
6.0	25	85	25	85	100	100
24.0	35	85	25	85	100	100
None	85	95	85	95	100	100

[a] *C. obtusifolia* treated with 375 g/ha.
[b] *S. spinosa* treated with 125 g/ha.
[c] *X. strumarium* treated with 32 g/ha.

Results, which were similar to those of Edmund and York,[5] showed that rainfall (0.6 cm) had only minor effects on imazaquin applied to both the soil and the foliage of any of these species (Table 12). Rainfall significantly affected plants which received only a foliar treatment. Control of *C. obtusifolia* and *S. spinosa* decreased from 85% with no rain to 25 to 35% with rainfall received anytime within the first 24 h (Table 12). *X. strumarium* was not as sensitive to rainfall. Control of this species was significantly reduced when rainfall occurred less than 3 h after treatment, but rainfall had no significant effect on control at 3, 6, or 24 h after treatment. These results again indicate that rainfall did not have a detrimental effect on the postemergence activity of imazaquin if root uptake was not prevented. When root uptake was prevented, rainfall reduced the control of certain species. However, species which have significant levels of foliar absorption only showed negative effects when rainfall occurred soon after treatment.

Herrick[8] conducted a similar study to examine the effects of rainfall on imazethapyr applied only to the foliage of *Anoda cristata*. As the interval between herbicide application and the occurrence of rainfall increased, the level of control increased, with the control at 8 h equal to that obtained in plants receiving no rain. These results correlate well with the above study on *X. strumarium*, and it appears that a rainfall occurring 24 h after application of an imidazolinone should not have a detrimental effect, even on species in which foliar absorption is the major route of entry.

B. THE EFFECT OF RAINFALL ON SOIL APPLICATIONS
After soil application, rainfall is required for good activity of most herbicides, including the imidazolinones. It has been widely observed with the imidazolinones that poor activity may result if inadequate rainfall is received after preemergence and, sometimes, preplant-incorporated applications. However, little information has been published on this subject. Basham et al.[9] found that early soil moisture enhanced the activity of imazaquin, but they did not determine how much moisture was required for enhanced activity nor the importance of rainfall timing. Further work needs to be completed to answer additional questions regarding rainfall and its effects on soil application of imidazolinones.

IV. EFFECTS OF SOIL MOISTURE

Observations from field trials indicate that soil moisture can play an important role in

TABLE 13

**Effect of Soil Moisture on Foliar Activity of Imazamethabenz-methyl
against *Avena fatua* and *Alopecurus myosuroides***

Species[a]	Rate (g/ha)	Soil moisture (% of field capacity)		
		50	100	150
		Reduction in fresh weight (%)		
A. fatua	12.5	5	38	34
	25	16	40	58
	50	44	67	77
	100	36	78	77
	Check (g)	(1.40)	(7.00)	(10.28)
A. myosuroides	25	42	17	0
	50	35	26	34
	100	74	62	56
	200	61	77	83
	Check (g)	(1.15)	(4.20)	(4.92)

[a] Plant stage at treatment was 3 leaves, 2 tillers.

the activity of the imidazolinone herbicides. To date, however, formal studies have only investigated soil moisture effects on imazamethabenz-methyl and imazaquin.

A. THE EFFECT OF SOIL MOISTURE ON IMAZAMETHABENZ-METHYL

Caseley and Pillmoor[1] have studied the effect of soil moisture on the herbicidal activity of imazamethabenz-methyl. Soil moisture levels of 50%, 100%, and 150% of field capacity were established 1 week before postemergence applications to *Alopecurus myosuroides* and *Avena fatua*. Plants were treated at the three-leaf, two-tiller growth stage. Applications were made only to the foliage, as soil was protected by plastic chips during spraying. The different soil moisture levels were maintained throughout the experiment, with fresh weights taken 3 weeks after treatment.

In this test, soil moisture did not have much effect on the activity of imazamethabenz-methyl against *A. myosuroides* (Table 13). With *A. fatua,* the reduction in fresh weight from herbicide treatment was less at the lowest soil moisture than at the two higher levels. However, growth of the untreated checks was also greatly reduced at the lowest moisture level, which could have influenced the results.

In later studies, Allen and Caseley[1] evaluated the effect of soil moisture regimens and growth stage at application on the activity of imazamethabenz-methyl against *A. myosuroides.* Three moisture regimens were evaluated: 50%, 75%, and 100% of field capacity. *A. myosuroides* was treated at three growth stages: preemergence, one leaf, and two leaves, one tiller. For postemergence treatments, plants were grown at 75% of field capacity, with moisture regimens established 5 d before treatment. In contrast to the previous test described with *A. myosuroides,* postemergence applications were to foliage and soil. Pregerminated seeds were used for preemergence applications. Dry weights were taken when the controls for each growth stage — moisture regimen combination reached the same size (about 10 tillers).

Allen and Caseley found that soil moisture did not significantly reduce dry weight of *A. myosuroides* when treated preemergence or at the later postemergence timing (two leaves, one tiller). When imazamethabenz-methyl was applied early postemergence (one leaf), injury increased as soil moisture increased (Table 14). Imazamethabenz-methyl at 0.1 and 0.2 kg/ha caused a 10% and 32% reduction, respectively, at 50% field capacity, compared to 46% and 66% reductions at 100% field capacity.

TABLE 14
Effect of Soil Moisture and Growth Stage at Treatment on Activity
of Imazamethabenz-methyl against *Alopecurus myosuroides*

| Growth stage | Rate (g/ha) | Soil moisture (% of field capacity) | | |
		50	75	100
		Reduction in dry weight (%)		
Preemergence	100	13	0	21
	200	35	51	45
One leaf	100	10	11	46
	200	32	58	66
Two leaves, one tiller	100	25	14	21
	200	42	51	51

Pillmoor and Caseley[1] studied the effect of waterlogged soils on the activity of imazamethabenz-methyl on *A. fatua* and *A. myosuroides*. In these studies "waterlogged" conditions consisted of growing plants in pots with no drain holes, and adding water until standing water was just evident on the soil surface. Three water regimens were evaluated: (1) soil maintained at field capacity; (2) soil waterlogged for 4 d, starting 3 d before herbicide treatment; and (3) soil waterlogged for 4 d, starting the day after treatment. After the waterlogged period, the soil was not watered until it returned to field capacity, which took about 7 d.

Imazamethabenz-methyl was applied postemergence when *A. fatua* and *A. myosuroides* had three leaves and one or two tillers. Soil was not protected, so applications were to the foliage and soil surfaces. Waterlogged-soil treatments had little effect on herbicidal activity against *A. fatua*. Imazamethabenz-methyl activity against *A. myosuroides* was greater under waterlogged conditions than when soil was maintained at field capacity, with the greatest activity when waterlogged soil was initiated after spraying. The importance of soil activity of imazamethabenz-methyl, even when applied postemergence, is much greater for *A. myosuroides* than for *A. fatua*, which could explain the difference between species.

Pillmoor and Caseley have also extensively evaluated the effect of waterlogged soil on imazamethabenz-methyl injury to wheat (cv. Fenman). Waterlogged-soil regimens were the same as described above. Wheat was treated postemergence at rates of 0.38 to 3.0 kg/ha. Waterlogged-soil treatments increased stunting of wheat, especially when waterlogged soil was initiated after spraying. Stunting was evaluated by measuring sheath height 3 weeks after treatment. However, there was little effect on fresh weight of wheat taken 3 weeks after treatment. Pillmoor and Caseley then evaluated the effect of waterlogged soils when wheat was treated at the one- to three-leaf stages. Waterlogged-soil conditions were initiated the day before herbicide treatment and maintained for 6 d. Fresh weights were taken 3 and 6 weeks after treatment.

When soil was maintained at field capacity, imazamethabenz-methyl caused little injury to wheat, even at rates of 3 kg/ha (over four times the use rate). The greatest injury was at the earliest growth stage, when 3 kg/ha of imazamethabenz-methyl caused a 27% reduction in fresh weight 3 weeks after treatment (Table 15). Waterlogged-soil conditions increased injury, with injury greatest for the youngest plants. However, the only severe effects were at rates of 1.5 and 3 kg/ha, with all other timings and rates causing less than 25% reduction 3 weeks after treatment. There was good recovery by 6 weeks after treatment, when the only treatments with over 10% reduction were at the 3 kg/ha rate.

Pillmoor and Caseley then evaluated the importance of foliar vs. soil activity of imazamethabenz-methyl for wheat injury under waterlogged conditions. Imazamethabenz-methyl was applied at high rates to wheat during the first-leaf through coleoptile stage, with ap-

TABLE 15
Effect of Waterlogged-Soil Conditions on Susceptibility of Wheat (Cv. Fenman) to Injury from Imazamethabenz-methyl

kg/ha	Field capacity				Waterlogged[a]			
	Z10[b]	Z11	Z12	Z13	Z10	Z11	Z12	Z13
	% Reduction in fresh weight 3 WAT							
0.38	15	1	3	0	1	9	0	4
0.75	15	1	4	4	13	11	6	8
1.50	19	10	3	4	40	18	3	8
3.00	27	11	8	8	50	23	18	16

[a]   Waterlogged-soil conditions were imposed one day before spraying and maintained for 6 d.

[b]   Growth stages are given using the Zadoc scale.[10] Z10 = first leaf through coleoptile; Z11 = one leaf expanded; Z12 = two leaves expanded; Z13 = three leaves expanded.

plications to both foliage and soil, to foliage only, or to soil only. The waterlogged-soil regimen was as described above. Imazamethabenz-methyl caused very little injury when applied only to the foliage, while applications to the soil or to the foliage and soil caused significant and similar injury. Therefore, injury to wheat under waterlogged-soil conditions is produced by the soil activity of imazamethabenz-methyl and not by foliar uptake.

These researchers further evaluated the effect of timing of waterlogged soil on wheat injury. Waterlogged conditions were maintained for 7 d, beginning either 1 d before herbicide treatment, 1 d after treatment, or 7 d after treatment. All waterlogged-soil treatments caused increased injury, even when waterlogging was delayed until 7 d after herbicide treatment. They also found that both waterlogged-soil conditions and "wet" soil (175% of field capacity) caused increased injury to wheat.

The effect of duration of waterlogged soil was also evaluated. Waterlogged-soil regimens were initiated on the day after herbicide treatment and maintained for 1, 7, or 14 d. There was no significant difference between durations, although the 14 d treatment caused the greatest injury to wheat. This is not surprising, as it took 8 d for waterlogged soils to return to field capacity, and wet soils also caused increased injury.

The mechanism of the increased injury to wheat under waterlogged-soil conditions was evaluated by applying ^{14}C-imazamethabenz-methyl to the soil surface. Waterlogged conditions were initiated 1 d after herbicide application, maintained for 6 d, and compared to field capacity treatments. Soil and wheat plants were evaluated to determine the fate of the imazamethabenz-methyl at 2, 6, and 12 d after treatment.

Soil was extracted with methanol and analyzed by thin-layer chromatography. Recovery of ^{14}C was 82%. Imazamethabenz-methyl was fairly stable in the soil, with 90% present as the parent 6 d after treatment under both moisture regimens (Table 16). By 12 d after treatment, waterlogged soil had slightly less of the parent (84% vs. 87%) and slightly more of the acid (9% vs. 5%) than did the soil at field capacity. Substantial growth inhibition was apparent by 6 d after treatment in the waterlogged-soil treatments; therefore, this injury could not be due to increased metabolism of imazamethabenz-methyl to the acid in the soil.

A greater amount of ^{14}C-imazamethabenz-methyl was taken up by wheat plants under waterlogged conditions than when grown at field capacity. The difference was greatest at 2 d after treatment but was still substantial 12 d after treatment (Table 17). Most of the radiolabel was found in the foliage, although a greater percentage was in the roots under waterlogged conditions as compared to field capacity conditions.

These results indicate that the increased injury to wheat seen under waterlogged-soil conditions is due to the increased herbicide uptake from the soil under these conditions.

TABLE 16
Metabolism of ^{14}C-Imazamethabenz-methyl in Waterlogged or Field Capacity Soils

Moisture regimen	Days after treatment	Parent ester	Acid	Other
			% of total	
Field capacity	2	92	3	5
	6	90	5	5
	12	87	5	8
Waterlogged	2	92	4	4
	6	90	5	5
	12	84	9	7
LSD (0.05)		2.1	0.6	2.2

TABLE 17
Effect of Waterlogged Soil on Uptake and Distribution of ^{14}C-Imazamethabenz-methyl by Wheat (Cv. Fenman)

Moisture regimen	Days after treatment	% of applied recovered	dpm[a]/plant		Distribution (%)	
			dpm	SQRT[b](dpm)	Foliage	Root
Field capacity	2	92	30	5.5	91	9
	6	96	445	20.8	80	20
	12	96	3063	55.1	86	14
Waterlogged	2	96	1932	43.7	62	38
	6	95	6360	79.3	77	23
	12	95	14003	118.2	78	22
LSD(0.05)		67	—	9.1	12.2	12.2

[a]   Disintegrations per minute.
[b]   Data was transformed to square root for statistical analysis. LSD applies to transformed data.

Pillmoor and Caseley also evaluated the effect of soil pH on wheat injury under water-logged vs. field capacity conditions. The pH of a sandy loam soil (6.4) was adjusted by addition of ground chalk to give a pH of 7.4 or by addition of qualine phosphate to give a pH of 5.8. However, when they evaluated soil pH again at the end of the experiment, they found that the pH of all soils had moved toward neutrality. By the end of the test, soil pHs were 6.4, 6.9, and 7.1.

When wheat was treated at the three-leaf, two-tiller stage, there was little injury with any treatment, even at 3 kg/ha of imazamethabenz-methyl. When wheat was treated at a younger stage (first leaf through the coleoptile but not completely unfolded), the researchers again found greater injury under waterlogged conditions. There was a trend of increasing injury with increasing pH under both soil moisture regimens, although injury was more pronounced under waterlogged conditions.

B. THE EFFECT OF SOIL MOISTURE ON IMAZAQUIN ACTIVITY

Cyanamid researchers evaluated uptake of imazaquin by corn in different soils that had different water-holding capacities. The soils ranged from a sandy loam to a clay loam with field capacities of 12.5 to 38.5 g water/100 g soil. In these tests all the soils were treated with ^{14}C-imazaquin to bring the concentration to 50 ppb on a dry weight basis. Twenty-five grams of each soil was added to test tubes containing enough water to bring each soil to its field capacity. Two corn seeds were planted in each tube. Thus, there were equal amounts

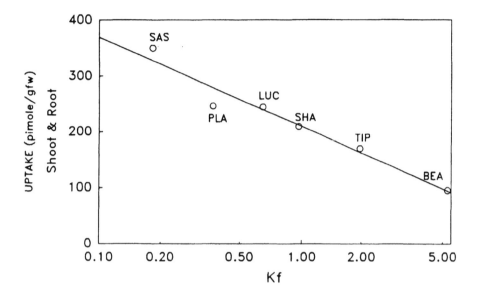

FIGURE 9.   ¹⁴C-imazaquin uptake by corn plants grown in different soils. Each of the six soils was maintained at field capacity.

of herbicide and soil in each unit, but different amounts of water were added, depending on the soil water-holding capacity of each soil.

Corn plants were grown for 6 d in all soils except the New Jersey sassafras sandy loam, in which plants were grown for 12 d. Plants grew slower in this soil, so they were grown until they reached the same size as plants in the other soils. Soils were maintained at field capacity throughout the test. At the end of the growing period, shoots, roots, and soil were separated, fresh and dry weights of plant material determined, and the amount of ¹⁴C was determined by combustion for roots, shoots, and soil.

In a parallel experiment the Freundlich absorption coefficient (K_f) of each soil for imazaquin was determined as well as the concentration of herbicide in the soil by extracting water from the soils with a Carver press. The K_f varied from 0.18 for the sandy loam soil to 5.28 for the clay loam soil.

The amount of ¹⁴C-imazaquin taken up (disintegrations per minute [dpm]/grams fresh weight) was highly and negatively correlated with soil water-holding capacity ($R^2 = .95$ for linear regression). As the water-holding capacity of the soil increased, the amount of imazaquin taken up decreased (Figure 9).

The corn used in the above test was susceptible to injury by imazaquin, and there were visible herbicidal effects on corn growth, particularly in the sassafras sandy loam. The imazaquin rate used (50 ppb) is sufficient to cause severe corn injury when plants are grown in the greenhouse for several weeks. It is possible that herbicide injury to the corn plant would also affect uptake. In order to remove any effect of herbicide injury, this test was repeated using a corn line tolerant to imazaquin. This tolerant corn has an altered AHAS binding site for imazaquin, so that imazaquin has no effect on the corn, even at very high rates.

Results obtained using tolerant corn were very similar to results with susceptible corn, with less herbicide absorbed in soils with higher K_f values for imazaquin (Figure 10).

The concentration of imazaquin in the soil solution decreases as the K_f value for a soil increases. Work done by Cyanamid researchers showed that as soil moisture decreases, there is a concomitant decrease in the concentration of imazaquin in the soil solution (Figure 11). These results suggest that the binding of imazaquin to the soil colloids increases as soil

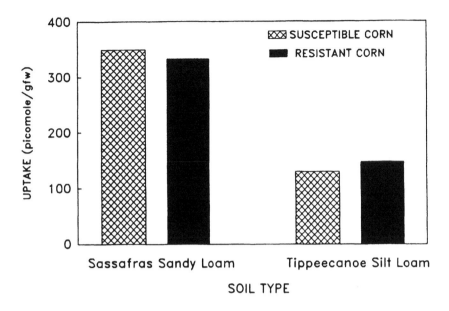

FIGURE 10. ¹⁴C-imazaquin uptake by imazaquin-resistant corn plants grown in different soils. The soils were maintained at field capacity.

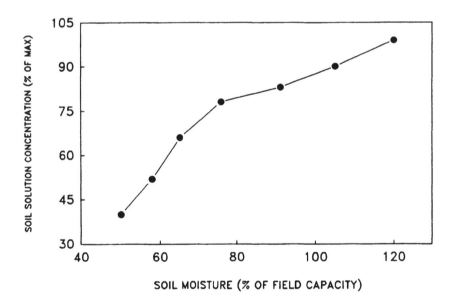

FIGURE 11. The concentration of imazaquin in the soil solution as influenced by soil moisture.

moisture decreases. If this is so, then the herbicidal efficacy of imazaquin should be less in a dry soil as compared to a moist soil.

Tests were conducted with a range of soil moisture levels in six soils to evaluate this hypothesis. Methods used were similar to those described above, except the six soils were each evaluated at six moisture levels, ranging from 25% to 200% of field capacity.

In three of the lighter soils, total uptake of ¹⁴C-imazaquin increased linearly as water content of the soil increased (Figure 12). A similar trend of increasing uptake of herbicide with increasing moisture levels was also seen with the other three soils, although the response

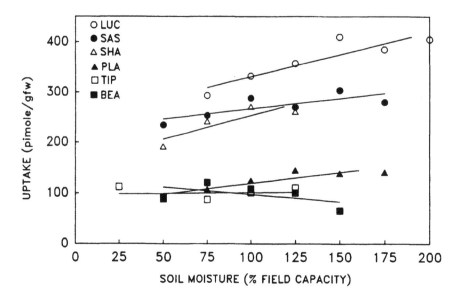

FIGURE 12.   Effect of soil moisture level on uptake of ^{14}C-imazaquin by corn. The soils studied were: Lucedale fine sandy loam (LUC); Sassafras sandy loam (SAS); Sharkey clay loam (SHA); Plano silt loam (PLA); Tippeecanoe silt loam (TIP); and Beardon clay loam (BEA).

was not as dramatic. Much less herbicide was taken up from the heavier soils than from the sandy loam at each moisture level (Figure 12).

These tests with varying soil moistures of the same soil are consistent with the measurements on the effect of soil moisture on the concentration of imazaquin in the soil solution. In drier soils, less herbicide is available to the plant for uptake than in moist or saturated soil. Consequently the herbicidal activity of imazaquin decreases under dry conditions as compared to moist conditions. These results are consistent with field observations where imazaquin provides erratic weed control under drought conditions.

Interactions between imidazolinones and soils are discussed in more detail in other sections of this book (Chapter 16).

The effect of multiple cycles of wetting/drying of soil on imazaquin uptake by corn was also studied. Two soils were used, the Sassafras sandy loam and the Tippeecanoe silt loam. Methods were similar to the previously described tests, except soils were put through five wetting/drying cycles. In each cycle, soil was brought to field capacity, corn was grown and harvested, ^{14}C-imazaquin uptake was determined, and the soil was then air-dried. One cycle lasted 14 d from planting to planting, with corn harvested after 8 d.

There was a clear trend of decreasing uptake with repeated wetting and drying when considering dpm/g dry weight, dpm/g fresh weight, or dpm taken up per test tube. However, when the loss of ^{14}C-imazaquin via plant uptake with each cycle was taken into account, the percent of remaining herbicide taken up by corn with each cycle remained fairly constant (Table 18). In each cycle, an average of 12.4% of the remaining herbicide was taken up from the sandy loam and about 5.9% from the silt loam.

V. EFFECTS OF RELATIVE HUMIDITY

The activity of postemergence herbicide applications is greater at higher humidities than at lower humidities.[11,12] This has also been demonstrated for a wide variety of herbicides of different chemistries. Herbicide solution droplets on the plant leaf surface dry slower at higher humidities. Most herbicides would not be expected to be taken up by plants once

TABLE 18
Uptake of ¹⁴C-Imazaquin by Corn through Five Wetting/Drying Cycles

Soil	Cycle	Uptake dpm/g fw	Uptake dpm/test tube	Total remaining[a] dpm/test tube	Uptake of remaining ¹⁴C (%)
Sassafras sandy loam	1	3129	4776	50,809	9.4
	2	2621	5688	32,544	17.5
	3	2210	4176	31,398	13.3
	4	1755	3504	29,949	11.7
	5	1250	2336	22,902	10.2
Tippeecanoe silt loam	1	1345	2560	51,594	6.9
	2	1274	3763	43,200	8.7
	3	943	2518	46,630	5.4
	4	649	1643	38,209	4.3
	5	549	1226	30,650	4.0

[a] Radiolabel remaining in the soil was determined by combustion of a small sample of soil at the beginning of each cycle.

they have formed crystals on the leaf surface. At higher humidities, the cuticle will be more hydrated, aiding movement of herbicides, particularly hydrophilic herbicides, through the cuticle.

Shaner et al.[13] evaluated the effect of relative humidity on uptake of ¹⁴C-imazameth-abenz-methyl by *A. fatua*. They found much greater uptake at higher humidity levels, with 60.5% absorption at 98% relative humidity compared with 20% absorption at 50% relative humidity.

Wills[6] evaluated the effect of temperature and relative humidity on uptake of imazaquin by *Ipomoea lacunosa*. Imazaquin was applied postemergence at 0.28 kg/ha to *I. lacunosa* in the three- to four-leaf stage. Plants were placed immediately into growth chambers each set at 40% and 100% relative humidity at three temperatures (18°C, 27°C, 35°C). Separate tests were conducted using ¹⁴C-imazaquin or unlabeled imazaquin. For the ¹⁴C-imazaquin experiments, treated plants were evaluated for herbicide uptake and translocation 2 d after treatment. Other treated plants were left under the environmental treatments for 3 d, and then returned to the greenhouse and visually evaluated for injury at 2, 4, and 6 weeks after treatment.

There was significantly greater injury at 4 and 6 weeks after treatment at 100% relative humidity than at 40% relative humidity at all temperatures evaluated (Table 10). Results with radiolabeled imazaquin indicated that absorption of imazaquin into the treated leaf and movement out of the treated leaf was significantly greater at 100% than at 40% relative humidity.

VI. SUMMARY AND CONCLUSIONS

Some overall trends can be seen regarding environmental effects on activity of the imidazolinones. The activity of acid imidazolinones such as imazaquin, imazethapyr, and imazapyr, is generally greater at cooler temperatures because the acid imidazolinones are more rapidly deactivated via metabolism at warmer temperatures.

The effect of temperature on imazamethabenz-methyl is not as defined and appears to be species dependent, although differences may also be affected by metabolism. Imaza-methabenz-methyl may be metabolized more rapidly at warmer temperatures, but one of its metabolites is the acid, which is herbicidally active. Thus, increased metabolism of this ester at warmer temperatures can increase or decrease herbicidal effects. The ultimate response depends on the relative effect of temperature on metabolism to the acid vs. metabolism to inactive compounds.

Rainfall tests confirm that uptake of the imidazolinones is fairly rapid, with sufficient uptake for control of sensitive species occurring within a few hours. The effect of rainfall on activity is species dependent. This can be partially explained by the relative importance of soil uptake for some species, even from postemergence applications. High relative humidity generally increases activity of these imidazolinones, as is true with most herbicides.

Soil moisture is also an important factor. Increased uptake of imazaquin by corn and imazamethabenz-methyl by wheat has been demonstrated with increased soil moisture, particularly under conditions of excess soil moisture. In greenhouse tests excess soil moisture has resulted in increased injury to wheat from imazamethabenz-methyl, but only at rates much higher than those used when applied at a very early growth stage, when this herbicide would not typically be applied.

Most of the results discussed in this chapter are from tests conducted under controlled growth-chamber conditions, where only one environmental factor was varied while others remained constant. In the field, all of these environmental factors are interdependent, and a change in one factor will often result in changes in many more. For example, rainfall can result in high humidity, excess soil moisture, and changes in temperature, while conditions of low soil moisture are the result of low relative humidity and high temperatures. In addition, many herbicide applications to soybeans and wheat (the major markets for imazaquin and imazamethabenz-methyl, respectively) are combinations of two or more herbicides. These multiple-factor interactions would be very difficult to evaluate in controlled-environment studies.

Even though experiments under controlled conditions have indicated that imidazolinone activity can be affected by environmental conditions, the imidazolinones have consistently performed very well when applied over a wide range of conditions. There is still a need, however, for further study of how this new class of herbicides is affected by environmental conditions.

REFERENCES

1. Extensive research has been done on the effect of environmental factors on imazamethabenz-methyl by J. C. Caseley and J. B. Pillmoor (1982), J. B. Pillmoor and J. C. Caseley (1982 to 1984), and R. Allen and J. C. Caseley (1985 to 1986). This research was conducted under contract for American Cyanamid at the Agricultural Research Council, Weed Research Organization, at Yarnton, Oxford, England. Results of this research were provided to American Cyanamid in a series of research reports. Some, but not all, of this research has been published.
2. **Pillmoor, J. B.,** Influence of temperature on the activity of AC 222,293 against *Avena fatua* L. and *Alopecurus myosuroides* Huds., *Weed Res.,* 25, 433, 1985.
3. **Shaner, D. L., Simcox, P. D., Robson, P. A., Mangels, G., Reichert, B., Ciarlante, D. R., and Los, M.,** AC 222,293: translocation and metabolic selectivity, in *Br. Crop Prot. Conf. — Weeds,* Vol. 1, British Crop Protection Council Publications, Croydon, England, 1982, 333.
4. **Malefyt, T., Robson, P., and Shaner, D. L.,** The effect of temperature on AC 252,214 in *Glycine max, Abutilon theophrasti,* and *Digitaria sanguinalis,* in *Aspects of Applied Biology,* No. 4, Association for Applied Biology, Warwick, England, 1983, 265.
5. **Edmund, R. M. and York, A. C.,** Effects of rainfall and temperature on postemergence control of sicklepod (*Cassia obtusifolia*) with imazaquin and DPX-F6025, *Weed Sci.,* 35, 231, 1987.
6. **Wills, G. D.,** Translocation of imazaquin in pitted morningglory (*Ipomoea lacunosa*), *Weed Sci. Soc. Am.,* 26 (Abstr.), 18, 1986.
7. **Malefyt, T. and Shaner, D. L.,** The effect of temperature on AC 263,499 in soybeans and alfalfa, *Weed Sci. Soc. Am.,* 26 (Abstr.), 71, 1986.

8. **Herrick, R. M.,** Factors Affecting the Herbicidal Activity of Imazethapyr and its Control of Spurred Anoda (*Anoda cristata* L. Schlect), Ph.D. thesis, Rutgers University, New Brunswick, NJ, 1987.
9. **Basham, G., Lavy, T. L., Oliver, L. R., and Scott, H. D.,** Imazaquin persistence and mobility in three Arkansas soils, *Weed Sci.,* 35, 576, 1987.
10. **Zadoks, J. C., Chang, T. T., and Konzak, C. F.,** A decimal code for the growth stages of cereals, *Weed Res.,* 14, 415, 1974.
11. **Bukovac, M. J.,** Herbicide entry into plants, in *Herbicides: Physiology, Biochemistry, Ecology,* Vol. 1, 2nd ed., Audus, L. S., Ed., Academic Press, New York, 1976, 335.
12. **Muzik, T. J.,** Influence of environmental factors on toxicity to plants, in *Herbicides: Physiology, Biochemistry, Ecology,* Vol. 2, 2nd ed., Audus, L. J., Ed., Academic Press, New York, 1976, 203.
13. **Shaner, D. L., Robson, P. A., and Umeda, K.,** Effect of mecoprop and dichlorprop on ^{14}C–AC 222,293 uptake, translocation, and metabolism in *Avena fatua, Proc. West. Soc. of Weed Sci.,* 37, 227, 1984.

Chapter 9

PHYSIOLOGICAL EFFECTS OF THE IMIDAZOLINONE HERBICIDES

Dale L. Shaner

TABLE OF CONTENTS

I. INTRODUCTION

Researchers at American Cyanamid Company have documented that imidazolinone herbicides kill plants by inhibiting the biosynthesis of the branched-chain amino acids through the inhibition of acetohydroxyacid synthase (Chapter 6). However, the physiological consequences of this inhibition are not obvious, nor is it apparent how this inhibition leads to the death of the plant. This chapter describes the physiological responses of plants to inhibition of AHAS by the imidazolinones and how these responses may eventually result in plant death.

II. INHIBITION OF GROWTH AND CELL DIVISION

In corn, root application of imazaquin inhibits leaf elongation within 6 h (Figure 1).[1] This growth retardation appears to result from an inhibition of cell division. Pillmoor and Caseley[2] found that imazamethabenz-methyl reduced the number of cells in division by 73% in *Avena fatua* and 55% in *Alopecurus myosuroides* within 24 h after application. In both species, cell division had stopped during interphase. Rost et al.[3] reported that imazapyr, imazaquin, imazethapyr, and imazamethabenz-methyl arrested cell division in pea roots in interphase.

Although imazapyr has little effect on respiration, photosynthesis, or lipid and protein synthesis, it does inhibit the rate of DNA synthesis by 63% within 24 h after application (Figure 2).[1] This inhibition is an indirect measurement of cell division. The inhibition of thymidine incorporation into DNA by imazapyr can be reversed by exogenously supplying corn with the branched-chain amino acids (Figure 3).[1] Rost et al.[3] demonstrated that supplementing pea root tips with all three branched-chain amino acids prevented inhibition of cell division by the herbicide. These results indicate a close relationship between inhibition of the synthesis of the branched-chain amino acids and DNA synthesis, cell division, and growth.

Although a relationship between synthesis of branched-chain amino acids and DNA synthesis is indicated, its mechanism remains unclear. The branched-chain amino acids could play a direct role in cell division. Although it has not been shown in plants, isoleucine appears to initiate DNA synthesis in Chinese hamster ovaries.[4,5] A similar phenomenon may occur in plants.

On the other hand, a protein vital for cell division may be rapidly synthesized and degraded. Any disruption of protein synthesis could affect the level of such a protein and interfere with cell division. More research is needed to define the relationship between branched-chain amino acids and cell division.

III. EFFECTS ON FREE AMINO ACID POOL SIZES

The imidazolinones also cause a change in the metabolite pool sizes of treated plants. Within 24 h after treatment, the levels of the free amino acid pools increased more than 30% in corn treated with imazapyr as compared to untreated corn, while the soluble protein pool in the root meristematic region decreased by 36% (Figure 4).[1] However, studies with radiolabeled amino acids show very little effect on protein synthesis rates.

The effects of imazapyr on the free amino acid pools and soluble protein pools suggest that imazapyr disrupts protein synthesis by changing the turnover rate of proteins. Rhodes et al.[6] found that the increase in free amino acid levels in *Lemna minor* caused by chlorsulfuron, another AHAS inhibitor, was due to turnover of preexisting proteins and not to new synthesis.

The changes in amino acid metabolism caused by the imidazolinone herbicides could be related to the plant's response to nitrogen starvation. The half-life of proteins in *L. minor*

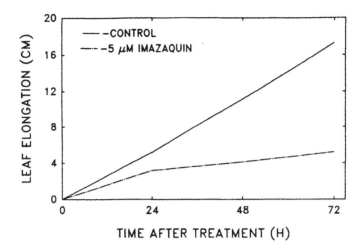

FIGURE 1. Effect of imazaquin on leaf elongation of maize. Plants were grown hydroponically and treated with 5 μM of imazaquin in the nutrient solution for 24 h. Elongation of the fourth leaf was measured.

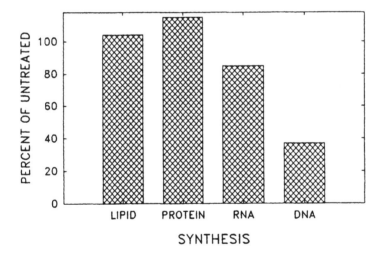

FIGURE 2. Effect of 150 μM of imazapyr on incorporation of ^{14}C-precursors into lipids, proteins, RNA, and DNA in corn root tips 24 h after initiation of treatment. The data are presented as percent of untreated tissue synthesis rates. (From Shaner, D. L. and Reider, M. L., *Pestic. Biochem. Physiol.*, 25, 248, 1986. With permission.)

decreases from 143 to 28 h if the plant is starved for nitrate.[7] The decrease in the synthesis of the branched-chain amino acids after imidazolinone treatment may signal the plant that nitrogen starvation has occurred, and this signal results in an increase in the rate of protein turnover.

This ability of the plant to scavenge amino acids from preexisting proteins may also be related to the observation that mature tissue appears to be relatively unaffected by the imidazolinones, since it has a large pool of proteins from which amino acids can be obtained; also, there is no shortage of energy since photosynthesis is unaffected. On the other hand, meristematic tissue, which is rapidly affected by the imidazolinones, requires a high level of amino acids for cell division and growth and has few protein reserves. Thus, inhibition

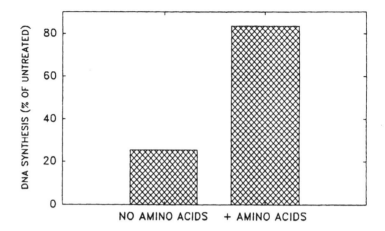

FIGURE 3.   Effects of supplementing corn root tips with 1 μM each of valine, leucine, and isoleucine on incorporation of ^{14}C-thymidine into DNA 24 h after treatment with 150 μM imazapyr. (From Shaner, D. L. and Reider, M. L., *Pestic. Biochem. Physiol.*, 25, 248, 1986. With permission.)

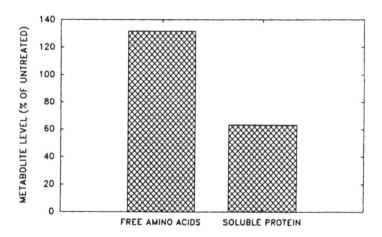

FIGURE 4.   Effects of 150 μM imazapyr on levels of free amino acids and soluble protein levels in corn root tips 24 h after initiation of treatment. (From Shaner, D. L. and Reider, M. L., *Pestic. Biochem. Physiol.*, 25, 248, 1986. With permission.)

of branched-chain amino acids in the meristematic region will have a profound and rapid effect, since the tissue cannot scavenge amino acids from other storage pools.

IV. EFFECT ON CARBOHYDRATE LEVELS AND PHOTOSYNTHATE TRANSPORT

The neutral sugar levels of corn leaves increase over 35% within 24 h after treatment with imazapyr.[1] This increase in sugars after imazapyr treatment is probably due to two phenomena. First, imidazolinones and other AHAS inhibitors do not inhibit photosynthesis,[1,8] so the plant continues to synthesize new photosynthate after herbicide treatment. Second, imidazolinones and sulfonylureas appear to inhibit the transport of photosynthate out of source leaves to the sinks in the plant.[9] The net result of these two phenomena is a build-up of sugars in the source leaves.

While it is still unclear why AHAS inhibitors retard photosynthate translocation, at least two different mechanisms could explain this inhibition. Work by Devine and co-workers[9,10] and Vanden Born et al.[11] suggested that chlorsulfuron disrupted photosynthate transport in excised leaves by inhibiting the transport out of the leaf. Thus, the AHAS inhibitors could affect the source leaf's ability to load photosynthate into the phloem. It is also possible that the inhibition of growth of the sinks within the plant reduces the sinks' demand for photosynthate. This decreased demand could result in less accumulation of photosynthate from the phloem so that the transport out of the source leaves is eventually halted. No data have been collected to support this second hypothesis.

This inhibitory effect on photosynthate transport can be reversed by exogenously applying the branched-chain amino acids to the plants,[9,10] indicating that the disruption of photosynthate transport is linked to the inhibition of synthesis of the branched-chain amino acids. However, the origin of this link is unclear.

One of the consequences of the disruption of photosynthate transport by imidazolinones is that these compounds inhibit their own transport and the translocation of other herbicides from source leaves to sinks. The transport of imazaquin is increased in sunflower when the plant is exogenously supplied with valine, leucine, and isoleucine (see Chapter 7). This increased transport is primarily to the apical meristem, which continues to grow in the presence of the branched-chain amino acids after imazaquin application.

Croon et al.[12] found that imazaquin and chlorimuron ethyl inhibited the translocation of haloxyfop-methyl out of the treated leaf to the growing points of sorghum. This inhibition could be overcome by exogenously supplying the branched-chain amino acids[13] or by applying one of the herbicides several days before the other. Since these AHAS inhibitors do not control most grass species, applying these herbicides separately probably allows the grass to metabolize them to nontoxic forms.

V. OTHER PHYSIOLOGICAL EFFECTS OF THE IMIDAZOLINONES

Imidazolinones inhibit transpiration in treated leaves within a few days after treatment. Imazaquin and imazethapyr decreased the water lost from *Xanthium strumarium* leaves to approximately 50% of the untreated plant's levels by 2 d after treatment (Figure 5). Imazapyr reduced water usage of *Imperata cylindrica* within 3 d after treatment. This reduction in plant water usage is due to two phenomena. First, growth inhibition in a treated plant will result in less water usage by that plant when compared to an untreated plant. Second, the rate of transpiration is reduced on a leaf-area basis, indicating that the stomata are also affected.

Inhibition of transpiration by herbicides is not unusual. Glyphosate, another inhibitor of amino acid biosynthesis, also causes a decrease in transpiration.[14] However, this loss occurs within a few hours after treatment rather than a few days, as is the case for the imidazolinones.

The mechanism of inhibition of transpiration by the imidazolinones is unknown but could be produced by disruption of many different processes in the plant since inhibition occurs 2 to 3 d after treatment.[15]

Interference with the hormonal status of the plant may be responsible for the decrease in transpiration and disruption of plant growth after treatment with an imidazolinone. Plants treated with imazaquin show leaf epinasty, proliferation of axillary buds, loss of apical dominance, chlorosis, and shortening of internodes.[16] These changes suggest that the hormonal balance in the plant has been altered in some way. Risley[16] found that imazaquin inhibited the release of auxin-induced ethylene from leaf discs in *X. strumarium* and soybeans, although the herbicide did not completely prevent the release. Such inhibition could be

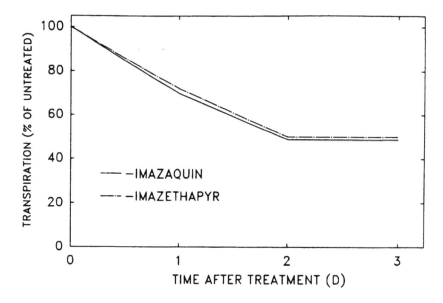

FIGURE 5.   Effects of 50 g/ha of imazaquin and imazethapyr on transpiration of *X. strumarium*. Herbicides were applied as a foliar spray with 0.25% DM 710 surfactant. Control plants were sprayed with surfactant alone. Transpiration was measured as loss in weight of pot plus plant and dividing this weight loss by the leaf area. The control plants transpired at a rate of 43 mg/cm²/day.

prevented if the leaf discs were supplemented with valine, leucine, and isoleucine, indicating that the inhibition was a secondary consequence of AHAS inhibition.

The loss of apical dominance could also be a secondary effect, since the production of auxin by the meristematic tissue enforces apical dominance. Treating a plant with an imidazolinone results in the death of the apical meristem and leads to a decrease in the rate of auxin production and the consequent release of apical dominance.

VI. THE RELATIONSHIP BETWEEN MODE OF ACTION OF THE IMIDAZOLINONES AND PHYSIOLOGICAL CHANGES

Treatment with the imidazolinones causes many different physiological changes in the plant: growth ceases because cell division and DNA synthesis are inhibited; transpiration and ethylene release decrease; and sugars and amino acids accumulate because protein synthesis and photosynthate transport are disrupted. Researchers have found that some of these changes relate in some way to the disruption of the synthesis of the branched-chain amino acids. These changes may also indicate a site of action of the imidazolinones other than AHAS inhibition.

However, results of work with imidazolinone-tolerant corn suggest that AHAS is the only site of action of the imidazolinone herbicides. In all cases of plant lines which have been selected for tolerance of the imidazolinones, the tolerant trait coincides with change in the AHAS enzyme in such a manner that it is no longer inhibited by the herbicide.[17] Line XA17 is highly tolerant of the imidazolinones and can survive levels more than three orders of magnitude higher than those required to kill a susceptible corn line.[17] Studies on the changes in amino acid levels, growth, and other processes have shown that none of these changes occur in XA17 after treatment with an imidazolinone (Figure 6). Therefore, these results indicate that these changes are secondary consequences of the inhibition of AHAS and are not attributable to other sites of action.

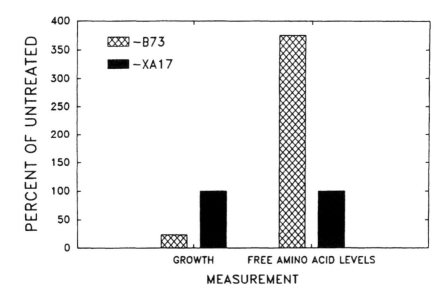

FIGURE 6.   Effect of 100 μM imazapyr applied as a soil drench on growth and free amino acid levels in susceptible (B73) and imidazolinone-tolerant (XA17) maize. The measurements were made two days after treatment. Growth was measured as the elongation rate of the fourth leaf.

VII. WHY DO PLANTS DIE AFTER IMIDAZOLINONE TREATMENT?

Work to date has shown that the mature tissue in the plant appears to be relatively unaffected by inhibition of AHAS. The tissue remains green for a long period of time and, in some cases, the plant can remain in an arrested state of growth for weeks after treatment before dying. Since photosynthesis is not affected, the green tissue is not devoid of energy.

However, disruption of photosynthate translocation can have a drastic effect on root growth, since the roots are completely dependent on energy supplied from the shoots. In fact, root growth inhibition has been shown to be a much more sensitive measure of imazaquin damage than shoot growth inhibition (Figure 7) because the effect of imazaquin on photosynthate transport results in cutting off the energy supply to the roots.

Likewise, disruption of metabolism of the growing points of the plant significantly affects the hormonal balance of the plant. Although it is not well understood how the balance of auxins, gibberellins, cytokinins, and abscisic acid control plant growth and development, inhibition of the growth and development of the growing points, which are the source and site of action of these hormones, will undoubtedly have a major effect on the plant. It is possible that disruption of branched-chain amino acid metabolism kills the plant only indirectly by interfering with the normal balance of hormones and energy distribution in the plant.

There is another physiological effect of inhibition of AHAS in plants. Rhodes et al.[6] found that α-aminobutyrate accumulates in *Lemna minor* after application of imazapyr. The accumulation of α-aminobutyrate occurs because of the transamination of 2-oxobutyrate, one of the precursors to AHAS for the synthesis of isoleucine. LaRossa et al.[18] have suggested that the toxicity of sulfonylureas, another class of AHAS inhibitor, is due to the accumulation of 2-oxobutyrate, which has been shown to interfere with methionine synthesis through inhibition of succinyl CoA.

2-Oxobutyrate arises from the deamination of threonine; a reaction catalyzed by threonine dehydratase. This enzyme is normally feedback-regulated by isoleucine. If accumulation of

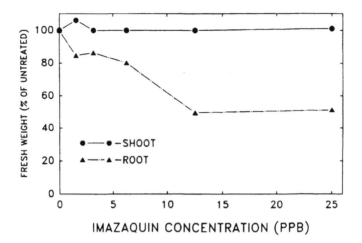

FIGURE 7.   Growth inhibition of corn shoots and roots by soil-applied imazaquin. Imazaquin was incorporated into a Sassafras sandy loam soil at the rates
indicated on a weight per weight basis. Measurements were taken 7 days after
planting the corn seeds.

either 2-oxobutyrate or α-aminobutyrate is playing a vital role in the phytotoxicity of imidazolinones, then isoleucine should provide some type of safening against these herbicides.
However, no work with plants has ever shown isoleucine to prevent the phytotoxicity of
any AHAS inhibitor. In light of this, it is difficult to propose a direct relationship between
the accumulation of either 2-oxobutyrate or α-aminobutyrate and the phytotoxicity of imidazolinones.

Understanding the sequence of events leading to the death of the plant after application
with an imidazolinone is very difficult because it involves the interaction of many different
pathways and metabolic routes. However, it also means that we now have a tool that can
be used to probe the relationships between branched-chain amino acid biosynthesis and other
metabolic processes.

REFERENCES

1. **Shaner, D. L. and Reider, M. L.,** Physiological response of corn *(Zea mays)* to AC 243,997 in combination
   with valine, leucine, and isoleucine, *Pestic. Biochem. Physiol.,* 25, 248, 1986.
2. **Pillmoor, J. B. and Caseley, J. C.,** The biochemical and physiological effects and mode of action of
   AC 222,293 against *Alopecurus myosuroides* Huds. and *Avena fatua* L., *Pestic. Biochem. Physiol.,* 27,
   340, 1987.
3. **Rost, T. L., Gladish, D., Steffen, J., and Robbins, J.,** Is there a relationship between branched-chain
   amino acid pool size and cell cycle inhibition in roots treated with imidazolinone herbicides, *J. Plant Growth
   Regul.,* 9, 227, 1990.
4. **Tobey, R. A. and Ley, K. D.,** Isoleucine-mediated regulation of genome replication in various mammalian
   cell lines, *Cancer Res.,* 31, 46, 1971.
5. **Tobey, R. A.,** Production and characterization of mammalian cells reversibly arrested in G1 by growth in
   isoleucine-deficient medium, *Meth. Cell Biol.,* 6, 67, 1973.
6. **Rhodes, D., Hogan, A. L., Deal, L., Jamieson, G. C., and Haworth, P.,** Amino acid metabolism of
   Lemna minor L. II. Response to chlorsulfuron, *Plant Physiol.,* 84, 775, 1987.
7. **Davies, D. D. and Humphrey, T. J.,** Amino acid recycling in relation to protein turnover, *Plant Physiol.,*
   61, 54, 1978.
8. **Ray, T. B.,** The mode of action of chlorsulfuron: a new herbicide for cereals, *Pestic. Biochem. Physiol.,*
   17, 10, 1982.

9. **Devine, M. D.,** Phloem translocation of herbicides, *Rev. Weed Sci.,* 4, 191, 1989.

10. **Devine, M. D., Bestman, H. D., and Vanden Born, W. H.,** Physiological basis for the different phloem mobilities of chlorsulfuron and clopyralid, *Weed Sci.,* 38, 1, 1990.

11. **Vanden Born, W. H., Bestman, H. D., and Devine, M. D.,** The inhibition of assimilate translocation by chlorsulfuron as a component of its mechanism of action, in *Factors Affecting Herbicide Activity and Selectivity,* Eur. Weed Res. Soc. Symp., in press.

12. **Croon, K. A., Ketchersid, M. L., and Merkle, M. G.,** Effect of bentazon, imazaquin and chlorimuron on the absorption and translocation of the methyl ester of haloxyfop, *Weed Sci.,* 37, 645, 1989.

13. **Hahn, K. L. and Coble, H. D.,** The effect of exogenously supplied amino acids on the antagonistic interaction between quizalofop and chlorimuron, *Weed Sci., Soc. Am.,* 29 (Abstr.), 86, 1989.

14. **Shaner, D. L.,** Effects of glyphosate on transpiration, *Weed Sci.,* 26, 513, 1978.

15. **Shaner, D. L.,** Absorption and translocation of imazapyr in *Imperata cylindrica* (L.) Raeuschel and effects on growth and water usage, *Trop. Pest Manage.,* 34, 388, 1988.

16. **Risley, M. A.,** Imazaquin Herbicidal Activity: Efficacy, Absorption, Translocation and Ethylene Synthesis, Ph.D. thesis, University of Arkansas, Fayetteville, 1986.

17. **Newhouse, K. E., Shaner, D. L., Wang, T., and Fincher, R.,** Genetic modification of crop responses to imidazolinone herbicides, in *Managing Resistance to Agrochemicals: From Fundamental Research to Practical Strategies,* Green, M. B., LeBaron, H. M., and Moberg, W. K., Eds., ACS Symp. Ser., No. 421, American Chemical Society, Washington, D.C., 1990, 474.

18. **LaRossa, R. A., VanDyk, T. K., and Smulski, D. R.,** Toxic accumulation of α-ketobutyrate caused by inhibition of the branched-chain amino acid biosynthetic enzyme acetolactate synthase in *Salmonella typhimurium, J. Bacteriol.,* 169, 1372, 1987.

Chapter 10

IMIDAZOLINONE-TOLERANT CROPS

Keith Newhouse, Theodora Wang, and Paul Anderson

TABLE OF CONTENTS

I. INTRODUCTION

Herbicide-tolerant crops offer a favorable alternative to the traditional approach of identifying and developing new crop herbicides.[1-3] Although the past success of the empirical screening method is apparent when one considers that selective herbicides are available for most cultivated crop species,[4] by making crops tolerant to currently registered herbicides, the search for novel, selective herbicides for the crop is virtually eliminated and the expense incurred in registering such herbicides is greatly reduced.

Registration of a herbicide for a specific crop may not be justified if the market size is limited, even if a selective herbicide were to be discovered. However, by making the crop tolerant to currently registered herbicides, new weed-control options become available for the crop. Alternatively, herbicide-tolerant crops may actually provide the impetus to register new commercial herbicides that have superior weed control characteristics but insufficient inherent crop selectivity. This situation could occur when use of the herbicide-tolerant crops provides enough additional market size to justify registration of the product.

Herbicide-tolerant crop varieties can also overcome the effects of residual herbicide activity on subsequent plants in a crop rotation. Residual activity will not harm a follow crop if that crop is tolerant of the herbicide. Therefore, the tolerance factor would permit herbicide application without concern for damage to the following crop.

Herbicide-tolerant crop plants can provide a multitude of new weed control and marketing opportunities. The decision to introduce novel tolerance factors into a crop would depend on the assessment of these opportunities for both the crop and the herbicide for which novel tolerance can be introduced.

II. RATIONALE FOR SELECTING IMIDAZOLINONE-TOLERANT CROPS

The imidazolinone herbicides have many favorable characteristics that make them ideal for use with herbicide-tolerant crops, including a wide spectrum of weed control activity, low environmental impact and low toxicity to mammals, flexibility of application, and registration of these herbicides in many countries for a variety of uses.

Corn was selected as the first target species for introducing imidazolinone tolerance into a crop. Trials in which imidazolinones were tested as soybean herbicides demonstrated their efficacy for control of the complex of weeds found in the midwestern U.S. The imidazolinones not only eliminated several weed species controlled by available corn herbicide products, but also provided control of *Sorghum bicolor* (L.) Moench, *Sorghum halepense*, and other weeds not adequately controlled by current products. Such activity could justify the development of an imidazolinone specifically for use on corn. Thus, imidazolinone-tolerant corn was viewed as a means by which the desirable weed control characteristics of the imidazolinone herbicides could be made available to corn producers.

The choice of canola as a target for imidazolinone tolerance was based on somewhat different considerations. Canola is in the mustard (Cruciferae) family and, therefore, related to several cruciferous weed species. This close relationship between the crop and its weeds makes the discovery of selective herbicides difficult to accomplish. In general, canola has low tolerance for most herbicides and is particularly sensitive to the imidazolinones. Discovery of an imidazolinone herbicide that is selective to canola, but that maintains desirable weed control, is an especially formidable task.

The imidazolinones are very effective in controlling cruciferous weeds. As such, imidazolinone-tolerant canola offers the potential for significantly better weed control than is provided by current canola herbicides. Also, the inherent sensitivity of canola could result in it being damaged by imidazolinone herbicides applied during a previous season. Thus, the development of an imidazolinone-tolerant canola would alleviate concerns for potential follow-crop damage.

Although the justifications for development of imidazolinone-tolerant corn differ from those for development of imidazolinone-tolerant canola, both crops are excellent candidates. Similar analyses have been conducted for other crops, to determine the feasibility and need for introduction of novel imidazolinone resistance.

III. METHODS FOR OBTAINING IMIDAZOLINONE-TOLERANT CROPS

At least three different types of *in vitro* selection have been successful in identifying imidazolinone tolerance in plant species. Callus culture selection was used to identify imidazolinone-tolerant corn (*Zea mays* L.);[5,6] suspension culture selection was used to identify imidazolinone-tolerant *Datura innoxia* P. Mill.;[7] and cultured microspore selection has been successful in identifying imidazolinone-tolerant canola (*Brassica napus* L.).[8,9] Imidazolinone-tolerant plants have also been obtained by screening the M_2 progeny of mutagenized seed.

Researchers at Plant Science Research Inc. (formerly known as MGI or Molecular Genetics Inc.) were among the first to regenerate plants routinely from corn tissue cultures. This ability was coupled with cell culture selection techniques in order to identify imidazolinone-tolerant corn.[5] Embryogenic corn cell cultures were initiated from immature embryos. Tissue from these cultures was placed on media containing sublethal doses of imidazolinone herbicides. Cultures with good growth were periodically transferred to fresh media with the same concentration of herbicide. Callus was subcultured every 15 to 30 d; the interval depended on the rate of growth. In some cases, the concentration of herbicide in the media was increased threefold after three to five subculture intervals. By using this protocol, vigorously growing tissue sectors could be identified. These sectors were subcultured to produce homogeneous imidazolinone-tolerant cell lines. The tolerant lines were characterized (Figure 1) and entered into a regeneration protocol. Regenerated plants expressed elevated tolerance of the imidazolinones at the whole-plant level. Thus, the procedure was successful in identifying imidazolinone-tolerant corn plants.

Minor inherent differences in tolerance for imidazolinone herbicides have been shown to occur among corn hybrids;[10,11] however, tolerance of the level and nature selected at Plant Science Research (PSR) Inc. had never been observed in corn elsewhere. Therefore, the selected lines were believed to represent novel mutational events in corn.

Suspension cultures were used to identify cell lines of *Datura innoxia* tolerant of imidazolinone herbicides.[7] Haploid cultures derived from anthers were mutagenized with ethyl methane sulfonate and grown in the presence of herbicide. The selection agents used in this particular study were sulfonylurea herbicides. Tolerant variants were isolated at the rate of one to three per 10^7 mutagenized cells. The selected variants differed widely in their resistance characteristics; some were highly tolerant of both imidazolinone and sulfonylurea herbicides, others were more tolerant of imidazolinones than sulfonylureas, and still others were more tolerant of sulfonylurea herbicides than of imidazolinone herbicides. None of these selections could be regenerated into whole plants.

Imidazolinone-tolerant *Brassica napus* L. plants were identified at Allelix, Inc. by yet a third *in vitro* selection system. Mutagenized microspores were cultured in the presence of normally lethal concentrations of imidazolinone herbicides. Susceptible microspores do not grow in the presence of the herbicides, whereas tolerant individuals grow and are thereby selected.[8,9] These selected microspores develop into embryos, from which tolerant plants grow. This procedure has also been used to identify variants tolerant of the imidazolinone as well as sulfonylurea herbicides.

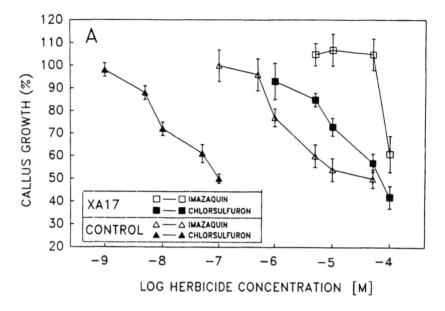

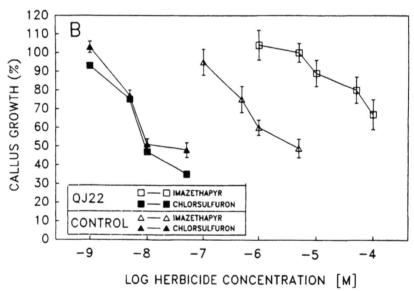

FIGURE 1. Panel A: effect of imazaquin and chlorsulfuron on growth of XA17 and unselected callus. Callus fresh weight was quantified following 14 d of growth, and data are presented as a percent of callus growth in the absence of herbicide. Values are the average and standard deviation of three replicates. Panel B: effect of imazethapyr and chlorsulfuron on growth of QJ22 and unselected callus. (From Anderson, P. C. and Georgeson, M., *Genome*, 31, 994, 1989. With permission.)

Imidazolinone-tolerant plants can be identified *in vivo* by screening progeny from mutagenized seed. Seed is treated with irradiation, chemical mutagens, or other mutagenizing agents. These treated seeds form the M_1 generation. The M_1 seeds are planted and M_2 generation seeds are harvested from the resulting plants. M_2 seedlings are screened for traits being selected. By screening at the M_2 rather than at the M_1 generation, it is possible to identify recessive as well as dominant mutations, with a much lower frequency of chimeric plants present. Therefore, plant selections from the M_2 generation are more likely to breed true for the selected phenotype than M_1 plant selections.

TABLE 1
Summary of Imidazolinone-Tolerant Corn Cell Lines Identified by *In Vitro* Selection

Cell line	Selection agent	Callus tolerance (fold)[a]			Regenerated plants	Resistant plants
		Imazaquin	Imazethapyr	Chlorsulfuron		
Class A[b]						
XA17	Imazaquin	100	300	1000	+	+
UV18	Imazaquin	100	100	100	−	−
AC17	Imazaquin	300	ND[c]	1000	+	−
QT15	Imazethapyr	ND	500	1000	−	−
Class B[d]						
QJ22	Imazethapyr	5	30—50	0	+	+
XS40	Imazaquin	10—30	100	0	+	+
XI12	Imazaquin	10—30	100	0	+	+
ZA54	Imazapyr	ND	30—50	0	+	+

[a] As compared to unselected parent cell lines.
[b] Class A lines have cross tolerance for chlorsulfuron.
[c] Not determined.
[d] Class B lines have little or no cross tolerance for chlorsulfuron.

From Anderson, P. C. and Georgeson, M., *Genome*, 31, 994, 1989. With permission.

Imidazolinone-tolerant wheat was identified by screening M_2 seed. M_2 seeds were produced by treating seeds of Fidel winter wheat with a 3 mM solution of sodium azide, planting them in the field, and harvesting the seeds from the resulting plants. Imidazolinone-tolerant variants were identified by treating the M_2 seeds with herbicide and selecting the resistant plants. Four plants were selected as having novel tolerance; it is not known whether the four selections represent the same or different mutational events. The selected tolerance trait was transmitted to the progeny of the selections.

The sulfonylurea herbicides possess the same mechanism of action as the imidazolinones.[12,13] Therefore, selection schemes that have successfully produced plants tolerant of sulfonylurea herbicides should also work for the imidazolinones. In addition to the selections made through suspension culture selection and microspore selection, plants tolerant of sulfonylurea herbicides have also been identified using tissue culture[14] and seed mutagenesis.[15-17] Also, the DNA fragments coding for tolerance have been isolated from sulfonylurea-tolerant plants and used to transform normally sensitive plants into sulfonylurea-tolerant plants.[3,18,19] Presumably, transformation with DNA fragments coding for tolerance of the imidazolinones would also be successful in creating imidazolinone-tolerant plants.

IV. CHARACTERISTICS OF IMIDAZOLINONE-TOLERANT CROPS

Imidazolinone-tolerant corn is the most advanced and best characterized of the imidazolinone-tolerant crops.[20] The tolerant corn will be described in this chapter to provide an example of characteristics associated with imidazolinone-tolerant crops. Other crops with imidazolinone tolerance are expected to exhibit similar characteristics, provided they employ the same mechanism of tolerance as the tolerant corn.

As described above, imidazolinone-tolerant corn cell lines were selected *in vitro* at PSR.[5] A number of independent variants were identified (Table 1). Concurrently, the mechanism of action of the imidazolinone herbicides was elucidated in independent experiments.[12,13,21] The imidazolinones were found to inhibit acetohydroxyacid synthase (AHAS). Following this discovery, the AHAS activity of protein extracts from tolerant cell lines was assayed.

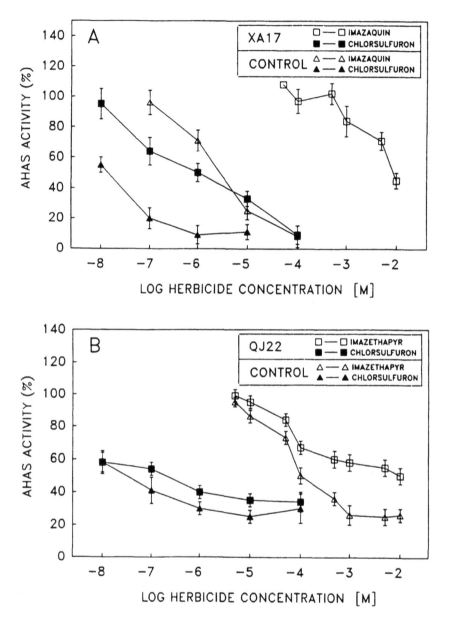

FIGURE 2.   Panel A: effect of imazaquin and chlorsulfuron on acetohydroxyacid synthase activity extracted from XA17 and unselected callus. Data are presented as a percent of enzyme activity in the absence of herbicide. Values are the average and standard deviation of three replicates. Panel B: effect of imazethapyr and chlorsulfuron on acetohydroxyacid synthase activity extracted from QJ22 and control callus. (From Anderson, P. C. and Georgeson, M., *Genome*, 31, 994, 1989. With permission.)

For each selection, the AHAS activity from the selected cell line was determined to be tolerant of inhibition by the imidazolinone herbicides (Figure 2). In contrast, AHAS activity from unselected tissue had normal inhibition by the imidazolinones.

Fertile plants with imidazolinone tolerance have been regenerated by PSR Inc. from five of the tolerant cell lines (Table 1). Nonsegregating tolerant progenies were obtained following two generations of self-pollination. These nonsegregating progenies were used to characterize the inheritance of the tolerance trait and the biochemical basis for tolerance.

TABLE 2
Genetic Analysis of Imidazolinone-Tolerant Corn[a]

Genotype	Observed ratio T:S	Expected ratio T:S	χ^2 fit of observed to expected	Probability of a larger χ^2 value	Number of progenies tested
Susceptible inbreds					
B73	All S	All S			
Mo17	All S	All S			
XA17 selection					
Homozygous	All T	All T			
F_1	All T	All T			
F_2	634:232	649.5:216.5	1.48	0.25—0.10	12
Testcross	800:859	829.5:829.5	2.10	0.25—0.10	20
XI12 selection					
Homozygous	All T	All T			
F_1	All T	All T			
F_2	823:285	831:277	0.30	0.75—0.50	15
Testcross	838:839	838.5:838.5	0.001	>0.995	15
QJ22 selection					
Homozygous	All T	All T			
F_1	All T	All T			
F_2	944:341	963.75:321.25	1.62	0.25—0.10	18
Testcross	622:704	683:683	1.29	0.50—0.25	18

[a]   Seedlings were treated postemergence at the 3-leaf stage with 150 g/ha of imazethapyr, and number of tolerant (T) and susceptible (S) individuals were determined at approximately 3 weeks after treatment. F_1 hybrids are the cross between susceptible inbreds and homozygous resistant selections, F_2 progenies were produced by self-pollination of the F_1 and testcross progeny are produced by crossing the F_1 hybrids with a susceptible inbred.

A. GENETIC ANALYSIS OF IMIDAZOLINONE-TOLERANT CORN

Nonsegregating tolerant progenies derived from the selections were crossed with a susceptible inbred line. For every selection, the resulting F_1 hybrids were uniformly tolerant of the imidazolinones, their F_2 progenies segregated approximately in a 3 tolerant: 1 susceptible ratio, and their testcross progenies segregated approximately in a 1 tolerant: 1 susceptible ratio (Table 2). These results are consistent with the hypothesis that the novel imidazolinone-tolerance is inherited as a single, dominant allele.

Lines with different tolerance alleles were intercrossed to determine whether more than one locus in the corn genome was conferring tolerance or whether the tolerance traits were alleles at the same locus. If the traits being tested were allelic or closely linked, the F_2 and testcross progenies from the intercrosses would be uniformly tolerant when sprayed with imidazolinone herbicides. If the traits were nonallelic and unlinked, the F_2 progenies would segregate in a 15 tolerant: 1 susceptible ratio and the testcross progenies would segregate in a 3 tolerant: 1 susceptible ratio.

The four alleles examined to date define two unlinked loci that can confer tolerance (Table 3), indicating that at least two loci for the AHAS enzyme are present in corn. The tolerance genes found in selections XA17 and XI12 appear to be alleles at one locus, and the tolerance genes from selections QJ22 and XS40 appear to be alleles at a second locus. The two loci have been mapped to chromosomes by using genetic marker stocks in which reciprocal translocations are linked with the waxy endosperm locus. These marker stocks facilitate assignment of a locus to a chromosome arm. The locus that XA17 and XI12 define appears to be on the long arm of chromosome 5. The locus that QJ22 and XS40 define is on chromosome 4 and may be close to the centromere, as associations with marker stocks for both the short and the long arm of chromosome 4 were observed.

TABLE 3

Chromosomal Location and Cross-Tolerance Spectrum for Five Imidazolinone-Tolerant Alleles in Corn Derived from Tissue Culture Selection

		Tolerance	
Resistance allele	Chromosome location	Imidazolinone	Sulfonylurea
XA17	5	Tolerant	Resistant
XI12	5	Tolerant	Susceptible
QJ22	4	Tolerant	Susceptible
XS40	4	Tolerant	ND[a]
ZA54	ND	Tolerant	ND

[a]   Not determined.

From Newhouse, K. E., Shaner, D. L., Wang, T., and Fincher, R., in *Managing Resistance to Agrochemicals: From Fundamental Research to Practical Strategies*, Green, M. B., LeBaron, H. M., and Moberg, W. K., Eds., ACS Symp. Ser. No. 421, American Chemical Society, Washington, D.C., 1990, 474. With permission.

TABLE 4

Increase in Tolerance for Three AHAS-Inhibiting Herbicides Provided by Three Imidazolinone-Tolerance Alleles[a]

	Tolerance Allele		
Herbicide	XA17	XI12	QJ22
	Increased level of resistance		
Imazethapyr	>1,000 ×	1,000 ×	300 ×
Imazaquin	>1,000 ×	60 ×	15 ×
Sulfometuron methyl	>1,000 ×	<2 ×	<2 ×

[a]   Tolerance determinations are based on postemergence herbicide applications that cause 50% growth inhibition.

From Newhouse, K. E., Shaner, D. L., Wang, T., and Fincher, R., *Managing Resistance to Agrochemicals*, Green, M. B., LeBaron, H. M., and Moberg, W. K., Eds., ACS Symp. Ser. No. 421, American Chemical Society, Washington, D.C., 1990, 474. With permission.

B. BIOCHEMISTRY AND PHYSIOLOGY OF IMIDAZOLINONE-TOLERANT CORN

The mechanism of action for the imidazolinones was still unknown when the program to select imidazolinone-tolerant corn began. Before starting the selection program, an assumption was made that a basic metabolic function was affected and that tolerance selected at the cellular level would be expressed at the whole-plant level. In practice, the magnitude and spectrum of tolerance measured in the tolerant cell lines are similar to expression at the whole-plant level. For example, the cell line XA17 has the greatest level of tolerance for imazethapyr in tissue culture, followed by XI12 and XS40; QJ22 has the lowest level of tolerance (Table 1). This same ranking for tolerance holds true at the whole-plant level (Table 4).

Tolerance at the whole-plant level has been investigated extensively for three of the mutants. Since the imidazolinones inhibit the same enzyme as the sulfonylurea herbicides,[12,13]

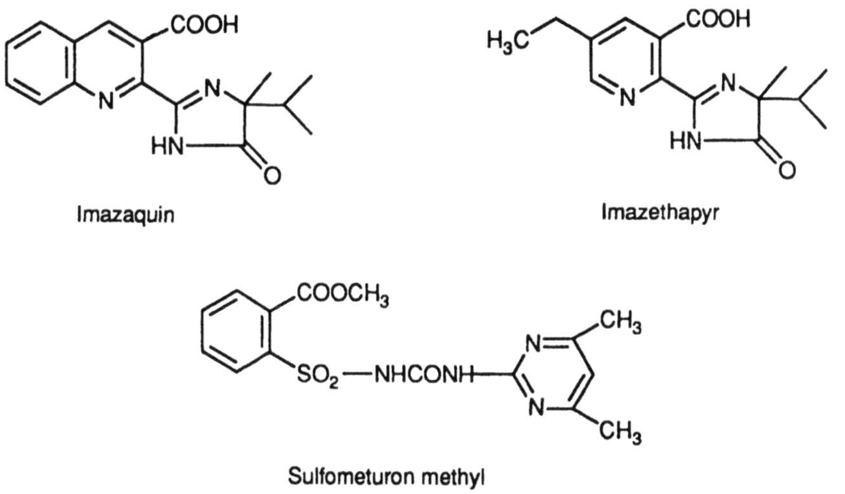

FIGURE 3. Structures of three AHAS-inhibiting herbicides: imazaquin and imazethapyr (imidazolinones) and sulfometuron methyl (a sulfonylurea).

tolerance of both classes of herbicides was investigated. Two imidazolinone herbicides — imazethapyr and imazaquin — and one sulfonylurea herbicide — sulfometuron methyl — were used. (Figure 3)

The tolerance from the selection XA17 confers a greater than 1000-fold increase in tolerance of the three herbicides tested (Table 4). A high level of tolerance for other imidazolinone and sulfonylurea herbicides was also noted. Plants with tolerance from the selection XI12 displayed high levels of tolerance for imazethapyr, somewhat less tolerance for imazaquin, and little or no tolerance for sulfometuron methyl. Plants containing resistance from the QJ22 selection are also tolerant of the imidazolinones and not of the sulfonylureas; however, the level of imidazolinone tolerance is only 20 to 25% that of XI12.

Protein extracts from tolerant plants contain AHAS activity insensitive to the imidazolinones, similar to what was observed for the cell cultures. AHAS activity profiles for extracts from the tolerant lines correspond to the responses at the whole-plant level.[22] AHAS activity from XA17 is highly insensitive to inhibition both by the imidazolinones and sulfonylureas; AHAS activity from XI12 and QJ22 is selectively insensitive to inhibition by the imidazolinones and sensitive to the sulfonylureas, but the level of insensitive AHAS activity from XI12 is greater than that from QJ22.

Aside from their novel tolerance of imidazolinone herbicides, plants containing the tolerant alleles appear to be unchanged. No morphological changes are observed in the absence of herbicide treatments. At the enzyme level, feedback inhibition by valine and leucine is unchanged for AHAS activity from the tolerant plants.[22] No deleterious effects attributable to herbicide-insensitive AHAS have been observed in tolerant plants.

V. COMMERCIALIZATION AND DEVELOPMENT OF IMIDAZOLINONE-TOLERANT CROPS

A. IMIDAZOLINONE-TOLERANT CORN

Pioneer Hi-Bred International is using the breeding method known as backcrossing to introduce the imidazolinone-tolerance trait into more than 100 inbred lines of corn. Genetic analyses have shown that imidazolinone tolerance is inherited as a single, semidominant gene. The backcrossing method is used to transfer simply inherited traits from one variety to another, thereby creating a nearly isogenic version of the original inbred line.

The backcrossing method[23] proceeds as follows: The tolerant donor is crossed to a susceptible elite line. The F_1 generation is all heterozygous and tolerant, receiving 50% of its genetic material from each parent. This generation is then crossed back to the elite (recurrent) parent. The gamete from the F_1 plant will, on the average, contain half of its genetic material from the donor and half from the recurrent parent due to independent assortment at meiosis. This gamete is united with one from the recurrent parent. Thus, progeny from this first backcross generation are now 75% recurrent parent and 25% donor parent. The process of crossing back to the recurrent parent is repeated until approximately 99% of the recurrent parent is recovered. The inbreds are then selfed to derive homozygous-tolerant seed.

Approximately 7 seasons are required to reach 99% recovery of the recurrent parent's genetic material, but by conducting the backcrossing program in a tropical environment, such as Hawaii, 3 seasons can be completed in 1 year. This reduces the number of years required for backcrossing and subsequent selfing to a minimum of 3 and probably a maximum of 4.

Once an inbred is converted, it must be tested in hybrid combinations to evaluate not only herbicide tolerance, but also performance for yield, lodging resistance, and other agronomic traits. Because of the potential for linkage of the herbicide tolerance allele to undesirable genes, a converted inbred may not necessarily substitute exactly for the original susceptible inbred in hybrid production. If performance of the hybrids produced with the converted inbred is equal to or better than that of the hybrids produced with the original inbred, then the imidazolinone-tolerant hybrid is ready for introduction. The field tests necessary to arrive at this decision can take up to 3 years depending on the philosophy of the company and its past experience with converted inbreds. Additionally, time must be allotted to produce the quantities of inbred seed that are needed for commercial hybrid production.

As Pioneer Hi-Bred International conducted the inbred conversion program, Cyanamid conducted efficacy and tolerance trials. Researchers treated tolerant hybrids with imidazolinone herbicides to determine the appropriate herbicide use recommendations for corn. Also, corn metabolism and residue studies were conducted in order to determine rates of absorption of the imidazolinone herbicides by the corn plants, accumulation sites, and the form in which the herbicide is present. Corn metabolism studies conducted with imazethapyr herbicide demonstrated a similar metabolism pattern for both susceptible (homozygous) and tolerant (heterozygous) corn. The pattern consists of similar metabolites to those found in soybeans (see Chapter 11): AC 263,499 (the parent compound), AC 288,511, and traces of the glucoside conjugate of AC 288,511.

Corn hybrids with excellent tolerance to the imidazolinones have been developed. Both heterozygous- and homozygous-tolerant corn hybrids exhibit field tolerance of imazethapyr. Yields have been measured in the absence of imidazolinone treatments for hybrids containing the imidazolinone-tolerance trait. To date, no indication of any deleterious effects attributed to or linked to the herbicide-tolerant allele has been observed. Likewise, no yield reductions have been observed when imidazolinone-tolerant corn is treated with commercial use rates of the imidazolinone herbicides.

B. OTHER IMIDAZOLINONE-TOLERANT CROPS

The strategies used to commercially develop other imidazolinone-tolerant crops will depend on the particular crop and how it is used. For example, using the tolerant crop only as a follow crop would require a different strategy than if application of the herbicide were made to the tolerant crop. Also, the herbicides used on the tolerant crop would depend on the level and spectrum of tolerance expressed in the crop being considered.

The ability to create imidazolinone-tolerant crops routinely through transformation procedures also will affect the direction that future programs for imidazolinone-tolerant crops will follow. Higher levels of tolerance may be achieved by using gene constructs with high expression levels or by incorporating multiple copies of the tolerance gene. The need for tedious backcrossing programs may, thus, be reduced or eliminated.

The experience gained during the commercial development of imidazolinone-tolerant corn and canola will also be important in guiding the development of other herbicide-tolerant crops and will help in introducing genetically modified crops to the marketplace.

VI. SUMMARY

The imidazolinone herbicides have desirable characteristics that could make them useful in several crop species. The decision whether to proceed with discovery and development of imidazolinone-tolerant crops must be considered separately for each crop in order to determine the challenges and opportunities that exist. In order for the project to be a success, herbicide-tolerant crop varieties should offer benefits to the farmer, the seed company, and the chemical company.

Imidazolinone-tolerant corn is in commercial development, and tolerant hybrids will be sold widely starting in 1992. Introduction of imidazolinone tolerance in other crops is in progress or under consideration.

Commercial development of imidazolinone-tolerant crop varieties requires a close working relationship between the seed supplier and the chemical company. The seed company must demonstrate that imidazolinone tolerance does not have deleterious agronomic consequences, either in the absence or presence of imidazolinone herbicides. Likewise, the chemical company must demonstrate that sufficient crop tolerance and safety margins exist for the rates and mixtures of herbicides needed for weed control. Close coordination of the seed company's program for herbicide-tolerant crop variety development with the chemical company's use recommendations and registration work ultimately culminates in the release of herbicide-tolerant crop combinations to the farmer.

REFERENCES

1. **Botterman, J. and Leemans, J.**, Engineering herbicide resistance in plants, *Trends Genet.*, 4, 219, 1988.
2. **Chaleff, R. S.**, Herbicide resistance, in *Biotechnology and Crop Improvement and Protection*, Day, P. R., Ed., British Crop Protection Council Publications, Croyden, England, 1986, 111.
3. **Mazur, B. and Falco, C.**, The development of herbicide resistant crops, *Annu. Rev. Plant Physiol. Plant Mol. Biol.*, 40, 441, 1989.
4. *Herbicide Handbook of the Weed Science Society of America*, 6th ed., Weed Science Society of America, Champaign, IL, 1989.
5. **Anderson, P. C. and Georgeson, M.**, Herbicide-tolerant mutants of corn, *Genome*, 31, 994, 1989.
6. **Anderson, P. C. and Hibberd, K.**, Evidence for the interaction of an imidazolinone herbicide with leucine, valine, and isoleucine metabolism, *Weed Sci.*, 33, 479, 1985.
7. **Saxena, P. K. and King, J.**, Herbicide resistance in *Datura innoxia*: cross-resistance of sulfonylurea-resistant cell lines to imidazolinones, *Plant Physiol.*, 86, 863, 1988.
8. **Swanson, E. B., Coumans, M. P., Brown, G. L., Patel, J. D., and Beversdorf, W. D.**, The characterization of herbicide tolerant plants in *Brassica napus* L. after *in vitro* selection of microspores and protoplasts, *Plant Cell Rep.*, 7, 83, 1988.
9. **Swanson, E. B., Herrgesell, M. J., Arnoldo, M., Sippell, D. W., and Wong, R. S. C.**, Microspore mutagenesis and selection: canola plants with field tolerance to the imidazolinones, *Theor. App. Genet.*, 78, 525, 1989.
10. **Renner, K. A., Meggitt, W. F., and Penner, D.**, Response of corn (*Zea mays*) cultivars to imazaquin, *Weed Sci.*, 36, 625, 1988.
11. **Sander, K. W. and Barrett, M.**, Differential imazaquin tolerance and behavior in selected corn (*Zea mays*) hybrids, *Weed Sci.*, 37, 290, 1989.

12. **Shaner, D. L., Anderson, P. C., and Stidham, M. A.,** Imidazolinones: potent inhibitors of acetohydroxyacid synthase, *Plant Physiol.,* 76, 545, 1984.

13. **Chaleff, R. S. and Mauvais, C. J.,** Acetolactate synthase is the site of action of two sulfonylurea herbicides in higher plants, *Science,* 224, 1443, 1984.

14. **Chaleff, R. S. and Ray, T. B.,** Herbicide-resistant mutants from tobacco cell cultures, *Science,* 223, 1148, 1984.

15. **Haughn, G. W. and Somerville, C. R.,** Sulfonylurea-resistant mutants of *Arabidopsis thaliana, Mol. Gen. Genet.,* 204, 430, 1986.

16. **Sebastian, S. A. and Chaleff, R. S.,** Soybean mutants with increased tolerance for sulfonylurea herbicides, *Crop Sci.,* 27, 948, 1987.

17. **Sebastian, S. A., Fader, G. M., Ulrich, J. F., Forney, D. R., and Chaleff, R. S.,** Semidominant soybean mutation for resistance to sulfonylurea herbicides, *Crop Sci.,* 29, 1403, 1989.

18. **Haughn, G. W., Smith, J., Mazur, B., and Somerville, C.,** Transformation with a mutant *Arabidopsis* acetolactate synthase gene renders tobacco resistant to sulfonylurea herbicides, *Mol. Gen. Genet.,* 211, 266, 1988.

19. **Lee, K. Y., Townsend, J., Tepperman, J., Black, M., Chui, C. F., Mazur, B., Dunsmuir, P., and Bedbrook, J.,** The molecular basis of sulfonylurea herbicide resistance in tobacco, *EMBO J.,* 7, 1241, 1988.

20. **Newhouse, K. E., Shaner, D. L., Wang, T., and Fincher, R.,** Genetic modification of crop responses to imidazolinone herbicides, in *Managing Resistance to Agrochemicals: From Fundamental Research to Practical Strategies,* Green, M. B., LeBaron, H. M., and Moberg, W. K., Eds., ACS Symp. Ser. No. 421, American Chemical Society, Washington, D.C., 1990, 474.

21. **Shaner, D. L. and Anderson, P. C.,** Mechanism of action of the imidazolinones and cell culture selection of tolerant maize, in *Biotechnology in Plant Science,* Zaitlin, M., Day, P. R., and Hollaender, A., Eds., Academic Press, Orlando, 287, 1985.

22. **Newhouse, K. E., Singh, B. K., Shaner, D. L., and Stidham, M. A.,** Mutations in corn conferring resistance to imidazolinone herbicides, manuscript in preparation.

23. **Fincher, R. R.,** Transfer of *in vitro* selected imidazolinone resistance to commercial maize hybrids, in *Prospects for Amino Acid Biosynthesis Inhibitors in Crop Protection and Pharmaceutical Chemistry,* Copping, L. G., Dalziel, J., and Dodge, A. D., Eds., British Crop Protection Council, Farnham, England, 69, 1989.

Chapter 11

PLANT METABOLISM

Anhorng Lee, Paul E. Gatterdam, Timmy Y. Chiu, N. Moorthy Mallipudi, and Ruth R. Fiala

TABLE OF CONTENTS

I. INTRODUCTION

The fundamentals of the metabolic reactions and pathways of modern herbicides in higher plants have been thoroughly reviewed.[1] The metabolic reactions, namely, ring closure, hydrolysis, aliphatic hydroxylation, and glucosidation of the imidazolinone herbicides, are similar to those of other herbicides. This chapter will cover the results from small-plot field experiments on the residue levels of the imidazolinones and their chemical nature. The experiments were designed to simulate field-use conditions.

Several greenhouse studies and *in vitro*, or cut-stem, techniques were also used to obtain samples for characterization and identification of the imidazolinone-derived residues. Based on these results, attempts will be made to propose partial metabolic pathways for imazapyr, imazaquin, imazamethabenz-methyl, and imazethapyr. These compounds represent different ring structures, that is, the pyridine ring in imazapyr, the alkyl-substituted pyridine ring in imazethapyr, the alkyl-substituted benzene ring in imazamethabenz-methyl, and the quinoline ring in imazaquin. Imazamethabenz-methyl is the only compound containing a carboxylic acid methyl ester, while the others contain a free carboxylic acid. The structural differences in these imidazolinones provide the plants with various sites for the enzymatic reactions leading either to bioactivation or to detoxification to take place (see Chapter 7 for additional discussion).

II. IMAZAPYR

($\pm$)-2-[4,5-dihydro-4-methyl-4-(1-methylethyl)-5-oxo-1*H*-imidazol-2-yl]-3-pyridinecar-
boxylic acid (CA)
($\pm$)-2-(4-isopropyl-4-methyl-5-oxo-2-imidazolin-2-yl)nicotinic acid (IUPAC)

A. ^{14}C- AND ^{13}C-LABELED IMAZAPYR
^{14}C/^{13}C-Carboxyl and ^{14}C/^{13}C-pyridine-ring-6-labeled imazapyr were used to define the metabolic profile in plants. The locations of the isotope labels are shown as follows:

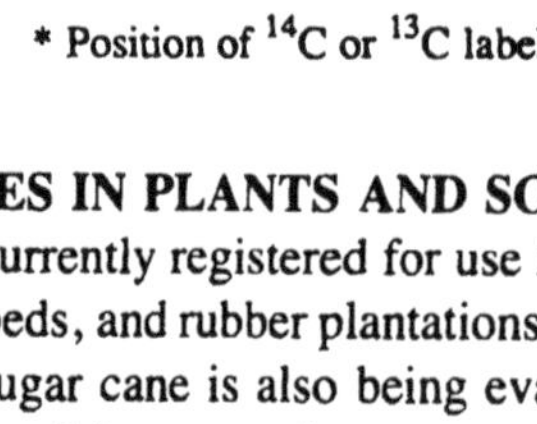

1 a 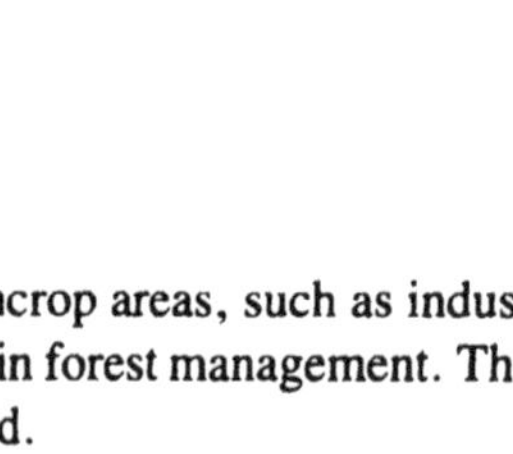   **1 b**

* Position of ^{14}C or ^{13}C label

B. ^{14}C RESIDUES IN PLANTS AND SOIL
Imazapyr is currently registered for use in noncrop areas, such as industrial sites, rights-of-way, railroad beds, and rubber plantations, and in forest management. The use of imazapyr in oil palm and sugar cane is also being evaluated.

A typical use of imazapyr is to spray the formulated material over the weeds (post-emergence treatment). After application, the compound is absorbed by the weeds and rapidly translocated throughout the plant. However, when the material is applied to a bare soil surface, the residue level and distribution pattern in plants may be considerably different from those seen with the postemergence treatment. Field and greenhouse metabolism studies using different application methods illustrate these differences.

A small-plot study was conducted with a 50:50 mixture of ^{14}C-ring- and ^{13}C-ring-labeled imazapyr formulated as the isopropylamine salt containing surfactant. The formulated material was sprayed uniformly at a typical use rate of 1.12 kg a.e./ha over a 1.22 × 1.22 m weed-covered sandy loam soil plot at the American Cyanamid Agricultural Research Center in Princeton, New Jersey.

Plant samples of *Digitaria sanguinalis* and *Ambrosia artemisiifolia* showed a rapid decline in total ^{14}C residues from the time of application to 24 h posttreatment, which then remained at a relatively constant level until the last sampling at 15 d (at which time the plants started to die).

The decrease in residues in the plant coincided with an increase of residues in the soil. Soil-residue levels in the top three inches increased from 0.008 ppm immediately after treatment to 0.029 ppm at 8 d after application. Since there was no rain during the first 8 d after application, the increased ^{14}C residues in the soil can be attributed to exudation from roots and the ^{14}C residue of the fine roots that could not be separated from the soil samples. The soil ^{14}C residues continued to increase and reached a peak of 0.23 ppm at 231 d posttreatment. After day 8, increasing soil residue levels are probably due to plant surface residues being washed off by rain and to the gradual release of imazapyr-related residue from the decomposition of dead plants and roots.

Studies were conducted in Costa Rica, Central America, and in St. Gabriel, Louisiana to determine the movement of ^{14}C-labeled imazapyr and its metabolites from soil into oil palm fruit and sugar cane. In these studies, ^{14}C-labeled imazapyr was applied to the bare ground beneath actively fruiting oil palm at a rate of 1 kg a.e./ha and to sugar cane as a preemergence application at a rate of 140 g a.e./ha.

Fruit samples collected from the oil palm at time of treatment and as they ripened at 7, 30, and 62 days posttreatment showed no detectable radioactivity in palm oil or other fruit fractions. Sugar cane harvested at maturity also showed no radioactive residues at a detection limit of 0.005 ppm. These results indicate that soil-applied imazapyr does not move rapidly to the root zone for uptake by the plants. Thus, negligible imazapyr-related residue would be found in the mature oil palm fruit and sugar cane.

In contrast to the results from soil applications, imazapyr in solution applied to coarse sand, which has very little adsorption capacity, showed that the compound is more readily available to the plants. In a greenhouse study, corn and soybean seedlings were transplanted into cups of coarse sand treated with ^{14}C-carboxyl-labeled imazapyr (**1b**) aqueous solution at levels of 8 and 35 μg per cup. Ten days after transplanting, the total ^{14}C residue was 0.87 μg per plant for soybean and 0.23 μg per plant for corn in the 8 μg per cup (0.01 ppm) treatment group. Residue levels in both soybean and corn did not increase proportionally with the increased treatment rate. In the 35-μg per cup (0.05 ppm) treatment group, 1.9 μg of ^{14}C residue was found in each soybean plant, while 0.51 μg was found in each corn plant. It is interesting to note that the residue in soybeans was mainly in the roots, while in corn the residue was more evenly distributed throughout the plant. The root-to-stem concentration ratio ranged from 11 to 30 in the soybean and from 0.7 to 2 in the corn samples.

C. CHARACTERIZATION OF THE IMAZAPYR-DERIVED RESIDUES

In the small-plot field study mentioned above, the radioactivity present in an external methanol wash of *Digitaria sanguinalis* and *Ambrosia artemisiifolia* was found to be in the unaltered parent compound. Eight days after treatment, about 95% of the total ^{14}C residues in the two plants was readily extractable with acidified methanol. Thin-layer chromatographic analyses of the extracts showed that about 78% of the extractable residue was imazapyr. Four minor metabolites, each accounting for less than 7% of the total extracted residues, were identified in *D. sanguinalis* and *A. artemisiifolia* as the hydrolysis product (**2**), 2-carbamoylnicotinic acid (**3**), 2,3-pyridinedicarboxylic acid (**4**), and an imidazopyrrolo-

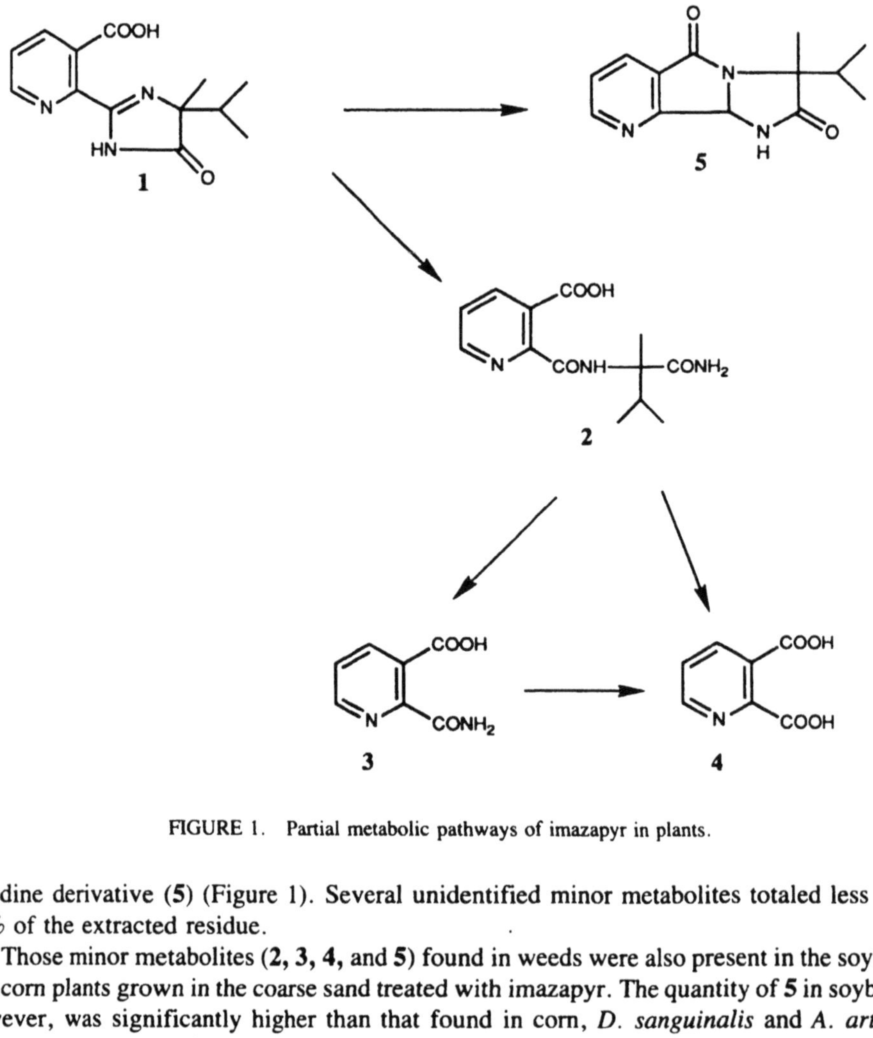

FIGURE 1.   Partial metabolic pathways of imazapyr in plants.

pyridine derivative (**5**) (Figure 1). Several unidentified minor metabolites totaled less than 10% of the extracted residue.

Those minor metabolites (**2, 3, 4,** and **5**) found in weeds were also present in the soybean and corn plants grown in the coarse sand treated with imazapyr. The quantity of **5** in soybean, however, was significantly higher than that found in corn, *D. sanguinalis* and *A. artemisiifolia*. This was confirmed by another experiment in which the stems of soybean, corn, *A. artemisiifolia*, and *Xanthium strumarium* were cut and submerged in an aqueous carboxyl $^{14}C/^{13}C$-labeled imazapyr (**1b**) solution to allow direct uptake of imazapyr into the plant systems. In 4 d, nearly 20% of the extractable residues in soybean was **5** as compared to 2.5% in corn and 0.5% in *A. artemisiifolia*; this metabolite was not detected in *X. strumarium*. The cyclic structure of metabolite **5** from the soybean cut-stem experiment was confirmed by comparing both positive- and negative-ion chemical ionization mass spectra of the isolate and the synthetic compound.

These results show that the major plant residue was imazapyr. The hydrolysis product (**2**) and two of its degradation products (**3** and **4**) were present in minute quantities. A ring-closure product (**5**) was also found as a minor metabolite in the plant systems.

D. SUGGESTED METABOLIC PATHWAYS

The metabolite profiles observed in *D. sanguinalis, A. artemisiifolia*, soybean, and corn suggest a slow metabolic transformation of imazapyr in plants. Hydrolysis of the imidazolinone ring to form product **2** is the first step leading to further degradation to **3** and **4**. Ring closure of the carboxyl group and the imidazolinyl nitrogen also takes place in plant systems as a minor metabolic route. The suggested partial metabolic pathways are shown in Figure 1.

III. IMAZAQUIN

($\pm$)-2-[4,5-dihydro-4-methyl-4-(1-methylethyl)-5-oxo-1H-imidazol-2-yl]-3-quinoline-carboxylic acid (CA)

($\pm$)-2-(4-isopropyl-4-methyl-5-oxo-2-imidazolin-2-yl)-3-quinolinecarboxylic acid (IUPAC)

A. ^{14}C- AND ^{13}C-LABELED IMAZAQUIN

^{14}C-Quinoline-4a,5,6,7,8,8a-labeled (**7b**) and ^{14}C-carboxyl-labeled imazaquin (**7a**) were used in several plant studies to define crop-residue levels and metabolic profiles. To aid mass spectroscopic analyses of metabolites, ^{13}C-labeled imazaquin was also used. The positions of the ^{14}C and ^{13}C labels are shown as follows:

7 a   7 b   7 c

* Position of ^{14}C label   ∧ Position of ^{13}C label

B. ^{14}C RESIDUES IN SOYBEANS

Imazaquin is registered for use in soybeans to control a wide spectrum of major broad-leaved weeds, grasses, and sedges by preplant-incorporated, preemergence, and postemergence applications. Soybean plants are capable of rapidly metabolizing imazaquin and thus exhibit exceptional tolerance to the herbicide.

Small-plot field investigations were conducted at several locations in the U.S. with ^{14}C-carboxyl- or ^{14}C-ring-labeled imazaquin at exaggerated dose rates ranging from 280 g a.e./ha to 560 g a.e./ha (two to four times the recommended use rate) to determine the residue levels and the nature of the residues in soybeans.

When imazaquin was applied postemergence to soybean plants, residues derived from both ^{14}C-ring-labeled and ^{14}C-carboxyl-labeled imazaquin declined rapidly to levels ranging from 0.1 to 0.66 ppm at maturity. Residue levels in the seeds ranged from 0.01 to 0.05 ppm. Imazaquin applied preplant incorporated resulted in lower plant residue levels (0.06 ppm to 0.19 ppm) at maturity than postemergence. However, total ^{14}C residue levels (0.01 to 0.04 ppm) in the seeds were similar to those treated postemergence.

Since these studies showed very low residues in soybean crops treated preplant incorporated or postemergence at two to four times the recommended use rates of imazaquin, it can be concluded that under normal use rates, the residue level in the soybean seeds would be negligible.

C. CHARACTERIZATION OF THE IMAZAQUIN-DERIVED RESIDUES

1. Extractable Residues in Soybean Foliage

^{14}C-Imazaquin-derived residues in green and mature, dry soybean plants were extracted by homogenizing plant tissue with aqueous acetone. This was followed by partitioning the extracts with dichloromethane to give organic- and water-soluble fractions. Plant residues from late growth stages and stalk required further extraction with acidified methanol or acetone mixtures to remove more tightly bound ^{14}C residues.

Extraction of the soybean plant samples from postemergence treatment at 280 g a.e./ha showed that organosoluble radioactivity represented 10% and 11% of the total plant residue at 2-week (2.31 ppm) and 1-month (0.52 ppm) intervals, respectively, and only 1% of the stalk ^{14}C residue (0.45 ppm) at maturity. The major portion of the recovered radioactive residue at all intervals consisted of water-soluble radioactivity. Imazaquin-derived residues in the unextracted fraction rose from 21% at 2 weeks to approximately 38% of the plant residue at maturity. Distribution of radioactivity in soybean foliage from plants treated preplant incorporated with ^{14}C-ring-labeled imazaquin was similar to those treated post-emergence at the same dose rate.

Several imazaquin metabolites in the extracts were identified based on high-performance liquid chromatography (HPLC) comparison with synthetic standards. An imidazopyrrolo-quinoline (8), a pyrroloquinoline acetamide (9), and the parent compound (7) were found in trace amounts in the methylene chloride fraction. The more polar metabolites (10, 11, 12, 13) and an unidentified metabolite were found in the aqueous fraction of the extracts (Figure 2).

Extraction of soybean plant residues with 70% aqueous methanol gave a very different metabolite profile with HPLC than those given by the aqueous acetone extracts. Two relatively apolar carboxylic methyl esters (14 and 15) as well as the carboxylic acids (10 and 11) were detected. In addition, metabolite 9 was never detected in the aqueous methanol extracts. It was later determined that 14 and 15 were extraction artifacts, because standard compound 9 readily forms two methanol derivatives, 14 and 15, when exposed to methanol. Compound 9 will also slowly hydrolyze to give 10 and 11 in water. However, when using 1% to 70% aqueous methanol gradient containing 1% formic acid as eluant, 9 is sufficiently stable in C-18 reverse-phase HPLC analysis to allow detection of this metabolite.

The cyclized metabolite (8) was detected only in trace quantities in soybean plant samples. For structure confirmation, metabolite 8 was isolated and purified from *in vitro* cut-stem experiments. Soybean shoots were "pulsed" with 40 ppm ^{14}C/^{13}C-carboxyl-labeled imazaquin mixture and extracted 1 day later with organic solvents. Thin-layer chromatography of the leaf extracts showed the presence of a nonpolar radioactive component corresponding to 8. Mass spectroscopic analysis confirmed the assigned structure because the

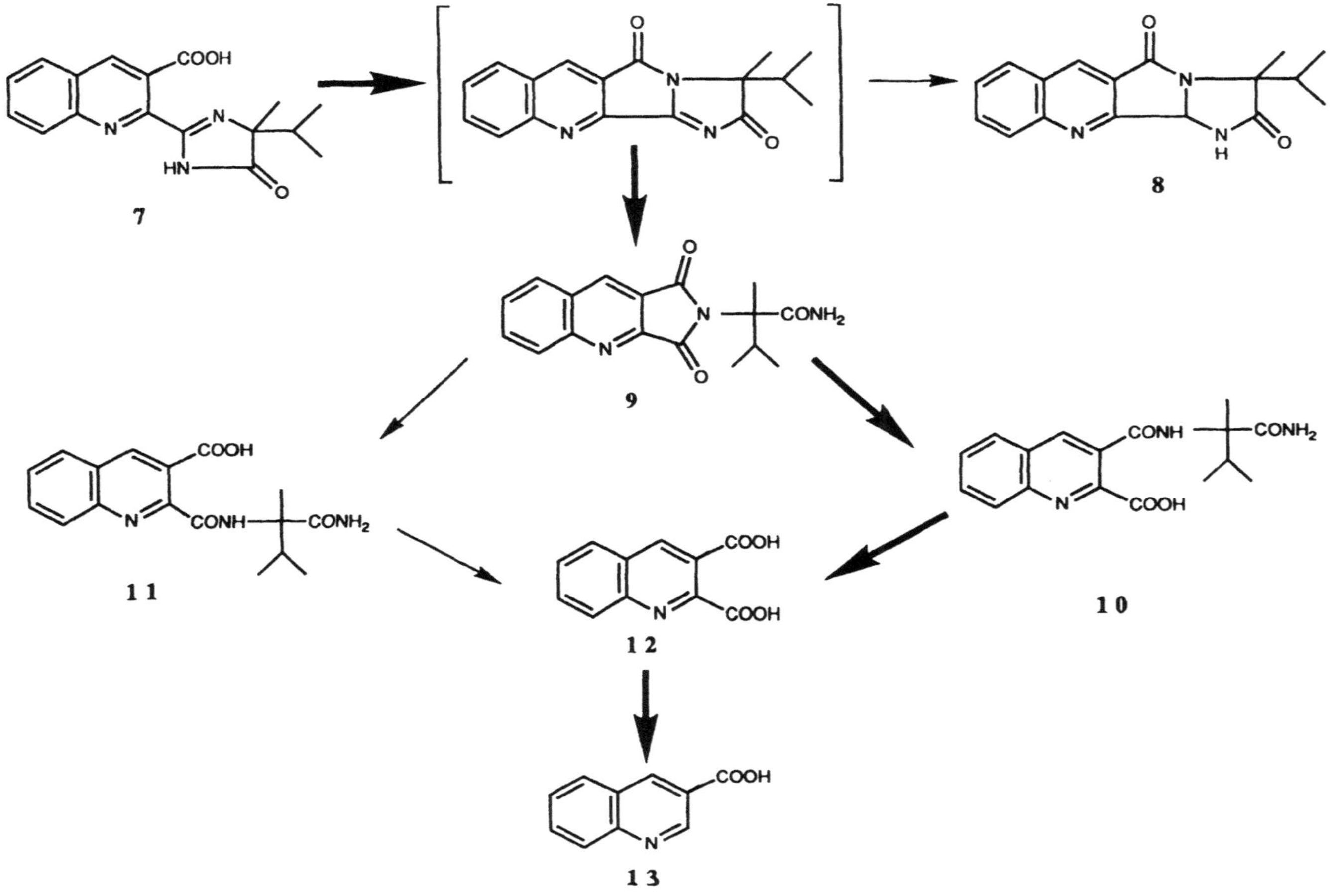

FIGURE 2.   Partial metabolic pathways of imazaquin in soybeans.

isolate exhibited ion doublets consistent with a molecular weight of 295, which corresponds to identical mass spectra of the authentic compound **8**.

2. Unextractable Plant Residues

As discussed earlier, a large portion of the soybean stalk residue was not extractable by conventional procedures and was considered to be bound residue. Extracted soybean stalk from plants treated postemergence at 420 g a.e./ha with ^{14}C-carboxyl-labeled imazaquin was refluxed in 10% sodium hydroxide by a procedure designed to isolate crude lignin and cellulose fractions.[2,3]

The isolated cellulose was hydrolyzed with 70% sulfuric acid and the resulting glucose was derivatized to glucosazone. Chemical ionization mass spectrometry of the generated osazone from soybean stalk confirmed the identify of the derivative by comparison with glucosazone spectra from starch. Initially, it was thought that the incorporation of the labeled carbon into cellulose must be the result of decarboxylation of the ^{14}C-labeled carboxyl group. However, similar distribution patterns of bound radioactive residues were also observed in the stalk of soybeans treated with ^{14}C-ring-labeled imazaquin. This finding suggests that imazaquin is not metabolized by pathways leading to direct decarboxylation of the imazaquin molecule. The pathways that result in the incorporation of the imazaquin metabolite into natural products have not been identified. Presumably, metabolite **13** could be oxidized further at the quinoline ring, and eventually the labeled carbon would be channeled to the general metabolic pool.

D. SUGGESTED METABOLIC PATHWAYS

The metabolite profiles observed in the field-grown soybean plant and cut-stem studies suggest that ring closure of the carboxyl group and the imidazolinyl nitrogen is the first step leading to the extensive degradation of imazaquin. The cyclized product is either reduced to form imidazopyrroloquinoline (**8**) or hydrolyzed to produce pyrroloquinoline acetamide (**9**). Metabolite **9** was not found in high concentration because it was rapidly hydrolyzed to form mainly **10** and a minor metabolite **11** which was also found as a hydrolytic product of imazaquin in buffered solution (c.f. Chapter 15). Direct hydrolysis of the imidazolinone ring is probably a minor metabolic pathway in soybean plants. The presence of quinoline-2,3-dicarboxylic acid (**12**) and quinoline-3-carboxylic acid (**13**) in the later growth stage suggests that carbamoyl groups of **10** and **11** were further hydrolyzed. There is one unknown metabolite present in concentrations similar to that of **10** at all sampling intervals. The origin of this metabolite has not been determined. Based on the metabolite profiles, the suggested partial metabolic pathways of imazaquin in soybean plants are shown in Figure 2.

A considerable amount of labeled imazaquin-derived radiocarbon was incorporated into natural constituents of soybean plants. The source of these radiocarbons has not been determined. Presumably, further oxidation at the quinoline ring eventually led to the incorporation of labeled carbon into plant constituents.

IV. IMAZAMETHABENZ-METHYL

Methyl (±)-2-[4,5-dihydro-4-methyl-4-(1-methylethyl)-5-oxo-1*H*-imidazol-2-yl]-4-methylbenzoate (CA), and
Methyl (±)-2-[4,5-dihydro-4-methyl-4-(1-methylethyl)-5-oxo-1*H*-imidazol-2-yl]-5-methylbenzoate (CA)

Methyl (±)-2-(4-isopropyl-4-methyl-5-oxo-2-imidazolin-2-yl)-*m*-toluate (IUPAC), and
Methyl (±)-2-(4-isopropyl-4-methyl-5-oxo-2-imidazolin-2-yl)-*p*-toluate (IUPAC)

A. ¹⁴C- AND ¹³C-LABELED IMAZAMETHABENZ-METHYL

Both ¹⁴C/¹³C *meta* isomer (16) and *para* isomer (17) were labeled at the carboxyl carbon and are shown as follows:

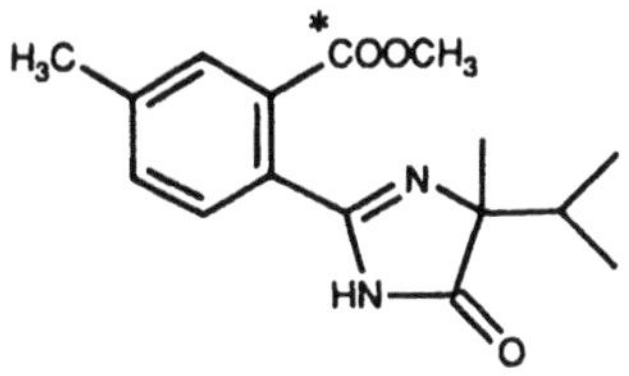

16 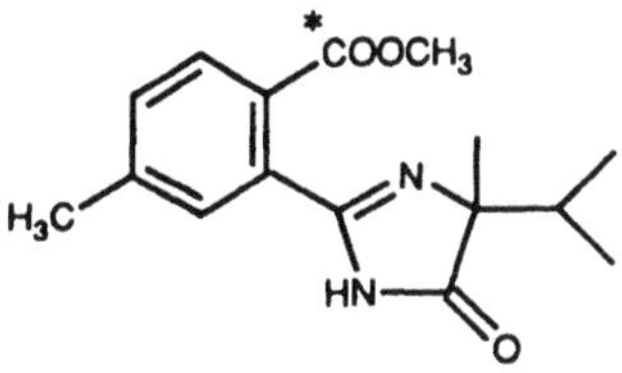 **17**

* Position of ¹⁴C or ¹³C label

B. ¹⁴C RESIDUES IN WHEAT

Imazamethabenz-methyl is a highly effective herbicide for control of *Avena* spp. and other weeds in wheat, barley, rye, sunflowers, and winter triticale. Field metabolism studies were conducted to determine the metabolic fate of **16** and **17** in spring wheat grown under realistic field conditions.

Spring wheat treated postemergence at the five-leaf stage with *meta* isomer or *para* isomer at 785 g a.i./ha in North Dakota showed only trace amounts of ¹⁴C residues in the grain at harvest. In the growing wheat plants, total ¹⁴C-*para* isomer-derived residues declined rapidly from 132 ppm immediately after application to 1.1 ppm at 2 weeks and 0.14 ppm at 2 months after application. Application of the *meta* isomer to wheat plants at the same growth stage and treatment rate showed similar results in ¹⁴C-residue level and dissipation pattern. The drastic decrease of radioactive residues within the first 2 months after application was presumably due to the dilution of radioactivity by the rapidly growing plants.

C. CHARACTERIZATION OF THE IMAZAMETHABENZ-METHYL-DERIVED RESIDUES

¹⁴C-Imazamethabenz-methyl-derived residues can be extracted from the wheat plants by homogenizing the plant tissues with methanol. In wheat plants sampled immediately after application of **16** or **17**, more than 99% of the residue could be extracted and shown by two-dimensional thin-layer chromatography to consist only of the unaltered parent compound. Residue extractability decreased to 32% and 51% of the 2 week and 1 month plant residues, respectively. However, metabolic profiles of extracted residues allowed the elucidation of the general metabolic pathways of imazamethabenz-methyl in wheat plants.

Although the wheat plants showed similar total ¹⁴C residue levels in *meta* and *para* regioisomer-treated plants, the two regioisomers showed significant differences in production rates of polar metabolites.

In the 2- and 4-week samples of *meta* isomer-treated wheat plants, the predominant extractable residue was identified as **21**, the glucose conjugate of *meta* alcohol (19), by acid and β-glucosidase hydrolysis and thin-layer chromatographic comparisons with the synthetic reference aglycone, **19**. Other extractable wheat plant residues included unaltered parent compound, unconjugated *meta* alcohol (19), *meta* acid (18), *meta* alcohol/acid (20) and **22**, which is a glucoside of *meta* alcohol/acid (20) (Figure 3).

By comparison, the major extractable residue in the *para* isomer-treated wheat plants was the unaltered parent compound. The degradation mechanism of the *para* isomer in wheat plants, however, is the same as that for the *meta* isomer, that is, hydroxylation at the aryl methyl group, hydrolysis of the carboxyl methyl ester, and glucosidation of the hydroxy-methyl group. Thus, the following metabolites were detected in the *para*-isomer-treated

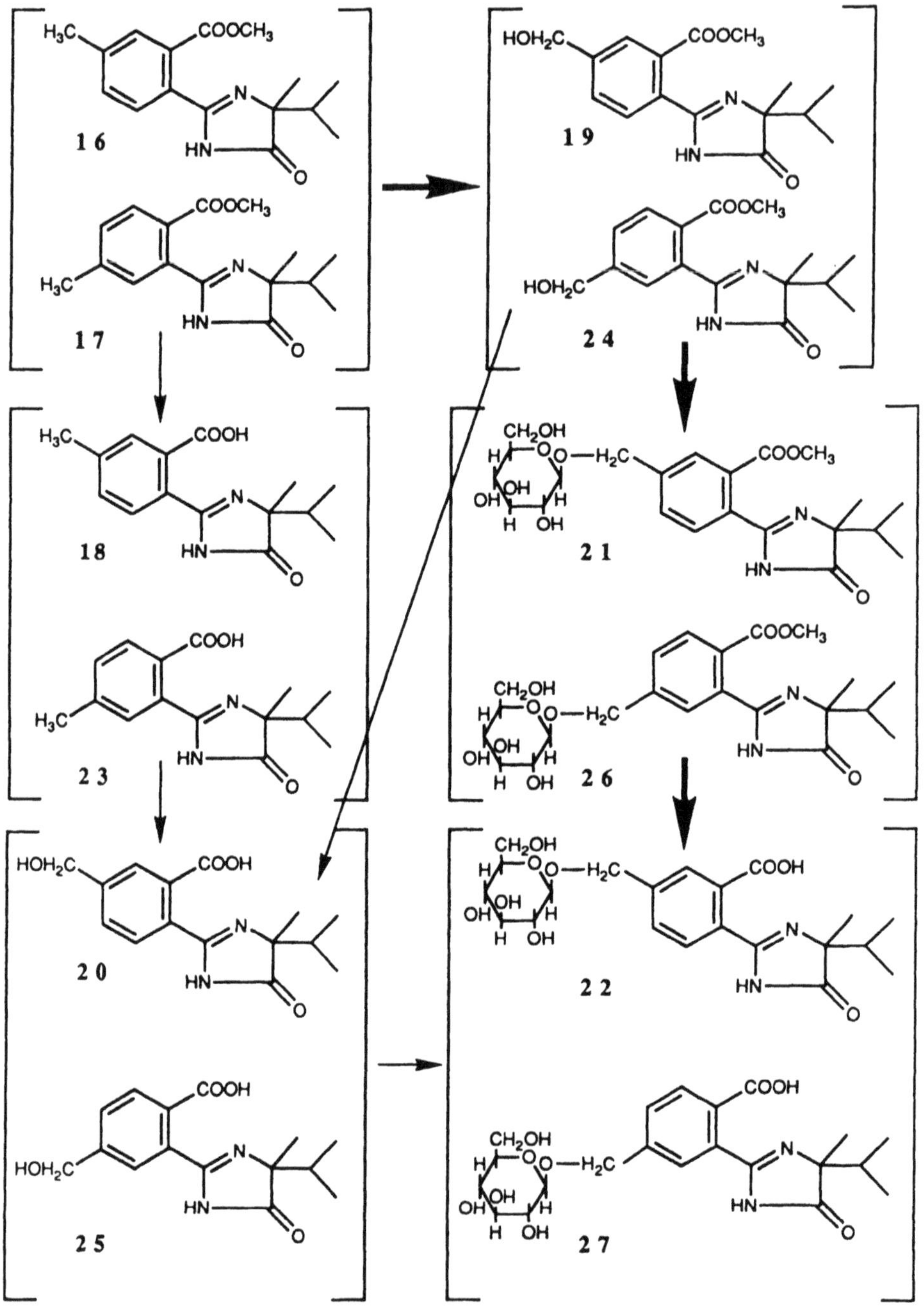

FIGURE 3.   Partial metabolic pathways of imazamethabenz-methyl in wheat.

wheat plant extracts: the *para* alcohol (**24**), the *para* acid (**23**), the *para* alcohol/ acid (**25**), the glucose conjugate of the *para* alcohol (**26**), and the glucose conjugate of the *para* alcohol/ acid (**27**).

The presence of the *meta* parent and ·the *para* parent, and the metabolites — *para* alcohol/acid, *meta* alcohol, and *meta* alcohol glucoside — have also been confirmed by the presence of the relevant ion doublets in negative ion mass spectroscopic analysis of the $^{13}C/^{12}C$ metabolite mixtures.[4]

D. SUGGESTED METABOLIC PATHWAYS

The metabolite profile in wheat plants suggested that oxidation of the aryl methyl group to the benzyl alcohol followed by glucose conjugation is the primary metabolic pathway. A significant amount of the glucosides of the *meta* and *para* alcohols was further hydrolyzed at the carboxylic acid methyl ester linkage as indicated by the presence of the glucosides of the *meta* and *para* alcohol/acids. However, the presence of two to four times more of the unaltered *para* parent compound in wheat-plant extracts proves that the *para* parent has greater metabolic stability than the *meta* parent. Young corn plants and *Avena* spp. have a metabolic pathway similar to that of wheat, with *Avena* spp. producing more of the phytotoxic carboxylic acid analog.[4] Based on metabolite profiles, the suggested partial metabolic pathways of imazamethabenz-methyl in wheat plants are summarized in Figure 3.

In summary, the most important metabolic pathway in the preharvest wheat plant is oxidation of the aryl methyl group to the benzyl alcohol group followed by glucose conjugation. The carboxylic methyl ester linkage of the glucose conjugates is further hydrolyzed to the corresponding carboxylic acid analog.

V. IMAZETHAPYR

($\pm$)-2-[4,5-dihydro-4-methyl-4-(1-methylethyl)-5-oxo-1*H*-imidazol-2-yl]-5-ethyl-3-pyridine-carboxylic acid (CA)

($\pm$)-5-ethyl-2-(4-isopropyl-4-methyl-5-oxo-2-imidazolin-2-yl)nicotinic acid (IUPAC)

A. ^{14}C- AND ^{13}C-LABELED IMAZETHAPYR

Both ^{13}C- and ^{14}C-imazethapyr were labeled in the pyridine ring at the C-6 atom adjacent to the nitrogen atom. The position of the label is shown as follows:

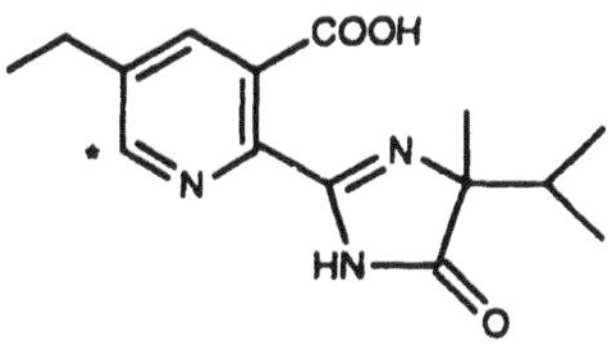

2 8

* Position of ^{14}C or ^{13}C label

B. ^{14}C RESIDUES IN AGRONOMIC CROPS

Imazethapyr is registered for use in soybeans *(Glycine max)* and other leguminous crops. The compound is being developed for use in alfalfa, peanuts, and corn.[5] Metabolism studies in soybeans, peas, beans, peanuts, and corn were conducted in small field plots, in the greenhouse, and *in vitro* to define metabolic profiles. The results of these studies are discussed in the following sections.

1. Soybean

As with imazaquin, the soybean plant is capable of rapidly metabolizing imazethapyr to more polar, less phytotoxic compounds. When ^{14}C-labeled imazethapyr was applied postemergence at a rate of 280 g a.e./ha (four times the maximum label rate) to soybeans at the fourth trifoliate stage, ^{14}C-residue levels in the plants declined rapidly from 27 ppm at two hours after application to 5.73 ppm at two weeks after application and to 0.34 ppm at harvest (4.5 months after application). Very little ^{14}C residue (0.02 ppm) was in the soybean seed at harvest. The preplant-incorporated application resulted in lower residue levels in the soybean plants than did the postemergence application — in the earlier growth stages, total ^{14}C residues were 0.28 ppm at 2 weeks after application, 0.36 ppm at 1 month after application, and 0.31 ppm in the mature straw at 5 months. Total ^{14}C-residue levels in the seed were the same as with the postemergence treatment. Results of the preplant-incorporated study indicate that within 1 month imazethapyr was readily absorbed from soil by the soybean seedlings and translocated to the growing parts of the plants. Thereafter, the decline of residues is probably due to the dilution effect of the radioactivity by the rapidly growing plants.

2. Peanuts

Florigiant-type peanut crops treated preplant incorporated or postemergence with ^{14}C-imazethapyr at 140 g a.e./ha showed less than 0.05 ppm imazethapyr equivalents in the peanut nutmeats. However, the dry hay at harvest showed significant amounts of ^{14}C residues ranging from 0.23 to 0.49 ppm.

3. Green Bean and Peas

At a rate of 105 g a.e./ha of imazethapyr applied preemergence or preplant incorporated very little residue (0.03 to 0.04 ppm) was present in the bean plants at 30 d after treatment and in tender bean pods (below the detection limit of 0.03 ppm) harvested at 60 d. Slightly more residue (0.08 ppm) was found in the pea vine at the same preplant-incorporated treatment rate. The postemergence application to peas resulted in an initial high residue of 2 ppm at 1 month but declined rapidly to 0.03 ppm at 75 d. Tender peas harvested at 75 d showed less than 0.03 ppm total ^{14}C residue.

4. Corn

Small-plot field studies were conducted with Pioneer 3747 field corn and two imidazolinone-tolerant corn varieties using preplant-incorporated and postemergence applications. The field corn was treated preplant incorporated with ^{14}C-labeled pyridine 6 imazethapyr at 34.7 g a.e./ha. Total residues in foliage, cob, and seed in mature plants were all below 0.006 ppm. The two imidazolinone-tolerant varieties were treated postemergence and pre-plant incorporated with ^{14}C-imazethapyr at 105 g a.e./ha. The ^{14}C residue rapidly declined in the foliage after postemergence application; total residues in the mature corn stalk from both treatments ranged from 0.02 to 0.26 ppm for the postemergence treatment and from 0.006 to 0.021 ppm for the preplant-incorporated treatment. Radioactive imazethapyr-derived residues in the seed of mature plants treated either postemergence or preplant incorporated ranged from 0.007 to 0.012 ppm. The residue levels in the stalk and seeds at harvest did not appear to be significantly affected by the method of treatment or the corn variety.

C. CHARACTERIZATION OF THE IMAZETHAPYR-DERIVED RESIDUES

Imazethapyr-related residues appear mainly in the vegetative portions of the plants. The metabolic profile of imazethapyr was developed by analyzing the plant samples at various growth stages.

[14]C residues from samples of soybean treated postemergence and preplant incorporated were readily extractable with 80% aqueous methanol. TLC analysis of methanol/water extracts revealed that **29** (the α-hydroxyethyl analog of imazethapyr) and **30** (the glucoside of **29**) were the predominant radiometabolites in green plants from both postemergence and preplant-incorporated treatments (Figure 4). A small amount of imazethapyr was also found in these samples. The metabolic profiles in the straw extracts of both samples were virtually identical, with compound **29** being the major metabolite in the methanol/water extractable fraction. Neither the parent compound nor the glucoside (**30**) was detectable in any straw sample.

Similar to soybeans, the ^{14}C residue in peanut hay was also readily extractable with aqueous acetone. HPLC analyses showed nearly half (48% to 56%) of the extractable residue

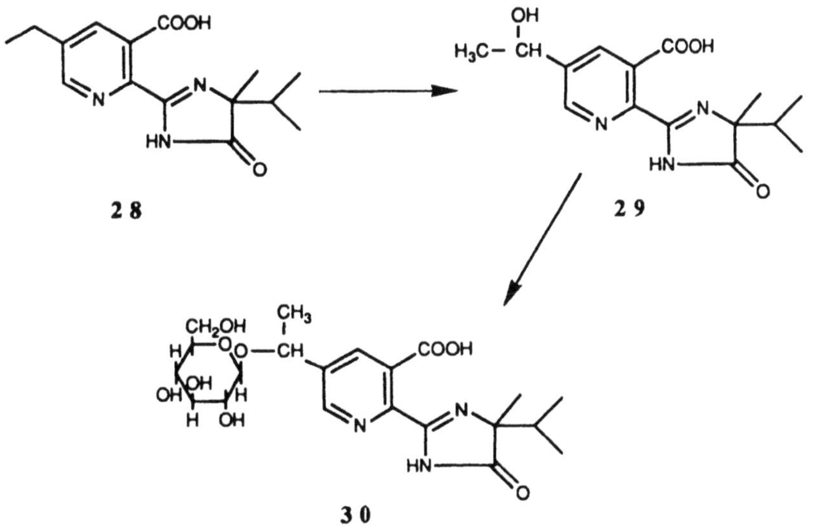

FIGURE 4.   Partial metabolic pathways of imazethapyr in legumes and corn.

was the glucoside of the α-hydroxyethyl imazethapyr (**30**) while 13% to 16% was the hydroxylated metabolite (**29**). The ratio of the glucoside to the hydroxylated metabolite was the same in the peanut hulls. The metabolite profiles did not appear to be significantly affected by application method or geographic location.

The metabolism of imazethapyr in corn was very rapid, with a complete conversion to one metabolite in 48 h in sweet corn cell suspensions.[6] The polar metabolite was identified as the α-hydroxylated compound (**29**). Similarly, field corn and imidazolinone-tolerant corn metabolize imazethapyr to **29** as the predominant residue. Trace amounts of the glucose conjugate (**30**) and imazethapyr were also detected in corn seeds and plants at various growth stages.

D. SUGGESTED METABOLIC PATHWAYS

Based on metabolic profiles of extractable residues from green soybean plants, oxidative hydroxylation at the α-carbon atom of the ethyl substituent on the pyridine ring is the primary site for the initial metabolic conversion of the parent compound into α-hydroxyethyl imazethapyr. This hydroxyethyl analog then reacts rapidly with glucose, yielding the glucose conjugate as the major metabolite. However, in the soybean straw at harvest, the hydroxyethyl analog (**29**) was the predominant metabolite. Other leguminous crops, such as peanuts, showed metabolic pathways similar to those in soybean, except that the glucoside was shown to be the terminal metabolite. The oxidative hydroxylation of the ethyl group on the pyridine ring was also shown to be the preferred route of metabolism by corn plants. The imidazolinone-tolerant gene did not alter the final metabolic profiles. Unlike leguminous crops, further glucosidation of the alkylhydroxylated imazethapyr (**29**) was not a major metabolic pathway in corn. The partial metabolic pathways of imazethapyr in leguminous crops and corn are summarized in Figure 4.

VI. CONCLUSION

Imidazolinone herbicides such as imazaquin, imazamethabenz-methyl, and imazethapyr are highly selective and effective in controlling weeds in leguminous and cereal crops. In general, a single soil or foliar application at very low rates early in the crop growing stages is sufficient to maintain weed control for the growing season. A slightly higher rate is required for imazapyr to control perennial weeds. After foliar application, the imidazolinones penetrate rapidly into the plants and are translocated throughout the plant system. Crop-root uptake of imidazolinone-derived soil residues in the field is relatively small compared to uptake following foliar application. Total plant residues always decline very rapidly over a few weeks because of the high growth dilution factor. Studies with carboxyl- and ring-labeled imidazolinones showed no evidence to suggest that the decline in residue level was due to the formation of volatile metabolites.

The imidazolinone-derived plant residues consist of the unaltered parent compounds and metabolites. Metabolism of imidazolinones in plants involves oxidative, hydrolytic, reductive, and ring-closure processes.

For compounds that contain an alkyl substituent, such as in imazamethabenz-methyl and imazethapyr, the general metabolic pathway in most plants is the oxidation of the alkyl moiety to the alcohol followed by glucose conjugation. The rates of alkyl alcohol formation vary among plant species or crop varieties. In corn, glucose conjugation of the α-hydroxyethyl analog of imazethapyr is significantly lower than in leguminous crops. Ester hydrolysis of the carboxylic acid methyl ester of imazamethabenz-methyl occurs as expected in weeds and to a very small extent in wheat. The imidazolinone ring of these alkyl-substituted compounds appears to be unaffected in young plants. Hydrolysis of the imidazolinone ring and the formation of tricyclic imidazopyrrolo aryl compounds have not been shown in the plant samples.

In contrast to the alkyl aryl imidazolinones, imazapyr and imazaquin contain no alkyl substituents on the pyridine or quinoline ring. Ring closure of the carboxyl carbon and the imidazolinyl nitrogen appears to be the first step in the degradation of imazaquin and imazapyr. The cyclized product is either reduced to form a minor tricyclic imidazopyrrolo aryl compound (**8**) or hydrolyzed to produce a pyrroloaryl acetamide structure (**9**) that is not found in high concentrations because it is rapidly hydrolyzed to the 3-substituted-2-arylcarboxylic acid or 2-substituted-3-arylcarboxylic acid. The acetamide structure and the 3-substituted-2-arylcarboxylic acid compound are not seen in the imazapyr-treated plants. Presumably, the aryl ring structure could be further oxidized and eventually lose the labeled carbon to the general metabolic pool, as suggested by the incorporation of small amounts of ^{14}C atoms into natural products such as cellulose, protein, and lignin. Hydrolysis products of the imidazolinone ring were shown in very small quantity in imazapyr- and imazaquin-treated plants. The hydrolysis product of imazaquin could also be derived from the hydrolysis of the acetamide.

ACKNOWLEDGMENTS

We wish to thank Dr. F. Corbin of North Carolina State University at Raleigh; Dr. M. Barrett of the University of Kentucky, Lexington; Dr. J. Nalewaja of North Dakota State University, Fargo; Drs. J. M. DiTomaso and P. Fay of Montana State University, Bozeman; Dr. D. Schilling of University of Florida, Gainesville; Dr. L. Kitchen of Louisiana State University, Baton Rouge; and R. Paniagua of Cyanamid Inter-American Corporation, San Jose, Costa Rica, for their help in conducting many field experiments on the imidazolinone herbicides. These field metabolism studies have contributed significantly to our understanding of the crop-residue dissipation and metabolic fate of the imidazolinone herbicides in various crops under realistic field growing conditions.

We would also like to thank our colleagues: Dr. C. Fung for the imazethapyr peanut metabolism data; Messrs. S. Stout and A. daCunha for the mass spectroscopic analyses; and Ms. Barbara Knoll, Mrs. Barbara Davis, Ms. Yvonne Roman, Ms. Joyce Washington, Mr. Paul Brugmann, and Mrs. Pamela Stanley-Millner for their skillful technical assistance in all phases of the metabolism study.

REFERENCES

1. **Hatzios, K. K. and Penner, D.,** *Metabolism of Herbicides in Higher Plants,* Burgess Publishing Company, Minneapolis, MN, 1982.
2. **Powell, W. and Whittaker, H.,** The chemistry of lignin. I. Flax lignin and some derivatives, *J. Chem. Soc.,* 125, 357, 1924.
3. **Honeycutt, R. and Adler, I.,** Characterization of bound residues of nitrogen in rice and wheat straw, *J. Agric. Food Chem.,* 23, 1097, 1975.
4. **Brown, M. A., Chiu, T. Y., and Miller, P.,** Hydrolytic activation versus oxidative degradation of Assert herbicide, an imidazolinone aryl-carboxylate, in susceptible wild oat versus tolerant corn and wheat, *Pestic. Biochem. Physiol.,* 27, 24, 1987.
5. **Peoples, T. R., Wang, T., Fine, R. R., Orwick, P. L., Graham, S. E., and Kirkland, K.,** AC 263,499: a new broad-spectrum herbicide for use in soybeans and other legumes, in *Br. Crop Prot. Conf. — Weeds,* Vol. 1, British Crop Protection Conference Publications, Croydon, England, 1985, 99.
6. **Robson, P. A., Stidham, M. A., and Shaner, D. L.,** AC 263,499: Laboratory Studies. I. Metabolism of 14C-AC 263,499 in Corn and Identification of the First Major Metabolite, Cyanamid Internal Rep. DIS-P4-14, American Cyanamid Co., 1984.

Chapter 12

ANIMAL METABOLISM

Phillip Miller, Chien Fung, and Barbara Gingher

TABLE OF CONTENTS

I. INTRODUCTION

Of the four imidazolinone herbicides being used commercially, imazaquin, imazamethabenz-methyl, and imazethapyr are selective compounds that are registered for use in legumes and cereals, while imazapyr is a herbicide used for total vegetation control. The herbicidal uses of these chemicals are discussed in other chapters (Chapters 17 through 20). The metabolism of these compounds in animals, specifically the laboratory rat, lactating goat, and laying hen, as discussed below, was determined using radiolabeled compounds to facilitate detection and analysis.

II. IMAZAMETHABENZ-METHYL

Imazamethabenz-methyl is a mixture of the regioisomers **1** (CL 239,589) and **2** (CL 252,767). In the commercial herbicide the ratio of the isomers is approximately 60:40, respectively. The structures of these isomers differ only in the position of the methyl group attached to the benzene ring as shown in structures **1** and **2**.

1   **2**

Animal metabolism studies have been conducted in the rat, the lactating goat, and the laying hen, using a ^{14}C-carboxyl label and a uniformly labeled benzene ring.

Carboxyl-labeled regioisomers **1** and **2** were independently administered to rats to determine the absorption, distribution, elimination, and metabolic fate of imazamethabenzmethyl. The radiolabeled carbon was excreted rapidly, primarily in the urine. No evidence of accumulation was observed in the liver, kidney, muscle, fat, or blood of rats treated with regioisomers **1** or **2**. The major metabolites found in the urine were the corresponding hydrolysis products, carboxylic acids **3** (CL 252,768) and **4** (CL 222,575).

3   **4**

Oxidation of the benzene ring methyl groups was observed to yield **5** (CL 280,079) and **6** (CL 280,162). Minor amounts of **7, 8, 9**, and **10** were also found. These metabolites were identified by thin-layer co-chromatography and chemical ionization mass spectrometry.

The major metabolites observed in the extracts of liver, kidney, muscle, and fat were the carboxylic acids **3** and **4**, and the oxidation products **5** and **6**. Metabolite **10** (CL 263,068), derived from isomer **2**, was also found in significant levels in liver and kidney tissue.

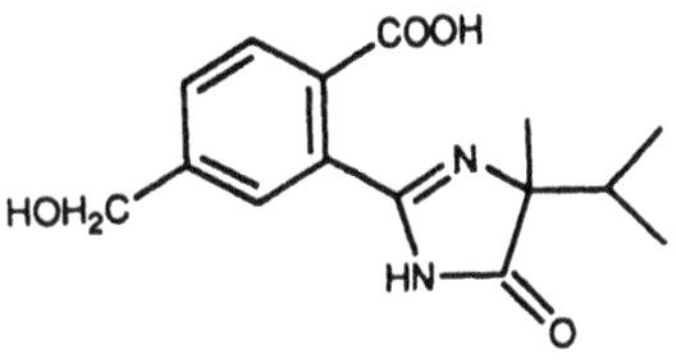

5

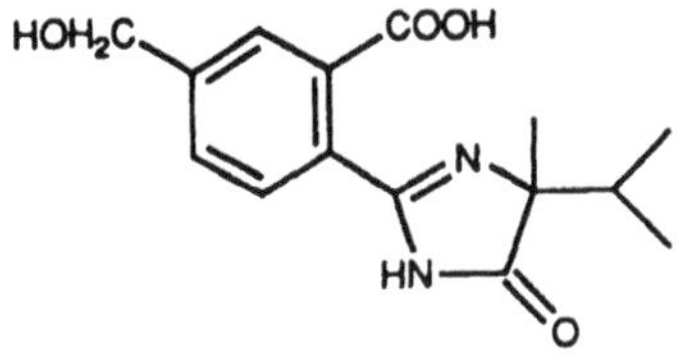

6

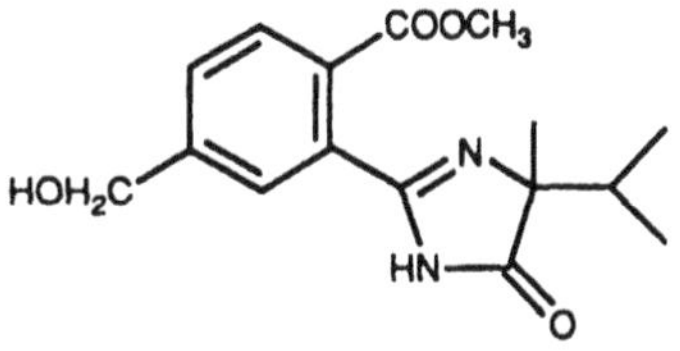

7

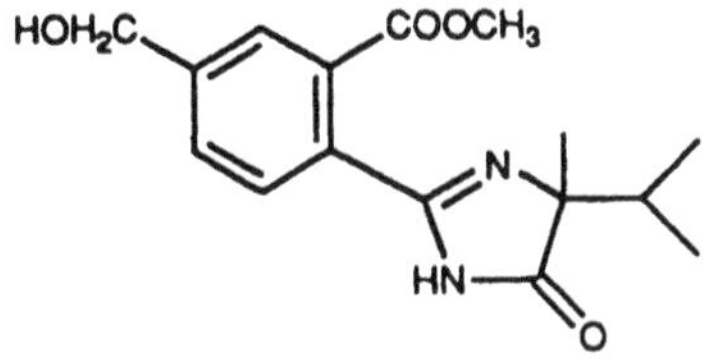

8

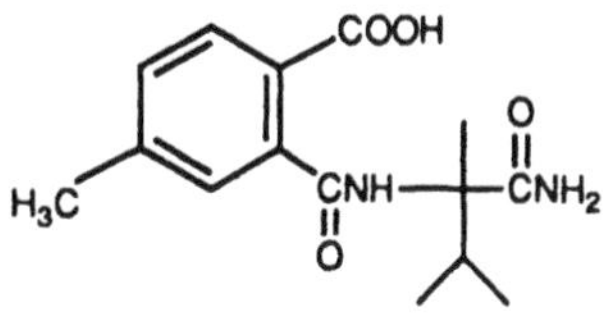

9

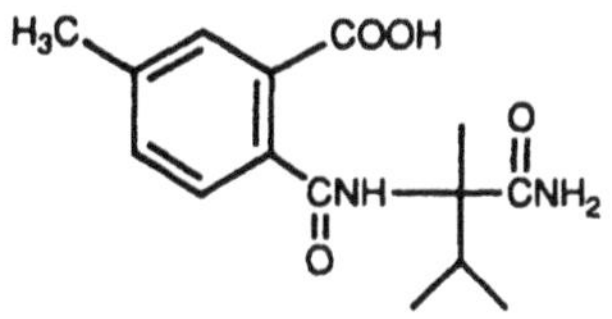

1 0

The metabolic profiles observed in the urine and tissues of rats suggest that hydrolysis of the ester function to the acid is the primary metabolic pathway (Figure 1). Oxidation of the benzene methyl group may also occur simultaneously or sequentially with hydrolysis. Opening of the imidazolinone ring appears to be an important process in tissue.

Goats were dosed with a 60:40 mixture of regioisomers **1** and **2** labeled either at the carboxyl carbon or uniformly labeled in the benzene ring. Imazamethabenz-methyl was administered by capsule once a day for 7 consecutive days. The highest dose administered was equivalent to 7.5 ppm in the diet. No detectable residues (less than 0.01 ppm) were found in the daily milk samples. At sacrifice all residues in liver, kidney, muscle, fat, and blood were nondetectable (less than 0.05 ppm).

Laying hens were given a 60:40 mixture of ^{14}C-labeled regioisomers **1** and **2** (^{14}C-carboxyl or benzene ring uniformly labeled) *ad libitum* in their feed for 7 consecutive days. No detectable residues (less than 0.05 ppm) were found in the eggs (yolks and whites) collected daily or in the liver, kidney, muscle, or skin with adhering fat or blood at sacrifice. The highest dose administered was 2.16 ppm (benzene ring, uniformly labeled).

III. IMAZETHAPYR

Imazethapyr was labeled with ^{14}C in the pyridine ring at the C-6 atom adjacent to the nitrogen atom and used to study the metabolism of the compound in the rat, goat, and laying hen.

Labeled imazethapyr was used to determine the absorption, distribution, excretion, and metabolism in rats. The compound was found to be excreted very rapidly, primarily in the urine, as unchanged imazethapyr. Metabolite **11** (CL 288,511), found in the urine and feces, was also observed at low levels in soybean plants, but not in the seed. As in the case of the other imidazolinones, no accumulation of residues was observed in the liver, kidney, muscle, fat, or blood.

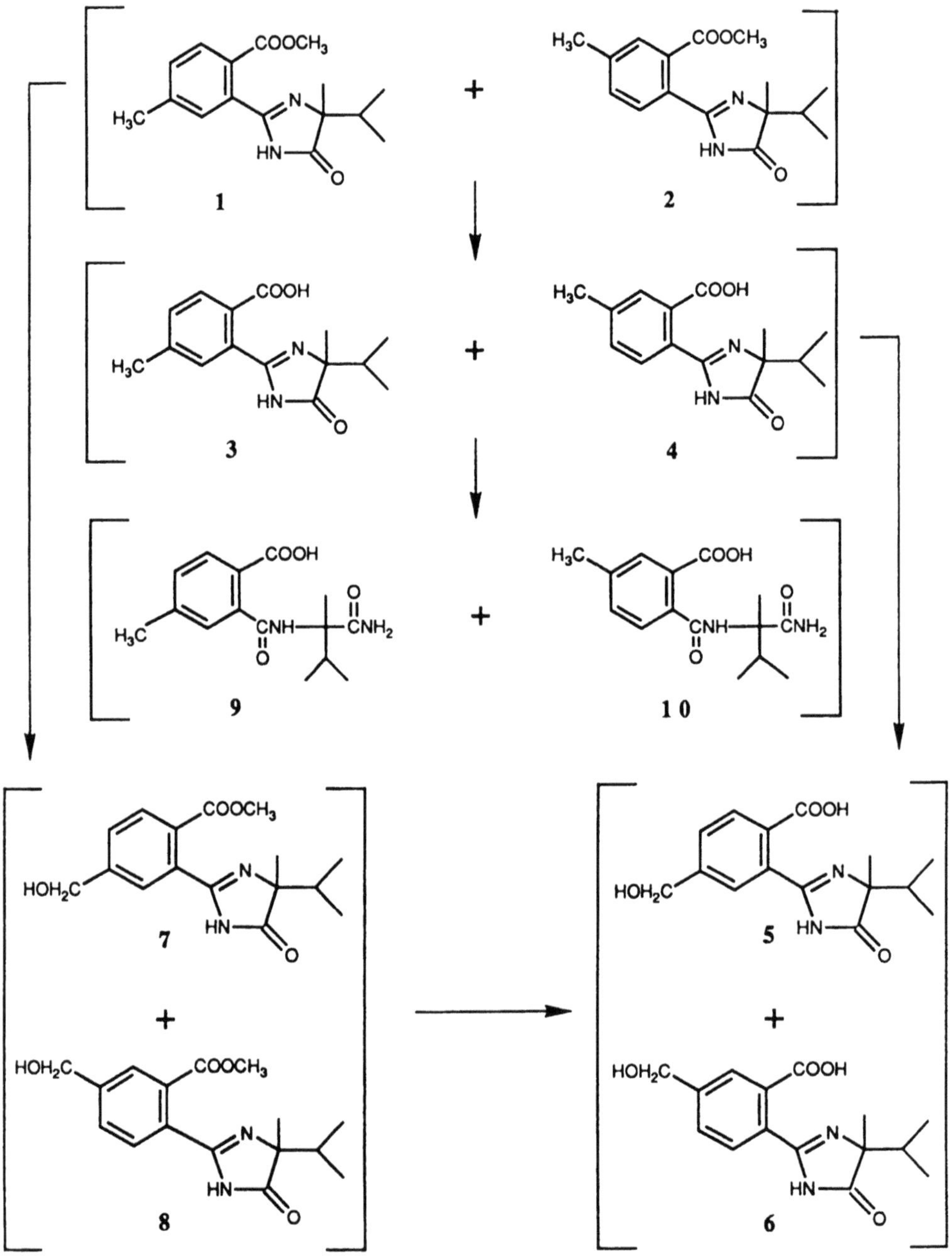

FIGURE 1.   Metabolic pathway of imazamethabenz-methyl in the rat.

Metabolite **11** (CL 288,511) was rapidly excreted unchanged when administered to goats and laying hens. No detectable residues were found in the edible tissues, milk, or eggs.

1 1

To determine the residue distribution in edible tissues and milk of lactating goats, ^{14}C-imazethapyr was administered in gelatin capsules at a daily dose equivalent to 0.25 or 1.25 ppm imazethapyr in the diet for 7 consecutive days. No detectable residues (less than 0.01 ppm) were observed in the daily milk samples or in the liver, kidney, muscle, fat, or blood (less than 0.05 ppm) at sacrifice.

Laying hens administered ^{14}C-imazethapyr in the feed at 0.25 or 1.25 ppm for 7 consecutive days showed no detectable residues (less than 0.05 ppm) in the eggs collected daily (yolks and whites) or in the liver, kidney, muscle, or skin with adhering fat or blood at sacrifice.

IV. IMAZAQUIN

^{14}C-Imazaquin labeled at the carboxyl group on the C-3 atom or uniformly labeled at the 4a, 5, 6, 7, 8, 8a carbon atoms of the quinoline ring was used to determine the fate of imazaquin in several animal species.

The ^{14}C-carboxyl-labeled imazaquin was used to determine the absorption, distribution, excretion, and metabolism of imazaquin in rats. The compound is excreted very rapidly in the urine as unchanged imazaquin. As in the case of imazapyr, residues did not accumulate in the liver, kidney, muscle, fat, or blood.

Goats receiving one capsule of the ring-labeled compound daily for 7 consecutive days at doses equivalent to 0.25 or 0.75 ppm in the feed showed no detectable residue (less than 0.01 ppm) in the daily milk samples. At sacrifice, liver, kidney, muscle, fat, and blood showed negligible residues (less than 0.05 ppm).

Laying hens were fed the ring-labeled compound *ad libitum* in their feed for 7 consecutive days at rates of 0.25 or 0.75 ppm. No detectable residues (less than 0.05 ppm) were found in eggs during the 7 d treatment period. Residue levels in the tissues and blood were also below the detection limit of the radioassay.

Thus, imazaquin and its derived residues have not been shown to accumulate in rat, goat, or poultry tissues or in goat's milk or chicken eggs.

V. IMAZAPYR

Imazapyr labeled at the carboxyl carbon with ^{14}C was used to determine its absorption, distribution, excretion, and metabolism in rats. The compound was rapidly excreted in both the urine and feces with no significant changes. Residues of imazapyr did not accumulate in the liver, kidney, muscle, fat, or blood.

VI. CONCLUSION

The imidazolinones are rapidly absorbed and excreted in the laboratory rat, primarily in the urine. With the exception of imazamethabenz-methyl, none of the imidazolinones are significantly metabolized. The major metabolites derived from imazamethabenz-methyl in the rat have been identified in the wheat plant. The single metabolite derived from imazethapyr in rat studies is the only metabolite found at trace levels in soybean plants. However, no detectable residues were found in soybean seeds.

Studies with imazamethabenz-methyl, imazaquin, and imazethapyr using the goat as a ruminant model have been conducted at dose rates representing significant safety factors with respect to the potential for exposure to residues in the human food chain. These studies show no evidence of significant residues or accumulation of residues in meat and milk.

Similarly, studies in which laying hens were given exaggerated doses of imazamethabenz-methyl, imazaquin, and imazethapyr showed no evidence of significant residues or accumulation of residues in edible tissue and eggs.

ACKNOWLEDGMENTS

The authors express their appreciation to the following scientists, without whose collaboration this chapter would not have been possible: C. Babcock, M. Brown, T. Chiu, J. Colavita, A. daCunha, T. Garces, P. Gingher, L. Goldenbaum, C. Gronostajski, C. Hebda, M. N. Hussain, D. Kirton, B. Knoll, T. M. Lee, L. Lucas, N. M. Mallipudi, Y. Roman, P. Stanley-Millner, S. Stout, and B. Taylor.

Chapter 13

RESIDUE ANALYSIS

James M. Devine

TABLE OF CONTENTS

I. INTRODUCTION

The residue methods for the various imidazolinone herbicides have been developed over many years. Although the chemical structures of these compounds are very similar, the chemical behavior as well as the biological activity differ substantially among the analogs. This situation, together with the rigorous means of extraction needed for certain types of samples, provided many challenges for developing the methods for each of the compounds. The use of solid-phase extraction cartridges, especially the ion-exchange cartridges, provided a simple and rapid means of cleanup for the various types of samples.

The earlier residue methods for the imidazolinones were not as simple nor as sensitive as the procedures used today. The methods presented here are the latest or recommended ones available at this time. Development of additional methods are still in progress for other applications and/or uses. Over the years many individuals have contributed to the residue knowledge of the imidazolinone herbicides, and their contributions are incorporated in the methods summarized in this section.

II. IMAZAPYR RESIDUE METHODS

No significant metabolites were found in metabolism studies conducted with soil or plants. Extensive degradation was observed, indicating complete breakdown of imazapyr. The residue methods developed, therefore, are only for the parent compound.

A. FORESTRY SAMPLES[1]

Residues of imazapyr are extracted from the finely ground samples of forest litter, pine twigs, or green leaves by shaking with an aqueous phosphate buffer (pH 6.5). Methanol is added to the extract, and the solution is partitioned with methylene chloride as a preliminary cleanup. The pH of the solution is then adjusted to 3.0, sodium chloride is added, and imazapyr is partitioned into methylene chloride. After removing excess salt by acetone precipitation, additional cleanup is obtained using a tandem C8/SCX aromatic sulfonic acid solid-phase extraction (SPE) cartridge arrangement. The C8 cartridge removes a significant amount of the sample coextractives while imazapyr passes through and is adsorbed by the SCX cartridge. The residue is removed from the SCX cartridge with a buffer solution which is then analyzed with an HPLC equipped with a UV detector set at 240 nm. An external standard is used for quantitation. The validated sensitivity of the method is 50 ppb for each type of sample. Recoveries averaged 87% for all three types of samples at fortification levels ranging from 50 to 20,000 ppb. Control samples usually had apparent residues of less than 5 ppb.

B. SOIL[2]

Residues of imazapyr are extracted from soil by shaking with 0.5 N sodium hydroxide in water. The pH of an aliquot of the extract is adjusted to pH 2.0 and the precipitated humic acids are filtered off. A C18 SPE cartridge is used to extract the imazapyr from the aqueous solution and further cleanup is accomplished by eluting the compound from the C18 cartridge onto an SCX aromatic sulfonic acid SPE cartridge. The imazapyr is eluted from the SCX cartridge with a buffer solution and partitioned into methylene chloride. Final determination is made with an HPLC using a UV detector set at 240 nm and comparing the sample peak size versus a standard. The validated sensitivity of the method is 5 ppb. Recoveries using five different types of soil averaged 91% over the fortification range of 5 to 1500 ppb. Control soil usually had apparent residues of less than 0.5 ppb.

C. WATER[3]

The water method for determining residue levels is similar to the soil method in that imazapyr residues are extracted from water by using a C18 SPE cartridge and cleanup and additional specificity are obtained by using an SCX cartridge. Similar to the previous two residue methods, quantitation is accomplished by liquid chromatography and a UV detector (240 nm). The validated limit of sensitivity is 1 ppb for water samples. Recoveries of imazapyr from water averaged 96% over the fortification range of 1 to 500 ppb. Control values were usually less than 0.2 ppb.

III. IMAZAMETHABENZ-METHYL RESIDUE METHODS

Metabolism studies have shown that the only significant metabolite of imazamethabenz-methyl in soil is the carboxylic acid hydrolysis product, CL 263,840. The acid metabolite is also a major metabolite of the compound in plants. Therefore, methods were developed for soil and crop samples to determine residues of both the parent and its acid metabolite. The crop methods use gas chromatography and determine a "total" residue (parent plus acid metabolite) while the soil method, which uses HPLC, quantitates the parent and acid metabolite separately.

A. WHEAT AND BARLEY GRAIN AND STRAW[4]

Homogenized plant samples are extracted by boiling a subsample with 0.5 N sodium hydroxide in 30% methanol/water. Any imazamethabenz-methyl residue is converted to the acid, CL 263,840, by the basic hydrolysis. The extract is cooled and filtered, and half of the extract is shaken with hexane as a preliminary cleanup step. For straw samples, Darco G-60 activated charcoal is added to the remaining aqueous/methanol solution which is then swirled for several minutes before filtering. The pH of the filtrate from straw samples or the remaining hexane-extracted aqueous methanol solution from grain samples is then adjusted to 3.5, and sodium chloride is added before extracting the CL 263,840 residues into methylene chloride. After removal of the methylene chloride, acetone is added to desalt the solution before further cleanup. The residue is dissolved in dilute acid and, after centrifuging, is passed through an SCX cartridge. After washing the cartridge to remove coextractives, the compound is eluted from the SCX cartridge with pH 6.5 buffer through an SAX quaternary amine SPE cartridge. After adding methanol to the eluate and adjusting the pH to 3.5, the CL 263,840 is partitioned into methylene chloride. The methylene chloride is removed and the residue is dissolved in methanol; trimethylanilinium hydroxide (TMAH) is added as a methylation (injection port) reagent. Final determination is made using gas chromatography with a nitrogen/phosphorus detector. Quantitation is done by comparing the sample versus a CL 263,840 standard to which TMAH is added. The validated sensitivity of the method is 0.10 ppm as a total residue (imazamethabenz-methyl plus CL 263,840). The average recovery in grain and straw for imazamethabenz-methyl was 81% and for CL 263,840 was 89% over the fortification range of 0.10 to 2.00 ppm. Control values were generally less than 0.02 ppm.

B. SUNFLOWER SEED[5]

Residues of imazamethabenz-methyl are extracted from homogenized sunflower seed by boiling the seed with 0.5 N sodium hydroxide, as previously described for wheat and barley. Half of the extract is concentrated on a rotary evaporator to a volume of 30 to 40 ml to remove the methanol. Acetone is added to precipitate coextractives, which are removed by filtration. The acetone is evaporated, pH 6.5 buffer is added, and the solution is partitioned with methylene chloride. The pH of the aqueous solution is adjusted to 3.5, and the CL 263,840 is partitioned into methylene chloride. The solvent is removed and the extract is passed through a C18 SPE cartridge with dilute acid. The compound is eluted from the C18 cartridge with 50% methanol/water and adsorbed onto an SCX cartridge. The SCX SPE cartridge is rinsed to remove coextractives and then eluted with pH 6.5 phosphate buffer. The eluate is shaken with methylene chloride, which is discarded. The pH is then adjusted to 3.5 before extracting the CL 263,840 into methylene chloride. The remainder of the procedure is the same as for wheat and barley in that TMAH and a gas chromatograph are used to determine the final results. The validated sensitivity of the method is 0.10 ppm as a total residue with control samples generally having less than 0.02 ppm apparent residue. The average recovery for imazamethabenz-methyl was 85% and for CL 263,840 was 84%. The fortification range was from 0.10 to 1.0 ppm.

C. SOIL[6]

Residues of imazamethabenz-methyl and CL 263,840 are extracted from soil with acidic methanol followed by 0.05 *N* sodium hydroxide. The acidic methanol extract is evaporated to a volume of approximately 10 ml using a rotary evaporator. The second basic extract is combined with this concentrated extract, and the pH is adjusted to 3.5 to precipitate the coextracted humic acids. Centrifugation is used to remove the humic acids, and the remaining solution is passed through a C18 SPE cartridge to adsorb the imazamethabenz-methyl and CL 263,840 residues onto the cartridge. A 50% methanol/water solution is used to elute the compounds from the C18 column onto an SCX cartridge, which is then rinsed with methanol. The residues are removed with a saturated potassium chloride/methanol solution. The methanol is removed, and pH 3.5 buffer is added before extraction with methylene chloride. The organic solvent is removed with a rotary evaporator, and the residue is dissolved in deionized water for analysis with an HPLC equipped with a UV detector set at 240 nm. The validated sensitivity of the method is 10 ppb for each compound. Recoveries averaged 85% and 91% for imazamethabenz-methyl and CL 263,840, respectively. Five different soils were used for the validation, and fortification levels ranged from 10 to 200 ppb. Control values were usually less than 1 ppb.

IV. IMAZAQUIN RESIDUE METHODS

In aerobic soil studies conducted with imazaquin no significant metabolites were formed. With soybeans, metabolism studies showed rapid and extensive degradation of imazaquin in the plants with no major metabolite identified. The residues were divided amoung 15 or more different compounds. Hence, the residue methods that were developed are for the parent compound only.

A. SOYBEANS[7]

Imazaquin residues are extracted from homogenized soybeans with acidic 80% methanol/water. Water is added to the extract and imazaquin is partitioned into methylene chloride, which is removed with a rotary evaporator, and the sample is partitioned between hexane and acetonitrile. The acetonitrile extracts are evaporated to dryness, and the residue is dissolved in acetone. A coagulation solution of ammonium chloride and phosphoric acid is added and, after 30 min, the solution is filtered with Celite 545 AW to remove the precipitated coextractives. The filtrate is then partitioned with methylene chloride, which is removed. The residue is dissolved in methanol and diluted with water before being passed through an SAX quaternary amine SPE cartridge. A dilute acid solution is used to elute imazaquin from the cartridge, and the eluate is then partitioned with methylene chloride. After evaporating the methylene chloride and dissolving the residue in methanol, TMAH is added as a methylation reagent. The determinative step is gas chromatography using a nitrogen/phosphorus detector. The validated sensitivity of the method is 50 ppb, with an average recovery of

95%, and the fortification range is from 50 to 500 ppb. Control seed contained apparent residues of less than 10 ppb.

B. WATER[8]

After filtration of the sample and pH adjustment to 3.0, imazaquin is isolated from water by the use of a C2 SPE cartridge. Acidic 40% methanol/water is used to elute the compound from the C2 cartridge through an SAX SPE cartridge. The eluate is then injected directly into an HPLC with a UV detector (240 nm). The validated sensitivity of the method is 1 ppb, and recoveries averaged 87% over the fortification range of 1 to 1,000 ppb. Apparent residues in control samples were usually less than 0.1 ppb.

C. SOIL[9]

Residues of imazaquin are extracted from soil by shaking with 0.5 N sodium hydroxide in 30% methanol/water. After filtration, an aliquot of the extract is adjusted to pH 2.0, and sodium chloride is dissolved before passing through a C18 SPE cartridge. The cartridge is rinsed with deionized water and imazaquin is eluted onto an SCX cartridge with 50% methanol/water. A pH 6.5 phosphate buffer is used to elute the imazaquin from the SCX cartridge. The eluate is adjusted to pH 2.0 and partitioned with methylene chloride. After evaporation of the methylene chloride, the residue is dissolved in pH 6.5 phosphate buffer in preparation for HPLC analysis with a UV detector (240 nm). Recoveries from five different types of soil averaged 81% over a fortification range of 5 to 200 ppb, and the validated sensitivity of the method is 5 ppb with control values generally less than 0.5 ppb.

V. IMAZETHAPYR RESIDUE METHODS

Metabolism studies with various legume crops (soybeans, peanuts, beans, and peas) showed no significant residues in the final edible crop. Degradation of imazethapyr in the plants was rapid and extensive; therefore, residue methods were developed only for the parent compound for the legume crops. Likewise, soil metabolism studies show that no significant metabolites were formed; therefore, the soil residue method only determines parent compound.

A. LEGUMINOUS VEGETABLES[10-12]

Residues of imazethapyr are extracted from leguminous vegetable samples by blending with acidic 60% methanol/water. The extract is adjusted to pH 2.0 and partitioned with methylene chloride. Residues are extracted into a pH 9 buffer and then back into methylene chloride after pH adjustment. After solvent evaporation, the sample is passed through an SAX quaternary amine SPE cartridge. A dilute acid solution is used to elute the imazethapyr from the SAX cartridge and after pH adjustment, methylene chloride is used to extract the eluate. The solvent is removed with a rotary evaporator, and methanol and TMAH are added. Gas chromatography with a nitrogen/phosphorus detector is used for the final analysis. The validated sensitivity of the method is 0.10 ppm. Recoveries averaged 84% for soybeans, 78% for peanuts, 94% for succulent beans and peas, and 89% for dry beans and peas. Fortification levels ranged from 0.10 to 1.0 ppm, and control samples usually contained less than 0.010 ppm.

B. SOIL[13]

Residues of imazethapyr are extracted from soil by shaking with 0.5 N sodium hydroxide solution. The pH of an aliquot of the extract is adjusted to pH 2.0 and the precipitated humic acids are removed by filtration. Residues are partitioned into methylene chloride, which is then evaporated on a rotary evaporator. The residue is dissolved in a dilute sodium hydroxide solution which is passed through a C18 SPE cartridge into a reservoir containing dilute acid. The resulting solution is then passed through an SCX cartridge where the imazethapyr is adsorbed. A pH 6.5 phosphate buffer and methanol are used to elute the SCX cartridge. After adjusting the pH of the eluate to 2.0, methylene chloride is used to extract the compound. The organic solvent is removed, and the final residue is dissolved in deionized water for HPLC analysis with a UV detector at 254 nm. The validated sensitivity of the method is 5 ppb. Recoveries were run with seven different types of soil and averaged 89% over the fortification range of 5 to 200 ppb, and control values were generally less than 2 ppb.

REFERENCES

1. **Picard, G.,** HPLC Method for the Determination of CL 243,997 Residues in Pine Twigs, Forest Litter, and Green Vegetation, American Cyanamid Company Method M-1612, 1986.
2. **Khunachak, A.,** HPLC Determination of CL 243,997 Residues in Soil, American Cyanamid Company Method M-1713.02, 1988.
3. **Khunachak, A.,** HPLC Method for the Determination of CL 243,997 Residues in Water, American Cyanamid Company Method M-1900, 1989.
4. **Picard, G.,** GC Method for the Determination of Total CL 222,293 and CL 263,840 Residues, Determined as CL 263,840, in Wheat and Barley Grain and Straw, American Cyanamid Company Method M-153, 1985.
5. **Khunachak, A. and Picard, G.,** GC Method for the Determination of Total CL 222,293 and CL 263,840 Residues, Measured as CL 263,840, in Sunflower Seed and Wheat and Barley Grain and Straw, American Cyanamid Company Method M-1761.1, 1988.
6. **Roman, M.,** HPLC Method for the Determination of CL 222,293 and CL 263,840 Residues in Soil, American Cyanamid Company Method M-1917, 1989.
7. **Roman, M.,** GC Method for the Determination of CL 252,214 Residues in Soybean Plant, Straw, and Seed, American Cyanamid Company Method M-1410, 1984.
8. **Picard, G.,** HPLC Method for the Determination of CL 252,214 Residues in Run-Off Water, American Cyanamid Company Method M-1591, 1986.
9. **Picard, G.,** HPLC Method for the Determination of CL 252,214 Residues in Soil, American Cyanamid Company Method M-1854, 1988.
10. **Babbitt, B.,** GC Method for the Determination of CL 263,499 Residues in Soybean Plants, Seed, and Straw, American Cyanamid Company Method M-1586, 1985.
11. **Potts, C.,** GC Method for the Determination of CL 263,499 Residues in Peanuts, American Cyanamid Company Method M-1762, 1987.
12. **Peterson, R.,** GC Method for the Determination of CL 263,499 Residues in Succulent and Dry Beans and Peas, American Cyanamid Company Method M-1855, 1988.
13. **Tondreau, R. and Guzman, B.,** Determination of CL 263,499 Residues in Soil by HPLC, American Cyanamid Company Method M-1719.03, 1988.

Chapter 14

TOXICOLOGY OF THE IMIDAZOLINONE HERBICIDES

James A. Gagne, Joel E. Fischer, Rajendar K. Sharma, Karl A. Traul,
Scott J. Diehl, Frederick G. Hess, and Jane E. Harris

TABLE OF CONTENTS

I. INTRODUCTION

The imidazolinone herbicides have been subjected to exhaustive toxicological studies to determine their toxicities to mammals and other nontarget organisms and to evaluate the potential hazard of handling and applying these herbicides. As demonstrated by the results of those studies, the imidazolinones have a low toxicologic potential, partially because they act by inhibiting a biosynthetic process at a site present only in plants. In addition, these herbicides are excreted rapidly by rats before they can accumulate in the tissues or blood. The results of these studies are presented below.

II. ACUTE TOXICITY TO MAMMALS

Imazapyr, imazamethabenz-methyl, imazethapyr, and imazaquin are no more than slightly toxic to mammals, as illustrated by the results of acute oral, dermal, inhalation, and sensitization studies summarized in Table 1.

III. SUBACUTE AND CHRONIC TOXICITY TO MAMMALS

Additional short- and long-term mammalian studies were conducted to evaluate the safety of the imidazolinone herbicides. The no-effect levels for these studies are summarized in Table 2.

TABLE 1
Acute Toxicity of the Imidazolinone Herbicides to Mammals
(Technical Material)

Study species	LD$_{50}$ (mg/kg body wt.) or result			
	Imazapyr	Imazamethabenz-methyl	Imazethapyr	Imazaquin
Oral				
Rat, male	>5,000	>5,000	>5,000	>5,000
Rat, female	>5,000	>5,000	>5,000	>5,000
Dermal				
Rabbit	>2,000	>2,000	>2,000	>2,000
Irritation				
Eye, rabbit	Irreversibly irritating	Mild[c]	Moderate[a]	None
Skin, rabbit	None	None	Mild[b]	Mild[b]
	LC$_{50}$ (mg/liter)[d] — analytical conc			
Inhalation				
Rat	>1.3	>1.4	>3.3	>5.7
Skin Sensitization				
Guinea pig	Nonsensitizer	Nonsensitizer	Nonsensitizer	Nonsensitizer

[a]   All signs of irritation subsided by 7 d after dosing.
[b]   Mild erythema, which disappeared by 72 h after treatment.
[c]   All signs of irritation subsided by 72 h after dosing.
[d]   4 h exposure.

TABLE 2
Subacute and Chronic Toxicity of the Imidazolinone Herbicides to Mammals (Technical Material)

Species; type of study	No-effect level[a]			
	Imazapyr	Imazamethabenz-methyl	Imazethapyr	Imazaquin
Rats				
13-Week dietary	10,000 ppm	1,000 ppm	10,000 ppm	10,000 ppm
24-Month dietary	10,000 ppm[b]	250 ppm	10,000 ppm	10,000 ppm
2-Generation reproduction	10,000 ppm	250 ppm[c]	10,000 ppm[f]	10,000 ppm[c]
			10,000 ppm[g]	
Teratology	300 mg/kg[d]	<250 mg/kg[d]	>1,125 mg/kg[d]	500 mg/kg[d]
	1,000 mg/kg[e]	250 mg/kg[e]	375 mg/kg[e]	500 mg/kg[e]
Dogs				
90-Day dietary	10,000 ppm	—	10,000 ppm	—
1-Year dietary	10,000 ppm	250 ppm	1,000 ppm	1,000 ppm
Mice				
18-Month dietary	10,000 ppm[b]	130 ppm	5,000 ppm	1,000 ppm
Rabbits				
21-Day Dermal	400 mg/kg	200 mg/kg	1,000 mg/kg	1,000 mg/kg
Teratology	400 mg/kg[d]	750 mg/kg[d]	300 mg/kg[d]	250 mg/kg[d]
	400 mg/kg[e]	250 mg/kg[e]	>1,000 mg/kg[e]	500 mg/kg[e]

[a] According to the U.S. Environmental Protection Agency (EPA).
[b] Studies in review at the EPA.
[c] Study involved 3 generations.
[d] Maternal toxicity.
[e] Fetal toxicity.
[f] Parental and reproductive toxicity.
[g] Pup toxicity.

IV. GENETIC TOXICITY

The imidazolinone herbicides have been tested in the following *in vitro* and *in vivo* genotoxicity studies:

In vitro studies
- Microbial mutagenesis in five strains of *Salmonella typhimurium*
- Microbial mutagenesis in one strain of *Escherichia coli*
- Mammalian cell mutagenesis in Chinese Hamster ovary (CHO) cells
- Chromosomal aberrations in CHO cells
- Unscheduled DNA synthesis in primary rat hepatocytes

In vivo studies
- Dominant lethal assay in rats
- Chromosomal aberrations in rat bone marrow cells (imazethapyr only)

All tests demonstrated that imazapyr, imazamethabenz-methyl, imazethapyr, and imazaquin are not genotoxic.

TABLE 3
Toxicity of the Imidazolinone Herbicides to Nontarget
Organisms (Technical Material)[a]

| | Acute toxicity | | | |
Test organism	Imazapyr	Imazamethabenz-methyl	Imazethapyr	Imazaquin
Birds				
Oral LD_{50} (mg/kg body wt.)				
Bobwhite quail, *Colinus virginianus*	>2,150	>2,150	>2,150	>2,150
Mallard duck, *Anas platyrhynchos*	>2,150	>2,150	>2,150	>2,150
8-d dietary LC_{50} (ppm)				
Bobwhite quail, *Colinus virginianus*	>5,000	>5,000	>5,000	>5,000
Mallard duck, *Anas platyrhynchos*	>5,000	>5,000	>5,000	>5,000
Fish				
96-h LC_{50} (mg/liter)				
Bluegill sunfish, *Lepomis macrochirus*	>100 (100)	420 (100)	420 (320)	410 (180)
Rainbow trout, *Salmo gairdneri*	>100 (100)	280 (100)	340 (<100)	280 (100)
Channel catfish, *Ictalurus punctatus*	>100 (100)	—	240 (180)	320 (140)
35-d MATC[b] (mg/liter)				
Fathead minnow, *Pimephales promelas*		0.32		
Arthropods				
Contact LD_{50} (μg/bee)				
Honey bee, *Apis mellifera*	>100	>100	>100	>100
48-h LC_{50} (mg/liter)				
Water flea, *Daphnia magna*	>100 (100)	220 (100)	>1000 (1000)	280 (56)
21-d MATC (mg/liter)				
Water flea, *Daphnia magna*		8.5		

[a]   The no-effect levels are indicated in parentheses.
[b]   MATC = Maximum acceptable toxicant concentration.

V. EFFECTS ON NONTARGET ORGANISMS

Many studies have been conducted to determine the potential hazard of the imidazolinones to birds, fish, and other nontarget organisms. The results of these studies are summarized in Table 3.

Chapter 15

BEHAVIOR OF THE IMIDAZOLINONE HERBICIDES IN THE AQUATIC ENVIRONMENT

Gary Mangels

TABLE OF CONTENTS

I. INTRODUCTION

The imidazolinone herbicides are not expected to reach bodies of water directly. The potential routes of entry of the imidazolinones, like those of other crop protection chemicals, into surface water bodies will be through spray drift or run-off. Volatilization of the imidazolinones from water bodies is not expected because of their high water solubilities and low vapor pressures (see Table 1), and thus their low Henry's law constants.

Four possible environmental degradation processes — hydrolysis, photolysis, and aerobic and anaerobic biodegradation — are described below.

II. HYDROLYSIS

Hydrolysis of the acid imidazolinones is extremely slow at environmentally relevant pHs and temperatures.

A. HYDROLYSIS OF IMAZAQUIN

The hydrolysis of imazaquin was studied at an initial concentration of 10 ppm using pH 5, 7, and 9 buffers and distilled water at 25°C for 30 d. There was no detectable degradation in the distilled water and pH 5 and 7 samples. Degradation in pH 9 buffer resulted in a calculated half-life of 169 d. The hydrolysis product was 2-[(1-carbamoyl-1,2-dimethylpropyl-carbamoyl]-3-quinolinecarboxylic acid.

B. HYDROLYSIS OF IMAZAPYR

The hydrolysis of imazapyr was studied at an initial concentration of 50 ppm using pH 5, 7, and 9 buffers and distilled water at 25°C for 30 d. There was no detectable degradation in the distilled water or pH 5 and 7 samples over 30 d. The half-life in pH 9 was calculated to be 325 d. The hydrolysis product formed is 2-[(1-carbamoyl-1,2-dimethylpropyl-carbamoyl]nicotinic acid.

C. HYDROLYSIS OF IMAZETHAPYR

Studies of the hydrolysis of imazethapyr were conducted at a concentration of 100 ppm using pH 5, 7, and 9 buffers and distilled water at 25°C for 30 d. There was no detectable degradation over 30 d in distilled water and at pH 5 and pH 7. Hydrolysis of the pH 9 samples was studied over 6 months, resulting in a calculated half-life of 9.6 months. The hydrolysis product is 2-[(1-carbamoyl-1,2-dimethylpropyl-carbamoyl]-5-ethylnicotinic acid.

The hydrolysis pathway for the imidazolinone acids is shown below, using imazaquin as an example.

D. HYDROLYSIS OF IMAZAMETHABENZ-METHYL

The hydrolysis of the regioisomers of imazamethabenz-methyl (AC 239,589 — the *para* isomer — and AC 252,767 — the *meta* isomer) was studied between the pHs of 5 and 9 and at a temperature range of 15 to 35°C.[1] In contrast to the slow hydrolysis of the acid imidazolinones, hydrolysis of the esters under basic conditions was rapid and resulted in the cleavage of the ester linkage to form the free acid. Also, small amounts of the acid-diamides were formed by adding water to the imidazolinone ring during ring opening.

TABLE 1
Physical Properties of the Imidazolinones

Compound	Water solubility (ppm)	Vapor pressure (mmHg)
Imazaquin	60 at 25°C	$<10^{-7}$ at 60°C
Imazapyr	9,740 at 15°C	
	11,300 at 25°C	
	13,500 at 35°C	$<10^{-7}$ at 60°C
Imazethapyr	1,415 at 25°C	$<10^{-7}$ at 60°C
AC 263,222	1,920 at 15°C	
	2,230 at 25°C	
	2,630 at 35°C	$<10^{-7}$ at 60°C
Imazamethabenz-methyl	860 (p-isomer)	1.13×10^{-8} at 25°C
	1370 (m-isomer)	1.96×10^{-7} at 37°C
	at 25°C	1.76×10^{-6} at 50°C
		2.41×10^{-5} at 61°C

TABLE 2
Half-Lives of AC 239,589 and AC 252,767 at Various pHs and Temperatures

		Half-life in days	
pH	Temperature (°C)	AC 239,589	AC 252,767
5	15	1340	730
	25	500	265
	35	185	94
6	25	430	220
7	15	420	240
	25	115	63
	35	28	16
8	25	16	10
9	15	6.2	3.5
	25	1.7	1.0
	35	0.5	0.4

The half-lives of regioisomers AC 239,589 and AC 252,767 at various pHs and temperatures are shown in Table 2.

The effects of pH and temperature on the hydrolysis rate constant, K_h (in days), is shown in Figure 1.

III. PHOTOLYSIS

In sharp contrast to the slow hydrolytic degradation of the imidazolinones, the compounds are rapidly degraded in water by light.

A. PHOTOLYSIS OF IMAZAQUIN

Using simulated sunlight from a borosilicate-filtered Xenon Arc lamp, American Cyanamid scientists studied the aqueous photolysis of ^{14}C-carboxyl-labeled imazaquin in dis-

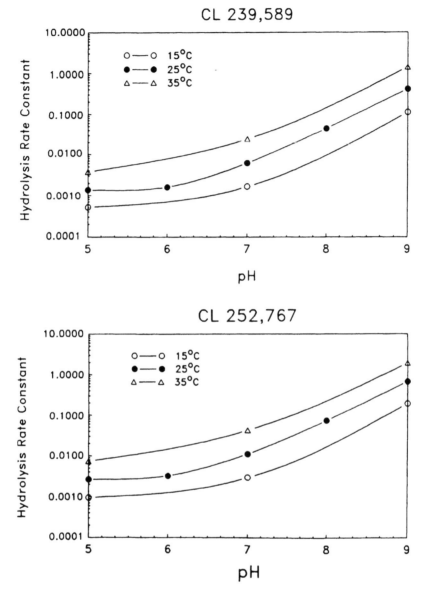

FIGURE 1. Effects of pH and temperature on the hydrolysis rate constant.

tilled water and pH 5, 7, and 9 buffers at 18 to 19 °C at an initial concentration of 9 ppm. The half-lives followed first-order kinetics and were calculated to be 16.7, 15.2, 21.0, and 7.8 h, respectively. Four photodegradation products were identified: 3-quinoline-carboxylic acid (**1**); 2,3-quinoline-dicarboxylic acid (**2**); 2-carboxamido-3-quinolinecarboxylic acid (**3**); and 2,3-dihydro-3-imino-1*H*-pyrrolo [3,4-*b*]quinoline-1-one (**4**).

American Cyanamid scientists also studied the aqueous photolysis of both ^{14}C-carboxyl-labeled and ^{14}C-quinoline ring-labeled imazaquin, using simulated sunlight from a borosilicate-filtered Xenon Arc lamp, in pH 7 buffer at 25°C at initial concentrations of 20 and 15 ppm, respectively. The half-lives followed first-order kinetics and were calculated to be 15 h under continuous irradiation. Three photodegradation products were identified from the ^{14}C-carboxyl-labeled imazaquin which accounted for more than 10% of the applied dose at any sampling interval: $^{14}CO_2$; 2-carboxamido-3-quinolinecarboxylic acid (3); and 2,3-dihydro-3-imino-1H-pyrrolo[3,4-b]quinoline-1-one (4). From the ^{14}C-quinoline ring-labeled imazaquin, three photodegradation products were identified which accounted for more than 10% of the applied dose at any sampling interval: 2-carboxamido-3-quinolinecarboxylic acid (3); 2,3-dihydro-3-imino-1H-pyrrolo [3,4-b]quinoline-1-one (4); and quinoline 2,3-dicarboxylic acid imide (5).

5

B. PHOTOLYSIS OF IMAZAPYR

Scientists at American Cyanamid Company studied the aqueous photolysis of imazapyr in distilled water and pH 5 and 9 buffers utilizing simulated sunlight from a borosilicate-filtered Xenon Arc lamp. The half-lives followed first-order kinetics and were calculated as 1.9 to 2.3, 2.7, and 1.3 d, respectively. Four degradation products were identified from the photolysis of ^{14}C-carboxyl-labeled imazapyr: quinolinic acid (6); quinolinimide (7); furo[3,4-b]pyridine-5(7H)-one (8); and 7-hydroxy-furo[3,4-b]pyridine-5(7H)-one (9).

6   7   8   9

C. PHOTOLYSIS OF IMAZETHAPYR

Photolysis of ^{14}C-pyridine-ring-C-6 imazethapyr was studied in distilled water and pH 5, 7, and 9 buffers. Solutions of the test compound were made in the buffers at an initial concentration of approximately 20 ppm and exposed to light from a Xenon Arc lamp. The light was filtered through borosilicate glass to mimic natural sunlight. Solutions were also maintained in the dark to serve as control samples to demonstrate that degradation did not result from hydrolysis or microbial activity. No radioactivity was lost from any of the samples during the study, indicating that no volatile compounds were formed. The control samples showed no degradation over the course of the study, indicating that degradation was caused by photolysis. The degradation of samples exposed to continuous irradiation followed first-order kinetics with half-lives of 46, 44, 50, and 57 h in distilled water, pH 5, 7, and 9 buffers, respectively. The two photodegradation products identified were 5-ethyl-2,3-pyridinedicarboxylic acid (10) and 5-ethyl-3-pyridinecarboxylic acid (11). At least six other minor degradation products were formed, none of which accounted for more than 10% of the applied dose.

10   **11**

D. PHOTOLYSIS OF IMAZAMETHABENZ-METHYL

The photochemical degradation of the regioisomers of imazamethabenz-methyl was studied under a variety of conditions with the ^{14}C-label placed in several positions on the molecule.[1] The photolysis of ^{14}C-4-imidazo-labeled imazamethabenz-methyl was studied in distilled water and pH 5, 7, and 9 buffers using a 450 W Hanovia* lamp as the light source. The light was filtered through borosilicate glass to remove light with a wavelength of less than 290 nm. The major photodegradation product containing the ^{14}C-label was determined by mass spectroscopy to be 3-butane-2-one. Photolysis was extremely rapid, with half-lives of approximately 13, 20, 13, and 6.5 h, respectively, in distilled water, pH 5, 7, and 9 buffers.

Photolysis of ^{14}C-imidazo-carbonyl-labeled imazamethabenz-methyl was studied in distilled water and in pH 5, 7, and 9 buffers using the same conditions as described above for the 4-imidazo-labeled study. The major photodegradation product which contained this ^{14}C-label was carbonate, as determined by trapping in NaOH and precipitating with a barium chloride solution. Photolysis was extremely rapid, with half-lives of approximately 13, 20, 13, and 6.5 h in distilled water and pH 5, 7, and 9 buffers, respectively. These half-lives are similar to the photolysis rates found with the 4-imidazo-labeled compound.

Photolysis of each of the ^{14}C-carboxyl-labeled regioisomers of imazamethabenz-methyl was also studied. The hydrolysis products previously described were found, along with the photodegradation products shown in the generalized photodegradation scheme listed below.

---

*   Hanovia is the trademark of Canrad Inc.

TABLE 3
The Effects of Location and Season on the Predicted Half-Lives of
the Regioisomers AC 239,589 and AC 252,767

Compound	Latitude (°)	Half-life (days)			
		Spring	Summer	Fall	Winter
AC 239,589	30	5.9	4.7	9.0	15.5
	40	7.8	5.4	14.5	35
	50	11.2	6.5	29.8	112
AC 252,767	30	4.8	3.8	7.3	13
	40	6.4	4.3	12.2	31
	50	9.6	5.3	26	106

The quantum yields of disappearance of the regioisomers of imazamethabenz-methyl in distilled water were measured on a standard photochemical revolving apparatus using 313 nm light isolated from a 450 W Hanovia mercury arc lamp inside a Pyrex* immersion well. The light intensity was measured using ferrioxalate actinometry. The quantum yields of disappearance of regioisomers AC 239,589 and AC 252,767 in distilled water were calculated to be 0.0215 and 0.016.

Using the procedures of Zepp and Cline[2] and their SOLAR computer program, the photolysis of the regioisomers of imazamethabenz-methyl was modeled in distilled water at 30°, 40°, and 50° latitude and at all four seasons in order to determine the effects of location and season on the rates of degradation. The predicted half-lives, in days, are shown in Table 3.

An outdoor experiment using sunlight was also conducted to compare the half-lives predicted with the SOLAR model and actual photolytic half-lives. Solutions of ^{14}C-carboxylic-acid-labeled AC 239,589 and AC 252,767 were placed in quartz test tubes and exposed to sunlight in the spring in the central New Jersey area. At various times, aliquots were assayed and degradation of the regioisomers was measured. The half-lives in the spring at 40° latitude were 7.9 d for AC 239,589 and 6.2 d for AC 252,767. These half-lives are in good agreement with predicted half-lives of 7.8 d and 6.4 d.

IV. MICROBIAL DEGRADATION

Imazapyr and imazethapyr were stable under aerobic and anaerobic aquatic conditions. In studies conducted by scientists at American Cyanamid Company, imazapyr did not degrade microbially over 1 year in an anaerobic sediment/water system taken from a lake. Similarly, there was no detectable degradation of imazapyr over a 28 d period in an aerobic sediment/water system. Imazethapyr also showed no degradation over 1 year in aerobic or anaerobic sediment/water systems taken from a river in Canada.

The fate of the primary photodegradation products from imazamethabenz-methyl, 3-hydroxy-6-methyl phthalide and 3-hydroxy-5-methyl phthalide, was studied under aerobic aquatic conditions with sediment using ^{14}C-labeled photodegradation products with the label in the carbonyl carbon. Approximately 90% of the applied dose of 3-hydroxy-6-methyl phthalide was recovered as ^{14}CO$_2$ in 9 d. After 28 d the water and sediment contained only 1.5% and less than 1% of the applied radioactivity, respectively. Approximately 85% of the applied dose of 3-hydroxy-5-methyl phthalide was recovered as ^{14}CO$_2$ in 28 d. After 28 d the water and sediment contained only 4.5% and 2.6% of the applied radioactivity, respectively. These studies demonstrate that the major photodegradation products from imazamethabenz-methyl are rapidly mineralized and would not pose risks to the environment.

* Pyrex is the trademark of Corning Glass Works.

V. SUMMARY

These studies demonstrate that the imidazolinones can be rapidly degraded in aquatic systems, primarily through photolysis. The regioisomers of imazamethabenz-methyl are also hydrolyzed rapidly under basic conditions.

REFERENCES

1. **Mangels, G.,** Photochemical Degradation of AC 222,293 (Assert Herbicide) in Aquatic Systems, Ph.D. thesis, Rutgers University, New Brunswick, NJ, 1985.
2. **Zepp, R. G. and Cline, D. M.,** Rates of direct photolysis in aquatic environment, *Environ. Sci. Technol.,* 11, 359, 1977.

Chapter 16

BEHAVIOR OF THE IMIDAZOLINONE HERBICIDES IN SOIL — A REVIEW OF THE LITERATURE

Gary Mangels

TABLE OF CONTENTS

I. INTRODUCTION

The imidazolinone herbicides are introduced into the environment by application to plants and/or soil. Once the compounds are placed in the soil they are subject to a variety of degradation and transport processes. Material that is present at the soil/air interface is exposed to possible interactions with sunlight, making photodegradation a possible degradation pathway. The microorganisms in the soil can also act to degrade these compounds by metabolizing them under aerobic conditions.

II. BINDING AND MOBILITY IN SOIL

Because of their structure, the acid imidazolinones have a unique set of acid-base properties. There are five distinct chemical species that are present at various pHs (Structure 1).

The acid-base dissociation constants of the various functional groups have been determined to be 2.8 to 3.1, 3.8 to 3.9, and 11.4. The effects of pH on the compositions of the various chemical forms is shown in Figure 1.

In the pH range of environmental interest, pH 5 to 9, species **4** is dominant. As the pH decreases below 5, species **2** becomes important and may theoretically exist with species **3**. As the pH decreases below 3 species **1** predominates. At pH values above 10, a fifth species, **5**, is present in significant quantities. An understanding of this equilibrium is key to understanding the factors affecting the adsorption of the acid-imidazolinones.

Molecular form **2** is the structure which is responsible for most of the binding of the imidazolinones to soil. This form is mostly uncharged, except for very weak positive charges on the nitrogens in the pyridine or quinoline ring and the imidazolinone ring. The adsorption behavior of this uncharged form is similar to that of nonionic compounds in that adsorption of this form is determined by the amount of organic matter in the soil. The uncharged form would be adsorbed onto nonionic resins, such as XAD-4 or C_{18}-solid phase media, and partitioned from water into organic solvents.

The anionic form of the compound, **4**, is the predominant species from pH 5 to 9. This form of the compound binds very weakly to organic matter or nonionic resins, but would bind well to anion-exchange resins and certain minerals. This form is not partitioned from water into organic solvents.

Studies conducted by scientists at American Cyanamid Company have demonstrated that

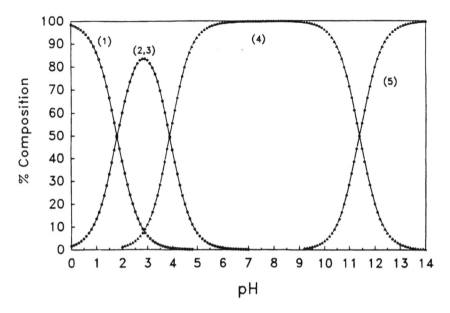

FIGURE 1. Effect of pH on the composition of various chemical forms.

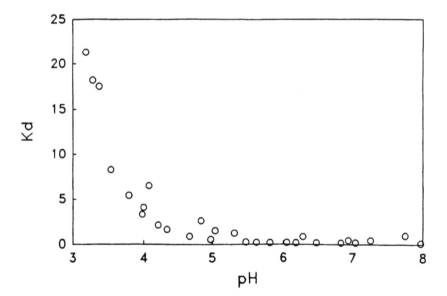

FIGURE 2. Effect of pH on soil/water partition coefficient (K_d).

both pH and organic matter significantly affect the behavior of the acid imidazolinones. The effects of pH on the adsorption of these compounds was studied by conducting batch equilibrium adsorptions and varying the pH by the addition of small amounts of HCl or NaOH. An example of the results are shown in Figure 2, which is a plot of the adsorption coefficient, K_d, versus pH.

These results are typical of those found using many different soils for each of the acid imidazolinones. In general, all soils studied showed very weak adsorption at pH values greater than 6. As the pH decreases below 6, the adsorption increases significantly as the pH approaches 3. Small pH differences, 0.2 to 0.4 pH units, can double the adsorption coefficient in the range of pH 3 to 6 because lowering the pH in small amounts greatly increases the amount of compound in the COOH form, (2).

III. FIELD DISSIPATION

Results from many rate-of-disappearance (ROD) studies in the field demonstrate that the imidazolinones dissipate from soil and do not leach under "real-world" conditions. Studies on the dissipation of imazethapyr serve as good examples of the behavior of this class of compounds.

Dissipation and downward migration of imazethapyr residues in soil were examined in 20 field ROD studies conducted in the U.S. (14 studies), Ontario, Canada (four studies) and the U.K. (two studies). In the U.S., dissipation was more rapid in the southern states (e.g., Georgia, Kentucky) than in the northern states (e.g., Iowa, Illinois, Nebraska). Imazethapyr residues were detected at greater than the limit of detection (5 ppb) in the 22.5- to 30-cm soil layer in only one of the U.S. trials. In all other U.S. studies, imazethapyr residues were less than the limit of detection (5 ppb) in samples from this layer at all times. In addition, imazethapyr residues were less than 5 ppb in samples from the 30- to 45-cm layer at all times in every study. Results from Canadian ROD studies were similar to those of the U.S. studies and demonstrated that imazethapyr did not show significant leaching below 15 cm.

Studies were conducted at Sion Hill and Riverside Fields, Manor Farm, Millby, N. Yorkshire, U.K. to determine the dissipation and downward migration of imazethapyr residues on bare ground plots following applications at 75 g a.e./ha. For the Sion Hill site, imazethapyr residues decreased from an average of 68 ppb at the time of application to 6 ppb by day 119 and were less than 5 ppb in samples taken 231 and 336 d posttreatment. Residues were below the validated sensitivity of the method (5 ppb) in all samples below 5 cm except for one of 6 ppb in the 5- to 10-cm layer at day 14. At the Riverside Fields site, imazethapyr residues dissipated from an average of 74 ppb at the time of application to 5 ppb by day 231 and were less than 5 ppb in samples taken 336 d posttreatment. Residues were below the sensitivity of the method in all samples below 10 cm except for one of 9 ppb in the 10 to 15 cm soil layer at day 28. Data from both sites further support the view that imazethapyr does not migrate downward to any significant degree.

In addition, a study was conducted specifically to investigate possible downward migration of imazethapyr in the field under conditions that would maximize leaching. A site was selected with soil of the highest leaching potential found in the U.S. soybean-growing area, and supplemental irrigation was provided to simulate the maximum area rainfall. Also, an unrealistically high preplant-incorporated application rate for imazethapyr was used: 140 g a.e./ha (0.125 lb a.e./A). Under this "worst-case" situation, imazethapyr soil residues in the upper 15 cm decreased from an average of 33 ppb on the day of application to less than 5 ppb by day 55. Furthermore, residues were not detected below 45 cm, and imazethapyr residues were detected on only two sampling dates in samples taken from the 30- to 45-cm layer (6 and 5 ppb on posttreatment days 14 and 21, respectively). Imazethapyr residues were less than 5 ppb, the limit of detection, in all subsequent sampling intervals from this layer.

Rotational crop studies in the field using ^{14}C-imazethapyr further demonstrated little downward movement of imazethapyr in the field under normal application conditions. In these studies, more than 95% of the recovered ^{14}C-activity was located in the top 15-cm layer, indicating that imazethapyr, as well as related degradation products, did not leach significantly.

Several university researchers have conducted a variety of studies on the dissipation of the imidazolinones. Basham et al.[1] studied the field persistence of imazaquin, which was applied preemergence at rates as high as 16 times the recommended rate of 0.14 kg a.i./ha in two locations in Arkansas in 1984 and 1985. Rapid dissipation occurred on Taloka silt loam under hot, dry field conditions in 1984, when there was no rainfall for 2 weeks after application. In 1985, phytotoxicity was greater and dissipation slower when the field was irrigated 7 d after treatment. Persistence and adsorption were greater on Sharkey silty clay than on Taloka silt loam. Imazaquin was weakly adsorbed and therefore readily bioavailable as well as mobile in the soil profile under cool, wet conditions. Imazaquin had higher persistence in soil with higher clay content and/or organic matter.

Basham et al.[1] also studied the mobility of ^{14}C-imazaquin in field plots under three different watering regimens, using H^{36}Cl as a reference compound. The plots were kept covered with polyethylene at all times except during rainfall. The watering regimens were:

1. 3 cm of water sprinkled over 2 d at a rate which did not exceed 0.7 cm/15 min to limit puddling and prevent run-off. The soil was allowed to dry for at least 3 h before additional water was applied.
2. 8 cm of water sprinkled over 10 d. The first 3 cm was applied as described above, and the following 5 cm were added comparably over the next 8 d.
3. Natural rainfall of 8 cm over 21 d, with the plots covered during periods of no precipitation.

Under regimen 1, 84% of the ^{14}C remained in the top 5 cm, and 98% remained in the top 10 cm. Only 2% of the H^{36}Cl remained in the top 5 cm and more than 25% leached beyond 15 cm, with amounts detectable to a depth of 26 cm. Under regimen 2, 78% of the ^{14}C remained in the top 5 cm, 19% was in the 5 to 10-cm layer and 3% was in the 10- to 15-cm layer. Under regimen 3, 25% of the ^{14}C remained in the top 5 cm and 90% remained in the upper 26 cm. The greater mobility in the covered plots was attributed to a reduction in the upward movement by evaporation. The conclusion of the study was that under normal field conditions leaching is not likely to be a major dissipation factor.

Renner and Meggitt[2] studied the field dissipation of preemergence and preplant-incorporated applications of imazaquin (0.28 kg/ha) to a clay loam soil. Soil samples were taken to a depth of 23 cm at 30, 60, 90, and 150 d after treatment. The soil samples were segmented and assayed by gas chromatography (GC). At 30 d after application, there was little movement in the soil with either type of application. Residue levels in the top 5 cm contained 80 ppb from the incorporated treatment and 50 ppb from the preemergence treatment. These amounts represented a 79% loss from the incorporated treatment and an 87% loss from the pre-emergence treatment. At 150 d after application the imazaquin residues remained predominantly in the top 5 cm of the soil profile, where there were 29 ppb from the incorporated treatment and 12 ppb from the preemergence treatment. These levels represented a 92% loss from the incorporated treatment and a 97% loss from the preemergence treatment. Imazaquin degraded rapidly during the growing season, but detectable residues were still present at 150 d after application in the top 5 cm at four times the recommended application rate.

In 1985 and 1986, Mills and Witt[3] conducted experiments to determine and compare the efficacy, phytotoxicity, and soil persistence of imazaquin, imazethapyr, and clomozone in no-till double-crop soybeans. The dissipation of each herbicide was described by first-order kinetics. Imazaquin and imazethapyr were more persistent in the soil than clomozone, with half-lives, averaged over both years, of 10, 43, and 60 d for clomozone, imazaquin, and imazethapyr, respectively.

Clomozone was not detected from 10 to 20 cm in the soil profile. Imazaquin levels in the 10- to 20-cm layer were consistent both years and decreased from 7 or 8 ppb at 30 d after application to 1 ppb at 120 d after application. More imazethapyr than imazaquin was detected from 10 to 20 cm in the soil profile in 1985, with levels of 81 ppb at 30 d, 15 ppb at 90 d, and 3 ppb at 120 d after treatment. In 1986, imazethapyr residues were at 4 ppb at 30 d and 7 ppb at 120 d after treatment.

Renner et al.[4] determined that preplant-incorporated applications of imazaquin at 280 g a.i./ha had significantly greater persistence than preemergence applications in 1985. In 1984 there were no differences between the treatments. Very little imazaquin was detected below 10 cm in the soil profile at any sampling for either year. A biphasic type of dissipation was noted both years for each type of treatment, with rapid dissipation occurring in the first 30 d after application. The half-lives in the first 30 d were calculated to be 30 to 36 d for both treatments in 1984, 48 to 52 d for preplant incorporated, and 20 to 23 d for preemergence in 1985. The dissipation rate decreased during the following 120 d.

Loux et al.[5] determined the availability and persistence of imazaquin, imazethapyr, and clomozine in a Cisne silt loam (1.3% organic matter [O.M.]) and a Drummer silty clay loam (5.8% O.M.). Availability of all three herbicides was greater in the Cisne soil than in the Drummer soil. In field experiments conducted in 1984, 1985, and 1986, all three herbicides were more persistent in the Drummer silty clay loam than in the Cisne silt loam. Imazaquin dissipated to undetectable levels in the upper 7.5 cm of Cisne soil within 123 d of application. Imazethapyr dissipated to undetectable levels in the Cisne soil within 60 d of application in 1985, but was detected through the final sampling date, approximately 150 d after treatment in 1986. The concentrations of imazaquin and imazethapyr in the top 7.5 cm of the soil did not change significantly from 80 d after application to the last sampling date at approximately 160 d in any of the 3 years. Imazaquin and imazethapyr residues were not detected by GC analysis in the Cisne soil 1 year after application, but residues of imazethapyr were detected in the Drummer soil 3 years following application. Herbicide residues found below 7.5 cm were greater in the Drummer than in the Cisne soil. Therefore, a relationship between availability and dissipation rate may exist since dissipation was most rapid in the soil where the availability was the greatest.

Ketchersid and Merkle[6] conducted dissipation studies of imazaquin on a Vamont fine montmorillonite clay and Lakeland fine sand. The pH of each soil was adjusted to 4.5, 6.0, and 7.5 using 1 M H_3PO_4 or commercial lime. Bioassays using cucumber plants showed that imazaquin persistence was greater in the clay than in the sand and tended to increase as the pH decreased. Leaching and persistence were monitored by extracting the soil and assaying the extracts by GC. Imazaquin was applied to each of the soils, the pH of the soils was adjusted, and the samples were incubated at 32°C. After 4 months, more than 45% of the applied dose remained in the soils. Mobility in the soil was determined using packed columns of each of the soils. Imazaquin was applied to the surface of the columns at a rate of 1.1 kg/ha. The columns were equilibrated for 24 h, then 9 cm of water was added. Approximately one third of the water applied was collected as leachate. Imazaquin leaching was greater in sand than in clay and increased with increasing pH, but was less than that of picloram under similar conditions.

Leaching under actual field situations, when compared to laboratory conditions, may be limited because crops and turf usually draw on the ground water reserve during the growing season. Thus, net water flow during the growing season is upward. Downward movement generally occurs before and after the growing season at those times when water in the soil is not frozen. Although many of the field ROD studies were sampled during these spring and fall periods of increased water movement, downward migration of imazethapyr residues was not detected. This result is consistent with the view that imazethapyr residues bind more tightly to the soil with time.

Based on the available field studies, we conclude that imazethapyr dissipates from the upper soil layers and will not leach from soil under normal field application conditions in spite of the fact that the adsorption/desorption studies conducted in the laboratory indicate the theoretical potential for low soil adsorption. For this reason, we believe there is little

TABLE 1
Mobility of Imazapyr and Other Herbicides in Soils

Soil	R_f			
	Imazapyr	Sulfometuron	Metribuzin	Atrazine
Decatur clay loam	0.95	0.95	0.65	0.39
Dothan loamy sand	0.68	0.97	0.73	0.49
Lucedale sandy clay loam	0.65	0.80	0.59	0.40
Eutaw clay	0.83	0.86	0.55	0.33
Sumter clay	0.83	0.73	0.46	0.29

From Wehtje, G., Dickens, R., Wilcut, J. W., and Hajek, B. F., *Weed Sci.*, 35, 858, 1987. With permission.

TABLE 2
Mobility of Imazethapyr and Other Herbicides in Soils

Soil	Imazethapyr	Paraquat	Diuron	2,4,5-T	2,4-D	Dicamba
Sandy loam, Toronto, Canada	1.00	0.00	0.16	0.71	0.87	1.00
Loam, Toronto, Canada	1.00	0.00	0.16	0.70	0.83	1.00
Silty clay loam, Toronto, Canada	1.00	0.00	0.12	0.60	0.77	1.00
Silt loam, Saskatchewan, Canada	0.95	0.00	0.08	0.54	0.70	1.00
Sandy loam, Manitoba, Canada	0.95	0.00	0.12	0.54	0.70	0.99
Beardon clay loam, ND	0.90	0.00	0.10	0.55	0.70	1.00
Sassafras sandy loam, NJ	0.88	0.00	0.42	0.79	0.89	1.00
Plano silt loam, WI	0.45	0.00	0.14	0.52	0.68	0.99

risk of groundwater contamination from imazethapyr and that significant downward migration of imazethapyr in soil is unlikely to occur.

While many field studies demonstrate that imazethapyr shows little movement through the soil under normal field conditions, laboratory studies indicate that this compound has the potential for high soil mobility. Soil thin-layer chromatography (TLC), adsorption/desorption, and column leaching studies have been conducted by several researchers with various results and interpretations. Some of these studies are discussed below.

IV. SOIL THIN-LAYER CHROMATOGRAPHY

The acid imidazolinones were shown to be mobile compounds when examined by soil TLC. Dickens and Wehtje[7] and Wehtje et al.[8] examined the mobility of radiolabeled imazapyr relative to three reference compounds in five soils representing the major soil types in Alabama. Imazapyr was more mobile than metribuzin and atrazine, but less mobile than sulfometuron. The results are shown in Table 1.

Goetz et al.[9] studied the mobility of imazaquin in five Alabama soils using soil TLC. Imazaquin was mobile in all soils with R_f values of 0.8 to 0.9. The mobility was Decatur silt loam = Sumter clay > Dothan sandy loam > Lucedale fine sandy loam > Eutaw clay. Stougaard et al.[10] determined the effect of pH on the mobility of imazaquin and imazethapyr on Sharpsburg silty clay loam, Holdrege silt loam, and Tripp sandy loam soils, which are representative soils from Nebraska. The pH of the soils was adjusted to 5, 6, and 7. Reducing the pH decreased the mobility and increased the adsorption of both herbicides. Imazaquin was more mobile on soil TLC plates than was imazethapyr.

American Cyanamid scientists studied the mobility of imazethapyr relative to five reference compounds of known mobility on eight soils. The results are summarized in Table 2 and indicate that imazethapyr is a mobile compound that can be classified as being in mobility class 4 or 5 in most soils, as classified by Helling.[11] On a Plano silt loam from Wisconsin, imazethapyr showed significantly less mobility and would be placed in mobility class 3.

Goetz and Lavy[12] measured the mobility of imazethapyr by soil TLC. The order of mobility was Taloka silt > Gallion silt > Conshatta silt > Crowley silt > Sharkey clay > iron subsoil. Lowering the pH of the Taloka silt from 8.0 to 4.8 reduced the R_f from 1.0 to 0.3. Stougaard et al.[10] determined that decreasing the pH decreased the mobility and increased the adsorption of imazethapyr, which was less mobile on soil TLC plates than was imazaquin.

These studies demonstrate that the acid imidazolinones are mobile on thin-layer soil plates and that the mobilities significantly decrease as the pH decreases.

V. ADSORPTION/DESORPTION

The adsorption/desorption properties of the imidazolinones have been studied by many researchers using batch equilibrium techniques. The general conclusion reached by the researchers is that the binding of imidazolinones to soil is generally weak. Several researchers have also demonstrated that adsorption increases as pH decreases and adsorption may also be correlated with various properties of the soil, such as organic matter.

Scientists at American Cyanamid Company studied the adsorption of imazapyr on eight different soils. The Freundlich adsorption coefficients (K_f) were determined to be

Buelah loamy sand (AR)	0.11	Sand (DE)	0.32
Plano silt loam (WI)	0.13	Tippecanoe silt loam (IN)	1.08
Beardon clay loam (ND)	0.21	Silt loam (Alberta, Canada)	1.35
Sassafras sandy loam (NJ)	0.24	Sharkey clay loam (AR)	1.48

Studies of the effect of pH on adsorption have demonstrated that adsorption increases significantly as the pH decreases to 6, as was shown in Figure 2. Dickens and Wehtje[7] conducted studies on the adsorption of imazapyr on five Alabama soils: Dothan loamy sand, Lucedale fine sandy loam, Decatur silt, Eutaw clay, and Sumter clay. In batch equilibrium studies, the adsorption of sulfometuron methyl was higher (average $K_d = 0.29$) than that of imazapyr (average $K_d = 0.09$), with minimal sorption on the two clay soils and maximum adsorption on the Lucedale fine sandy loam. Similar results were obtained with the soil solution recovery technique. Temporary reduction of soil water content or of soil pH increased the sorption of both herbicides. The effect of pH was more pronounced with imazapyr than with sulfometuron methyl.

Wehtje et al.[8] studied adsorption of radiolabeled imazapyr relative to three reference compounds in five soils representing the major soil types in Alabama. The K_d values of imazapyr were 0.07 on Decatur clay loam, 0.17 on Dothan loamy sand, and 0.19 on Lucedale sandy clay loam. Eutaw and Sumter clays showed no detectable adsorption. Clay soils showed the least adsorption while the sandy clay loam showed the maximum adsorption. Similar results were obtained in a soil solution recovery technique (Table 3). Reducing the pH of the soils increased the adsorption of imazapyr.

Sorption was enhanced on some soils by a temporary reduction in soil water content to 25% to 50% of field capacity for 48 h, followed by re-equilibration at field capacity for 48 h.

Researchers at American Cyanamid also studied the adsorption of imazaquin on four soils. The Freundlich adsorption coefficients were determined to be 0.327 on a Sassafras sandy loam, 0.589 on a Plano silt loam, 0.24 on a Beardon clay loam, and 3.57 on a muck soil.

Basham et al.[1] studied the adsorption of imazaquin on Taloka silt loam, Crowley silt loam, and Sharkey silty clay using a modified soil-slurry technique. The Freundlich ad-

TABLE 3
Effect of pH on Amount of Imazapyr in Solution

Soil	pH	Applied dose in solution (%)
Lucedale fine sandy loam	5.8	33
	6.0	52
	6.7	54
Decatur silt loam	5.1	56
	5.4	81
	5.8	100
Dothan sandy loam	6.3	43
	6.4	47
	6.6	66

TABLE 4
Effect of pH on Amount of Imazaquin in Solution

Soil	pH	Applied dose in solution (%)
Lucedale fine sandy loam	5.8	47
	6.3	48
	6.6	100
Decatur silt loam	4.7	38
	5.2	60
	5.5	75
Dothan sandy loam	6.0	50
	6.4	56
	6.6	65

From Goetz, A., Wehtje, G., Walker, R., and Hajek, B., *Weed Sci.*, 34, 788, 1986. With permission.

sorption coefficients were determined to be 0.14 on Taloka silt loam, 0.17 on Crowley silt loam, and 0.30 on Sharkey silty clay. The investigators concluded that imazaquin adsorption was significantly higher in the Sharkey soil, which contained more clay and organic matter than the other soils.

Goetz et al.[9] examined the sorption of imazaquin using batch equilibrium and soil solution recovery techniques on five soils from Alabama. No measurable adsorption on four of the five soils studied was detected, but these results may be related to the small amounts of radioactivity (10,000 dpm total) used in the study. Soil solution recovery was used to evaluate sorption as a function of soil type. For the five soils studied, the order of decreasing adsorption was: Lucedale fine sandy loam > Decatur silt > Dothan sandy loam > Sumter clay > Eutaw clay. Soil solution recovery was used to evaluate the influence of wetting and drying and pH on sorption. Temporary drying and rewetting of the soil to field capacity generally increased imazaquin adsorption but was soil dependent. Reducing the moisture level of each soil to 25 or 50% of field capacity before rewetting resulted in maximum adsorption. The increased sorption obtained by temporary drying can be attributed to the reduction in water-film thickness coating the soil minerals, which concentrates the compound near the sorption surface. This phenomenon is maximized at approximately 25% of field capacity rather than in air-dry soil because the compound is not in solution and cannot diffuse to sorptive surfaces. Lowering the pH enhanced sorption in the three soils studied. The influence of pH is shown in Table 4.

Soil sorption appeared to be governed by pH-dependent charge surfaces from Al and Fe oxyhydroxides, specifically hematite, gibbsite, and kaolinite.

Goetz et al.[13] found that the amount of labeled imazaquin extracted from five soils was not affected by the imazaquin concentration in the range of 0.01 to 1 ppm. At 1 ppm the amount increased, which was attributed to saturation of binding sites. The percentage of the total applied dose found in the soil solutions after a 1 ppm treatment were: Dothan sandy loam 63%, Lucedale fine sandy loam 41%, Sumter clay 73%, Eutaw clay 63%, and Decatur silt loam 66%. When the soil pH was raised or lowered by approximately 1 pH unit, allowed to equilibrate for 48 h, and then extracted, the percentage of applied imazaquin in soil solution increased with increasing pH. The highest percentage of imazaquin in the soil solution was obtained when the soils were air dried prior to being returned to field capacity. Various drying regimens were used to determine the effects of drying on desorption. Samples were treated at field capacity, and the moisture reduced to air dry, 25 or 50% of field capacity, or maintained at field capacity, equilibrated for 48 h at 30°C, and then returned to field capacity and assayed. The amount of imazaquin in solution decreased in Decatur and Lucedale soils with an increase in soil moisture content. In Dothan, Eutaw, and Sumter soils the response was more variable, decreasing at 25% of field capacity and increasing at the two higher moisture levels.

McKinnon and Weber[14] conducted column leaching and adsorption studies to determine the adsorption and relative mobility of imazaquin and metolachlor on Norfolk sandy loam, Rion sandy clay loam, Cape Fear sandy clay loam, and Webster clay loam. In adsorption studies the relative order of mobility for imazaquin was (Rion = Norfolk) > (Cape Fear = Webster), and for metolachlor was Rion > Norfolk > Cape Fear > Webster, which indicates that both imazaquin and metolachlor were less mobile in soils with higher levels of organic matter and montmorillonite clay. The Cape Fear soil had the lowest pH (4.7), which may have contributed to the higher adsorption on this soil. These results are comparable to those reported by Liu and Weber,[15] where there was moderate adsorption by H-saturated cation-exchange resins but not by Ca-saturated exchange resins. Adsorption of imazaquin on the Cape Fear soil increased as the pH of the solution decreased. Goetz et al.,[9] Liu and Weber,[15] and Renner et al.[16] have also reported greater adsorption of imazaquin at lower pH values because the amount of imazaquin in the COOH form (**2**) increases as pH decreases, and there is minimal adsorption of the anionic form **4** of imazaquin. The amount of imazaquin which was desorbed decreased as the pH decreased for the same reasons as found for adsorption. Soil organic and humic matter contents were highly correlated with adsorption, but correlation with clay content was weak.

Loux et al.[17] examined the adsorption of imazaquin and imazethapyr on a variety of soils, sediments, and clays. Batch equilibrium studies were conducted on 22 soil samples and six sediment samples, and the data were fitted to Freundlich isotherms. The molar Freundlich K_f values ranged from 0 to 3.79 for imazaquin and 0 to 9.27 for imazethapyr, with mean values of 1.88 and 1.01, respectively. These values indicate weak-to-moderate binding. Adsorption of imazethapyr was less than that of imazaquin on 13 of the 27 samples. Adsorption of imazaquin was positively correlated with organic carbon content and negatively with pH, while adsorption of imazethapyr was positively correlated with clay content and cation-exchange capacity (CEC). Increases in adsorption with CEC are reflective of increases in clay and/or organic carbon content. Multivariate regression analysis of Freundlich K_f values and soil and sediment properties yielded a model for imazaquin adsorption which included soil pH, organic carbon, and clay as significant independent variables. Clay content and pH were significant independent variables for the imazaquin adsorption model and yielded the following model:

$$Y = 7.844 - 2.94(pH) + 2.188(OC) + 0.111(clay)$$
$$+ 0.254(pH^2) - 0.273(pH*OC) - 0.015(pH*clay)$$

$$R^2 = 0.909$$

TABLE 5
Effect of pH on Freundlich Adsorption
Coefficient (K_f)

Soil	pH	K_f Imazaquin	K_f Imazethapyr
Sharpsburg	5	38.25	232.22
	6	7.84	41.59
	7	5.20	12.49
Holdrege	5	1.09	49.72
	6		11.48
	7		7.34
Tripp	5		12.53
	6		2.83
	7		0.65

From Stougaard, R. N., Shea, P. J., and Martin, A. R., *Weed Sci.*, 38, 67, 1990. With permission.

The adsorption model for imazethapyr was

$$Y = 16.442 - 6.035(pH) + 0.397(clay) + 0.511(pH^2) - 0.048(pH*clay)$$

$$R^2 = 0.716$$

Adsorption of imazaquin and imazethapyr was greater at pH values less than 6 than at higher pH, but adsorption was not dependent on pH alone. Adsorption is even greater with high organic carbon or clay content at low pH than at high pH. The reason is the increased amounts of imazaquin or imazethapyr in the nonionized forms, which bind better than the ionized form. Adsorption of both compounds on Ca- and H/Al-kaolinite and Ca-montmorillonite was very low (approximately 5% adsorbed) but increased significantly on H/Al-montmorillonite (50% to 87% adsorbed).

Renner et al.[16] found that adsorption of ^{14}C-imazaquin and imazethapyr to Hillsdale sandy loam soil increased as soil pH decreased from 8 to 3. Adsorption of imazethapyr was greater than imazaquin at the three pH levels studied — 3.0, 5.5, and 8.0. The K_d values at pH 3.0, 5.5, and 8.0 were calculated to be 2.76, 0.134, and 0.049, respectively, for imazaquin and 3.75, 0.349, and 0.072, respectively, for imazethapyr.

Stougaard et al.[10] studied the adsorption of imazaquin and imazethapyr on Sharpsburg silty clay loam, Holdrege silt loam, and Tripp sandy loam soils after adjusting the pH to 5, 6, and 7. Reducing the pH decreased the mobility and increased the adsorption of both herbicides. Imazethapyr was more strongly adsorbed on all three soils than was imazaquin. Imazaquin was not significantly adsorbed at any pH in the sandy loam soil nor at pH 6 and 7 in the silt loam. The Freundlich K_f values are shown in Table 5.

Researchers at American Cyanamid studied the adsorption of imazethapyr on four soils. The Freundlich adsorption coefficients were determined to be 0.75 on a Sassafras sandy loam, 0.54 on a Plano silt loam, 0.82 on a Beardon clay loam, and 0.46 on a sand. Studies of the effect of pH on adsorption have demonstrated significant increases in adsorption as the pH decreases to below 6, as shown in Figure 2.

Goetz and Lavy[12] measured the adsorption of imazethapyr by batch-equilibrium experiments and fitted the data to the Freundlich equation. The order of adsorption, based on Freundlich K_f values, was iron subsoil > Sharkey clay > Crowley silt > Conshatta silt > Gallion silt > Taloka silt. Lowering the pH of the Taloka silt from 8.0 to 4.8 increased the value of K_f from 0.01 to 0.111.

TABLE 6
Effective pH on Adsorption to Ion-Exchange Resins

Resin	pH	% Adsorbed	
		Imazaquin	Imazethapyr
Dowex 50W	2	96	92
	7	0	0
IR-45	2	79	63
	7	96	97
XAD-4	2	80	79
	7	10	7

From Loux, M. M., Liebl, R. A., and Slife, F. W., *Weed Sci.*, 37, 712, 1989. With permission.

In adsorption/desorption studies of imazamethabenz-methyl at American Cyanamid Company, five soils were tested using batch equilibrium. The Freundlich adsorption coefficients for CL 239,589 (*meta* isomer) and CL 252,767 (*para* isomer) were determined to be 0.18 and 0.32 on a Sassafras sandy loam, 0.54 and 0.67 on a Plano silt loam, 8.5 and 10.5 on a silt loam from Alberta, 1.4 and 1.6 on a sandy loam from Manitoba, and 3.2 for both isomers on a silt loam from Saskatchewan.

VI. ADSORPTION ON ION-EXCHANGE RESINS AND NONSOIL MEDIA

Liu and Weber[15] determined that imazaquin existed primarily in the anionic form at alkaline pH levels, was ionically bound to Cl^- or OH^--exchanged resins in high amounts, and was only bound to cation-exchange resins in very small amounts. These results indicate that imazaquin is anionic at neutral or alkaline pH and is ionically adsorbed through anion-exchange forces. Imazaquin was more readily displaced from the exchange resins by NaCl solution than by water. Adsorption/desorption by H^+-exchange resins confirmed that imazaquin has acidic properties because only molecular species were adsorbed onto H^+-exchange resins, and water was as effective as NaCl in displacing the physically adsorbed species. Adsorption of imazaquin to Ca-organic matter was seven times higher than for chlorsulfuron. There was negligible adsorption by Ca-montmorillonite. The adsorption to Cape Fear soil, Ca-montmorillonite, and Ca-organic matter was studied at several pH values. Adsorption increased as the pH decreased from 7 to 3. The percentage of imazaquin adsorbed on the Ca-montmorillonite increased from 1% to 71% when the pH decreased from 5 to 3. In the Cape Fear soil and Ca-organic matter, more compound was desorbed by water than by NaCl at all pH values suggesting that binding between imazaquin and clay or organic matter is extremely weak and of a nonionic type.

Loux et al.[17] studied the adsorption of imazaquin and imazethapyr on ion-exchange resins. There was very low adsorption (less than 5%) of both compounds on Ca- and H/Al-kaolinite and Ca-montmorillonite, but significantly higher amounts were adsorbed on H/Al-montmorillonite (50 to 87%). Imazaquin and imazethapyr were similarly adsorbed by ion exchange resins, as shown in Table 6.

The high degree of adsorption on the Dowex 50W resin, a cation-exchange resin, at pH 2 results either from cationic bonding, hydrogen bonding, or weak physical bonding between the herbicide and the resin. There was no binding to the cation-exchange resin at pH 7 because of repulsion of the anionic form **4** of the herbicide and the negatively charged resin. The high level of adsorption on the IR-45 resin, an anion-exchange resin, at pH 7 is believed to be an anion-exchange interaction between the positively charged resin and the negatively charged anionic forms of the herbicides. Adsorption to XAD-4 resin, a nonionic resin, was greatest when the herbicides were in the nonionic form **2** (pH 2), although there was a small amount of adsorption of the herbicides when they were present in the ionic form **4** (pH 7). These results indicate that binding of protonated herbicide at low pH could be a result of physical forces, hydrogen bonding, or cationic binding. At soil pH levels above the pK_as of imazaquin and imazethapyr, ionic bonding of herbicide anions with positively charged soil components could also occur.

McKinnon and Weber[14] studied the influence of organic matter and montmorillonite clay on the adsorption and bioavailability of imazaquin in model soil media. There was no measurable adsorption of imazaquin on Ca-montmorillonite clay, but there was moderate adsorption on calcium-saturated humic matter. The bioavailability of imazaquin decreased with increasing levels of organic matter added to soil media, which the researchers attributed to the retention of imazaquin by the organic matter.

VII. COLUMN LEACHING

As in soil TLC studies, the imidazolinones are mobile when examined in column-leaching studies.

Liu and Weber[15] studied the mobilities of imazaquin, chlorsulfuron, SD 95481, and prometryn in column-leaching studies on Norfolk loamy sand and Cape Fear sandy loam soils. On Norfolk loamy sand and Cape Fear sandy loam, 85 and 54%, respectively, of the applied imazaquin was leached through 35-cm-long columns with the addition of 540 mm of water, while 95 and 78%, respectively, of the chlorsulfuron leached through the soils. Mobilities in the Norfolk and Cape Fear soils were: chlorsulfuron > imazaquin ≫ SD 95481 ≥ prometryn.

McKinnon and Weber[14] conducted column leaching studies to compare the relative mobilities of imazaquin and metolachlor on Norfolk sandy loam, Rion sandy clay loam, Cape Fear sandy clay loam, and Webster clay loam soils using 35-cm-long soil columns. A total of 20 in. of water was added to the columns over 10 d. The amount of imazaquin which leached through the columns was 93.3, 95.1, 55.1, and 52.4% for the Norfolk, Rion, Cape Fear, and Webster soils, respectively, while the amount of metolachlor leached through the columns was 13, 46, 7, and 3% for the Norfolk, Rion, Cape Fear, and Webster soils, respectively. Imazaquin was more mobile than metolachlor. The mobility of imazaquin in the soils was Norfolk = Rion > Cape Fear = Webster. Both imazaquin and metolachlor were less mobile in soils with higher levels of organic matter.

McKinnon and Weber[14] studied the movement of imazaquin in Rion sandy loam and Cape Fear clay loam soil columns under four flow conditions. The flow conditions were (1) water added to the soil surface, (2) subirrigation, (3) water added to the surface followed by evaporation with bottom watering, and (4) water added to the surface followed by evaporation without bottom watering. Imazaquin was much more mobile in Rion sandy loam than in Cape Fear sandy clay loam in both leaching and capillary-flow studies. There was a significant difference in the distribution of imazaquin in columns where there was leaching of water versus leaching followed by 12 d of evaporation in a greenhouse. In the Rion sandy loam the amount of imazaquin (as percent of the recovered dose) in the top 15 cm was 27.9% in the leaching treatment (flow condition 1), 41.6% with evaporation without bottom watering (flow condition 4), and 89.3% with evaporation with bottom watering (flow condition 3). Columns that were subject to evaporation with a water supply at the bottom of the column showed greater redistribution towards the soil surface than those without that water supply (flow condition 2). Movement of imazaquin through capillary flow with the evaporating water could help explain why imazaquin remains near soil surfaces in fields with high water tables, wet soils, and soils with low organic matter. These conclusions are similar to those found by Basham et al.[1]

Lolas and Galopoulas[18] conducted column-leaching studies of imazaquin, metazachlor, and cinmethylin using 35-cm-long columns containing a sandy loam, a sandy clay, or an organic soil. The test compounds were applied at rates of 0.25, 0.75, and 1.0 kg a.i./ha respectively for imazaquin, metazachlor, and cinmethylin. The test compounds were dissolved in 25 ml water and added to the columns. The columns were allowed to equilibrate for 24 h, then 25 ml of water were added each day for 30 d. The columns were segmented and bioassayed. Imazaquin was uniformly distributed in the sandy loam columns. In the sandy clay, most of the imazaquin moved to the 20- to 30-cm layer, yet imazaquin remained in the 0- to 10-cm layer in the organic soil. The relative mobilities were imazaquin > metazachlor > cinmethylin.

American Cyanamid scientists conducted a column-leaching study of imazethapyr on a Sassafras sandy loam soil. Approximately 90% of the applied dose was found in the eluent after adding 20 in. of water.

Aged-column-leaching studies were also conducted with isomers of imazamethabenzmethyl on Sassafras sandy loam and Beardon clay loam. Samples were aged for 1 month in a greenhouse under artificial lighting, then leached with 20 in. of water. The leachates from the clay loam soils contained 12.4% and 9.4% of the applied radioactivity from CL 239,589 and CL 252,767 treatments, respectively, while the leachates from the sandy loam soils contained 34% and 23% of the applied radioactivity from CL 239,589 and CL 252,767 treatments, respectively. The majority of the material in the leachates from both soils were the corresponding acids of the applied compounds.

The dependence of adsorption of the acid imidazolinones on soil pH and organic matter content is typical of the adsorption of acid herbicides such as chlorsulfuron[19] and picloram.[20] Chlorsulfuron adsorption increases as the pH decreases from 7.8 to 4.2. Changes in pH below 6 have the greatest effect on adsorption.[21,22]

VIII. AEROBIC SOIL METABOLISM

The aerobic metabolism of the imidazolinones is characterized by a relatively slow, but extensive, degradation. Studies performed by a number of researchers have indicated that both the acid group and the carbons in the ring attached to the acid are mineralized. There is slightly greater evolution of $^{14}CO_2$ from the acid-labeled carbons than from ring-labeled carbons but the difference is only several percent of the applied dose. The mineralization of several sites in the imidazolinones indicates that there is extensive degradation of the molecules.

Scientists at American Cyanamid Company studied the aerobic soil metabolism of ^{14}C-carboxyl-labeled imazapyr at a temperature of 22 to 24°C for 1 year in both sandy loam and clay loam soils. Over 12 months, 14% of the applied dose was evolved as $^{14}CO_2$ from the sandy loam and 15% from the clay loam soil. The half-lives were calculated to be 17 and 37 months, respectively. The degradation products were determined to be $^{14}CO_2$, CL 252,974, CL 9,140, 2-carbamoylnicotinic acid, and 4-isopropyl-4-methyl-2-(2-pyridyl)-2-imidazolin-5-one. The aerobic soil metabolism of ^{14}C-ring-labeled imazapyr was studied on a sandy loam soil at 25 and 35°C, and after 12 months, 7 and 7.7%, respectively, of the applied dose was trapped as $^{14}CO_2$. The half-lives were determined to be 5.9 years at 25°C and 7.5 years at 35°C. Studies of the aerobic soil metabolism of ^{14}C-carboxyl-labeled imazaquin on a sandy loam soil at 25°C and 75% of 1/3 bar moisture showed that the half-life was approximately 7 months. About 15% of the applied dose was evolved as $^{14}CO_2$ over 12 months.

Basham and Lavy[23] studied the microbial degradation of ^{14}C-imazaquin by measuring $^{14}CO_2$ evolution over 7 months from five soils: Crowley silt loam, Taloka silt loam, Amagon silt loam, Pulaski silt loam, and Sharkey silty clay. The soils were treated with ^{14}C-carboxyl-

TABLE 7
Dissipation of Imazaquin in Various Soils

Temp (°C)	Water potential (−kPa)	Half-life (months)		
		Sharkey silty clay	Crowley silt loam	Taloka silt loam
18	100	11.7	7.2	2.2
	33	9.8	2.7	2.3
35	100	3.0	2.5	2.3
	33	2.2	1.3	2.3

From Basham, G. and Lavy, T., *Weed Sci.*, 35, 865, 1987. With permission.

or ^{14}C-ring-labeled imazaquin at a rate of 0.5 ppm and incubated at −33 kPa water potential at a temperature of 24 ± 2°C. After 11 weeks, a sucrose/potassium nitrate solution was added to each flask at 0.5% (w/w). The evolution of $^{14}CO_2$ was found to be soil dependent with the greatest amount of $^{14}CO_2$ evolution from Crowley silty loam, from which 10% of the carboxyl label and 7% of the ring label evolved. The Taloka silt loam and Pulaski silt loam both liberated approximately 7.5% of the carboxyl label and 6.5 to 5.5% of the ring label, while the Amagon silt loam generated approximately 5.5% from both labels. The Sharkey silty clay evolved only 3% of the carboxyl label and 2% of the ring label. The relative order of $^{14}CO_2$ evolution was: Crowley > Taloka = Pulaski = Amagon > Sharkey. The $^{14}CO_2$ evolution from both the carboxyl- and ring-labeled samples was similar except in the Crowley soil where there was more $^{14}CO_2$ evolution from the carboxyl-labeled samples. This similarity suggests that microbial populations found in the soil are effective in degrading ^{14}C-imazaquin irrespective of the location of the ^{14}C-label. The Sharkey silty clay, which contained the highest amount of organic matter and clay, evolved the lowest amount of $^{14}CO_2$ during the study. This soil also had the highest amount of adsorption in a previous study. There were no differences in the amount of $^{12}CO_2$ evolved from control soils or soils treated with 0.5 ppm imazaquin. Thus microbial populations of the soils are probably not adversely affected. The addition of the sucrose/potassium nitrate increased the evolution of $^{14}CO_2$ but did not affect the evolution of $^{12}CO_2$.

Basham and Lavy[23] also incubated samples of Taloka silt loam, Crowley silt loam, and Sharkey clay loam. These soils were placed in plastic bags, treated with two concentrations of imazaquin, and incubated for various times at 18 and 35°C and soil water potentials of −33 kPa and −100 kPa. Dissipation was measured by bioassay and the results are shown in Table 7.

Imazaquin degraded most rapidly in the Taloka silt loam regardless of incubation conditions. Dissipation in both the Sharkey and Crowley soils was more rapid when incubated at the higher temperature and/or moisture levels.

In studies conducted by Flint and Witt,[24] ^{14}C-ring- and ^{14}C-carboxyl-labeled imazaquin were applied at 125 ppb to microbially active and sterile Maury silt loam soils which were then incubated in the dark at 30°C. Evolved CO_2 was trapped in KOH. Less than 1% of the applied radioactivity was evolved from the sterile soil 32 weeks after treatment, but 10% of the ring label and 12% of the carboxyl label were released over 32 weeks from microbially active soil. In a separate experiment, the effects of soil moisture on the degradation of ^{14}C-carboxyl-labeled imazaquin were studied at moisture levels of 5, 25, 50, 75, and 100% of field capacity. There was little $^{14}CO_2$ released at the 5% moisture level, but the amount of $^{14}CO_2$ evolved increased as the moisture level increased to 75% of field capacity. There was no difference in the amount of $^{14}CO_2$ released between soil at 75 and 100% of field capacity over 8 weeks of treatment. The effects of temperature were studied by incubating soils treated with ^{14}C-carboxyl-labeled imazaquin at 15 and 30°C. Evolution of $^{14}CO_2$ was 2.5% of the applied dose at 15°C, and 11.8% of the applied dose at 30°C. These results indicate

TABLE 8
Effect of Location of ^{14}C-Labeled Imazaquin on Evolution of ^{14}CO$_2$

Soil	Label	Applied dose as ^{14}CO$_2$ (%)	Applied dose as parent (%)
Cisne	Carboxyl	23.7	42.1
	Ring	19.3	44.7
Drummer	Carboxyl	8.9	69.8
	Ring	8.5	68.4

that microbial activity is important in soil dissipation of imazaquin and that degradation may be limited by low moisture and low temperature. In a parallel study, unlabeled imazaquin was added to soils as described above and bioassayed over time. Herbicidal activity in the sterile soil decreased by 7% after 32 weeks of incubation, while nonsterile soil lost 83% of the herbicidal activity 4 weeks after treatment. There was a linear decrease in herbicidal activity from 4 to 32 weeks.

Studies by Cantwell et al.[25] on the degradation of imazaquin indicated that after 12 weeks of incubation at 28°C, 94% of the applied radioactivity could be recovered as the parent compound from both Cisne silt loam and Drummer silty clay loam soils which were sterilized by gamma radiation. Carboxyl-labeled imazaquin evolved 24% of the applied dose from Cisne soil and 9% from Drummer soil, while ring-labeled imazaquin evolved 19% from Cisne soil and 8% from Drummer. After 12 weeks of incubation, the parent compound accounted for 42 to 45% of the applied dose in the Cisne soil and 68 to 70% of the applied dose in the Drummer soil. The amount of nonextractable radioactivity was 20 to 27% from the Cisne soil and 9 to 12% from the Drummer soil. Some metabolite production was observed from the ring-labeled treatment of the Cisne soil. Evolution of ^{14}CO$_2$ from carboxyl-labeled imidazolinones was greater than from ring-labeled; however, the differences were small, only several percent. These results follow those of Basham and Lavy.[23] Evolution of ^{14}CO$_2$ from the soil exhibited first-order kinetics. Evolution of ^{14}CO$_2$ from each of the compounds over 3 months is summarized in Table 8. The greater ^{14}CO$_2$ evolution in this study relative to the results reported by Basham and Lavy[23] may be attributable to the use of different soils, different temperatures of incubation, quantity of biomass in soils, or ^{14}CO$_2$-trapping efficiency. These data indicate that complete degradation of the herbicides may be expected. In the Cisne soil, both the carboxyl- and the ring-labeled imazaquin were degraded more rapidly than in the Drummer soil. Loux et al.[5] measured the adsorption of imazaquin on the same soil types and determined the Freundlich K_f value to be 2.2 in a Drummer soil and 0.4 in a Cisne soil.

Scientists at American Cyanamid Company studied the aerobic metabolism of pyridine-ring-labeled imazethapyr at 25°C for 1 year in loamy sand, sandy loam, silt loam, and clay loam soils. After 12 months of incubation, 2%, 2.8%, and 13.8% of the applied dose were evolved from the respective soils. The half-lives were calculated to be 8.5, 12, 5.6, and 3.2 years, respectively.

Degradation of ^{14}C-labeled imazethapyr was studied under conditions similar to those for imazaquin.[24] Less than 1% of the applied radioactivity was evolved from sterile soil 32 weeks after treatment, but 10% of the ring label and 12% of the carboxyl label were released over 32 weeks in the Maury silt loam. In a separate experiment, the effects of moisture on the degradation of ^{14}C-carboxyl-labeled imazethapyr were studied at moisture levels of 5, 25, 50, 75, and 100% of field capacity. The amount of CO$_2$ evolved increased as the moisture level increased to 75% of field capacity and stabilized between 75 and 100% of field capacity. The effects of temperature were studied by incubating soils treated with ^{14}C-carboxyl-labeled imazethapyr at 15 and 30°C. Evolution of ^{14}CO$_2$ was 2.5% of the applied dose at 15°C and 17.7% of the applied dose at 30°C.

TABLE 9
Effect of Location of ^{14}C-Labeled Imazethapyr on Evolution of ^{14}CO$_2$

Soil	Label	Applied dose as ^{14}CO$_2$ (%)	Applied dose as parent (%)
Cisne	Carboxyl	44.0	39.3
	Ring	37.4	36.7
Drummer	Carboxyl	15.3	65.8
	Ring	13.4	63.9

Studies by Cantwell et al.[25] on the degradation of imazethapyr indicated that after 12 weeks of incubation at 28°C, 95% of the applied radioactivity could be recovered as the parent compound from both Cisne silt loam and Drummer silty clay loam which were sterilized by gamma radiation. Carboxyl-labeled imazethapyr evolved 44% of the applied dose from Cisne soil and 15% from Drummer soil, while ring-labeled imazaquin evolved 37% from Cisne soil and 13% from Drummer. After 12 weeks of incubation, the parent compound accounted for 37 to 39% of the applied dose in the Cisne soil and 64 to 66% of the applied dose in the Drummer soil. The amount of nonextractable radioactivity was 8% from the Cisne soil and 6 to 8% from the Drummer soil. There was some metabolite production observed from the ring-labeled treatments. Evolution of ^{14}CO$_2$ from carboxyl-labeled imazethapyr was greater than from ring-labeled; however, the differences were small, only several percent. Evolution of ^{14}CO$_2$ from the soil exhibited first-order kinetics. Evolution of ^{14}CO$_2$ from each of the compounds over 3 months is summarized in Table 9.

These data indicate that complete degradation of the herbicides can be expected. Imazethapyr was degraded more rapidly in the Cisne soil than in the Drummer soil. Loux et al.[5] measured the adsorption of imazethapyr on the same soil types and determined Freundlich K_f value for imazethapyr to be 2.1 in a Drummer soil and 0.1 in a Cisne soil.

Evolution of ^{14}CO$_2$ from treated Cisne soil containing serial dilutions of activated charcoal demonstrated that adsorption was negatively correlated with degradation. Therefore, microbial degradation of the imidazolinones is regulated by the amount of compound in soil solution, which is determined by the properties of the soil. Basham et al.[1] and Renner et al.[16] also associated increased imidazolinone persistence with adsorption to soil.

IX. ANAEROBIC SOIL METABOLISM

The anaerobic soil metabolism of imazaquin, imazapyr, and imazethapyr was studied on a variety of soils. There was no significant degradation of any of the compounds over 2 months of incubation, indicating that the imidazolinones are stable to anaerobic degradation.

X. SOIL PHOTOLYSIS

While the photolysis of the imidazolinones in water is fairly rapid (see Chapter 15), the photolysis on soils is slower. In general, the half-lives of the photolysis of the imidazolinones on dry soil are approximately 4 months. There are no significant (more than 10% of the applied dose) photodegradation products formed by any of the imidazolinones, but a variety of minor products are formed.

The photolysis of ^{14}C-carboxyl- and ^{14}C-ring-labeled imazaquin on thin-layer soil plates was studied using a borosilicate-filtered Xenon Arc lamp as the light source. The half-life of imazaquin under these conditions was determined to be 60 d with continuous irradiation or 120 d with a 12 h light cycle. There were several minor unknown degradation products, none of which exceeded 5% of the applied dose.

Basham and Lavy[23] studied the volatilization and photodecomposition of imazaquin by applying an acetone/^{14}C-imazaquin solution to glass slides or slides coated with soil. Crowley,

Taloka, and Sharkey soils were used. After the solvent had evaporated the slides were placed in incubation chambers at either 18 or 42°C under UV lights or in the dark. At various sampling intervals slides were crushed, placed in a liquid scintillation cocktail, and radioassayed. An additional set of slides of Taloka soil was exposed to sunlight for up to 56 h in October. Samples were assayed as previously described. There were no significant losses of radioactivity from the glass plates maintained at 42°C, but there were decreases in recoveries based on extraction with a liquid scintillation cocktail.

Both the ^{14}C-labeled incubation study and the storage study suggest that soil microbial activity is important in the breakdown of imazaquin in soil. Photodecomposition would not be a factor in dissipation when the compound is applied as an incorporated treatment or when rainfall incorporates the compound into the soil. Microbial degradation will occur once imazaquin is incorporated into warm, moist soil. McKinnon and Weber[14] studied the photodecomposition of aqueous solutions of imazaquin, imazapyr, imazethapyr, imazamethabenz-methyl, and AC 263,222 by exposing samples to a light source (Oriel ARC lamp) for various time intervals up to 16 h. Results from spectrophotometric determinations showed that all of the imidazolinones were degraded. Bioactivity, as determined by a bioassay with cotton, decreased with exposure time. The imidazolinone family is not subject to losses by vaporization because of the low vapor pressures of these herbicides. Basham et al.[1] reported rapid dissipation of imazaquin applied preemergence under hot, dry field conditions. The actual breakdown from photodecomposition in the field would be expected to be less than the decomposition in an aqueous solution.

XI. SUMMARY

The movement of the acid imidazolinones is largely influenced by many properties of the soil, the most important of which are pH and organic matter, with clay content playing a less important role. Binding of the acid imidazolinones increases as pH decreases. Because most soils are acidic, and their surfaces become more acidic as moisture levels decrease, mobility in the field is limited. The limited mobility is also influenced under most field conditions by movement of the compounds to the soil surface through capillary action and with evaporation.

Microbial degradation under aerobic conditions is the primary degradation mechanism, with a small contribution from photolysis. Conditions which tend to favor microbial activity such as warm, moist soils are also the conditions under which the imidazolinones are most rapidly degraded. As with many other compounds, dissipation of the imidazolinones in the field is generally more rapid than shown in laboratory studies.

REFERENCES

1. **Basham, G., Lavy, T., Oliver, L., and Scott, H. D.,** Imazaquin persistence and mobility in three Arkansas soils, *Weed Sci.,* 35, 576, 1987.
2. **Renner, K. A. and Meggitt, W. F.,** Corn variety response to imazaquin residue in the soil, *Proc. 39th North Cent. Weed Control Conf.,* 1984, 105.
3. **Mills, J. A. and Witt, W. W.,** Efficacy, phytotoxicity, and persistence of imazaquin, imazethapyr, and clomazone in no-till double-crop soybeans, *Weed Sci.,* 37, 353, 1989.
4. **Renner, K. A., Meggitt, W. F., and Leavitt, R. A.,** Influence of rate, method of application, and tillage on imazaquin persistence in soil, *Weed Sci.,* 36, 90, 1988.
5. **Loux, M. M., Liebl, R. A., and Slife, F. W.,** Availability and persistence of imazaquin, imazethapyr and clomazone in soil, *Weed Sci.,* 37, 259, 1989.
6. **Ketchersid, M. L. and Merkle, M. G.,** Effect of soil type and pH on persistence and phytotoxicity of imazaquin (Scepter), *Proc. So. Weed Sci. Soc. Am.,* 39, 420, 1986.

7. **Dickens, R. and Wehtje, G.**, Mobility and soil solution characteristics of imazapyr (Arsenal) and sulfometuron methyl (Oust) in Alabama soils, *Proc. So. Weed Sci. Soc. Am.*, 39, 368, 1986.

8. **Wehtje, G., Dickens, R., Wilcut, J. W., and Hajek, B. F.**, Sorption and mobility of sulfometuron and imazapyr in five Alabama soils, *Weed Sci.*, 35, 858, 1987.

9. **Goetz, A., Wehtje, G., Walker, R., and Hajek, B.**, Soil solution and mobility characterization of imazaquin, *Weed Sci.*, 34, 788, 1986.

10. **Stougaard, R. N., Shea, P. J., and Martin, A. R.**, Effect of soil type and pH on adsorption, mobility, and efficacy of imazaquin and imazethapyr, *Weed Sci.*, 38, 67, 1990.

11. **Helling, C.**, Pesticide mobility in soils. II. Applications of soil thin-layer chromatography, *Proc. Soil Sci. Soc. Am.*, 35, 737, 1971.

12. **Goetz, A. and Lavy, T. L.**, Mobility and sorptive properties of imazethapyr in Arkansas soils, *Proc. So. Weed Sci. Soc. Am.*, 41, 337, 1988.

13. **Goetz, A., Wehtje, G., and Walker, R. H.**, Mobility and soil solution characteristics of imazaquin in Alabama soils, *Proc. So. Weed Sci. Soc. Am.*, 38, 476, 1985.

14. **McKinnon, J. and Weber, J. B.**, Leaching and capillary movement of imazaquin in soils, *Proc. So. Weed Sci. Soc. Am.*, 41, 337, 1988.

15. **Liu, S. L. and Weber, J. B.**, Retention and mobility of AC 252,214, chlorsulfuron, prometryn and SD 95481 in soils, *Proc. So. Weed Sci. Soc. Am.*, 38, 465, 1985.

16. **Renner, K. A., Meggitt, W. F., and Penner, D.**, Effect of soil pH on imazaquin and imazethapyr adsorption to soil and phytotoxicity to corn *(Zea mays)*, *Weed Sci.*, 36, 78, 1988.

17. **Loux, M. M., Liebl, R. A., and Slife, F. W.**, Adsorption of imazaquin and imazethapyr on soils, sediments, and selected adsorbents, *Weed Sci.*, 37, 712, 1989.

18. **Lolas, P. C. and Galopoulas, A.**, Soil bioactivity, persistence, and leaching of cinmethylin, imazaquin and metazachlor, *Zizaniology*, 1, 221, 1985.

19. **Shea, P.**, Chlorsulfuron dissociation and adsorption on selected adsorbents and soils, *Weed Sci.*, 34, 474, 1986.

20. **Hamaker, J., Goring, C. A. I., and Youngson, C.**, Sorption and leaching of 4-amino-3,5,6-trichloropicolinic acid in soils, in *Organic Pesticides in the Environment Symposium*, Advances in Chemistry Series 60, American Chemical Society, Washington, D.C., 23, 1966.

21. **Mersie, W. and Foy, C. L.**, Phytotoxicity and adsorption of chlorsulfuron as affected by soil properties, *Weed Sci.*, 33, 564, 1985.

22. **Thirunarayanan, K., Zimdahl, R. L., and Smika, D. E.**, Chlorsulfuron adsorption and degradation in soil, *Weed Sci.*, 33, 558, 1985.

23. **Basham, G. and Lavy, T.**, Microbial and photolytic dissipation of imazaquin in soil, *Weed Sci.*, 35, 865, 1987.

24. **Flint, J. L. and Witt, W. W.**, Soil temperature and moisture effects on imazaquin and imazethapyr degradation, *Proc. So. Weed Sci. Soc. Am.*, 41, 338, 1988.

25. **Cantwell, J. R., Liebl, R. A., and Slife, F. W.**, Biodegradation characteristics of imazaquin and imazethapyr, *Weed Sci.*, 37, 815, 1989.

Chapter 17

IMAZAPYR HERBICIDE

Richard A. Beardmore, Ronaldo Hart, Richard Iverson, Mark A. Risley, and Mark Trimmer

TABLE OF CONTENTS

I. INTRODUCTION

Imazapyr is a broad-spectrum herbicide which controls most annual and perennial grasses and broadleaved weeds, woody brush, and deciduous trees in noncrop areas such as railroads, utility, pipeline and highway rights-of-way, utility plant sites, petroleum tank farms, pumping installations, fence rows, storage areas, and nonirrigation ditchbanks. Imazapyr is used to establish and maintain wildlife openings, to prepare sites for conifer planting, and to release conifers from competing vegetation. Imazapyr may be used for the release of *Cynodon dactylon* and *Paspalum notatum* in unimproved turf, such as roadsides. Imazapyr has also been developed for use in sugar cane and plantation crops, such as rubber and oil palm.

Imazapyr herbicide may be applied either preemergence or postemergence to weeds; however, postemergence application is preferred in most situations, particularly for control of perennial weeds. For maximum herbicidal activity, imazapyr should be applied when the weeds are growing vigorously. The residual activity of imazapyr provides control of most newly germinating weed species following a postemergence application. Granular applications provide preemergence control of annuals and also of established perennials.

II. REGULATORY STATUS

Table 1 lists countries and uses where imazapyr is registered.

III. FORMULATIONS

Imazapyr herbicide is formulated either as an aqueous solution to be mixed with water and applied as a spray or as a granular formulation to be applied to the ground. For postemergence applications, a nonionic surfactant is required for optimum imazapyr activity. The surfactant is either included in the formulation or is added to the spray solution.

Imazapyr is marketed in the U.S. under the following trademarks: Arsenal, Arsenal Applicators Concentrate, Chopper, Arsenal 0.5G, and Arsenal Forestry Granule (Table 2).

Commercial formulations marketed outside the U.S. are listed in Table 3. Imazapyr labels should be consulted for specific recommendations.

A package-mix, soluble-concentrate formulation containing 12.5 g a.e./l of imazapyr and 300 g a.i./l of atrazine is sold in the U.K. under the tradename Arsenal XL.

TABLE 1
Countries Where Imazapyr Is Registered

Country	Use	Country	Use
Argentina	Noncrop	Japan	Noncrop
Australia	Noncrop	Luxembourg	Noncrop
Belgium	Noncrop	Malaysia	Noncrop
Brazil	Noncrop		Rubber
	Rubber	Mexico	Noncrop
Bulgaria	Noncrop	Morocco	Noncrop
Chile	Noncrop	New Zealand	Noncrop
Colombia	Noncrop	Nicaragua	Noncrop
Costa Rica	Noncrop	Norway	Noncrop
	Rubber		Forestry
Czechoslovakia	Noncrop	Panama	Noncrop
	Forestry	Paraguay	Noncrop
Dominican Republic	Noncrop	Philippines	Rubber
El Salvador	Noncrop	Poland	Noncrop
East Germany	Noncrop		Forestry
France	Noncrop	Puerto Rico	Noncrop
French West Africa	Rubber	South Africa	Noncrop
Guatemala	Noncrop		Forestry
	Rubber	Spain	Noncrop
Honduras	Noncrop	Thailand	Noncrop
	Rubber		Rubber
Hungary	Noncrop	U.K.	Noncrop
	Forestry	U.S.	Noncrop
Indonesia	Noncrop		Forestry
	Rubber	Venezuela	Noncrop
Ireland	Noncrop		Sugar cane
	Forestry	Yugoslavia	Noncrop
Israel	Noncrop	Zimbabwe	Noncrop
Ivory Coast	Rubber		

TABLE 2
Commercial Formulations of Imazapyr in the U.S.

Trade Name	Formulation	Amount of active ingredient	Use category
Arsenal	Aqueous solution	240 g a.e./l	Noncrop
Arsenal App. Concentrate[a]	Aqueous solution	480 g a.e./l	Forestry
Chopper	Aqueous solution	240 g a.e./l	Forestry, noncrop
Chopper RTU	Aqueous solution with a penetrant	3% w/w	Forestry, noncrop
Arsenal 0.5G	Granule	0.5%	Noncrop
Arsenal Forestry Granule	Granule	5%	Forestry

[a] Formulation does not contain surfactant.

Marble is a package-mix, soluble-concentrate formulation available in South Africa and in Belgium under the tradename Marbre. It contains 83.3 g a.e./l of imazapyr and 416.7 g a.e./l of diuron.

IV. INDUSTRIAL/NONCROP WEED CONTROL

As a result of the wide range of weeds controlled and long residual activity of imazapyr, the compound is well suited for use in noncrop areas, were eradication of all existing

TABLE 3
Commercial Formulations of Imazapyr in Countries outside the U.S.

Tradename	Formulation	Amount of active ingredient	Amount of surfactant (g/l)
Arsenal	Aqueous solution	250 g a.e./l	312.5
Arsenal 250A			
Assault 250A			
Arsenal 250LC	Aqueous solution	250 g a.e./l	375
Arsenal 100A	Aqueous solution	100 g a.e./l	375
Arsenal 100A	Aqueous solution	100 g a.e./l	125
Assault 100			
Arsenal 75A	Aqueous solution	75 g a.e./l	—
Arsenal 50A	Aqueous solution	50 g a.e./l	75
Arsenal 25% AS	Aqueous solution	25%	—
Chopper 100AS	Aqueous solution	100 g a.e./l	—
Chopper 250AS	Aqueous solution	250 g a.e./l	—
Arsenal 0.5G	Granule	0.5%	—

vegetation and maintenance of bare ground is required, including rights-of-way (railway, highway, utility, and pipeline), industrial sites (petroleum tank farms, pumping installations, utility plant sites, storage areas, and military bases), and aquatic uses (ditchbanks and drainage channels).

A. WEED SPECTRUM/METHOD OF APPLICATION

For noncrop use, imazapyr has shown a very broad spectrum of activity, proving to be effective on annual and perennial weeds of many botanical families. Imazapyr will provide postemergence activity with residual control of newly germinating annual and perennial seedlings of the target vegetation species. In general, annual weeds may be controlled by preemergence or postemergence applications of imazapyr herbicide; whereas, for established biennials and perennials, postemergence applications are recommended. The speed of kill varies according to species, with deeply rooted perennials, established weeds, and woody species generally being more difficult to control.

1. United States

Weeds controlled by imazapyr in the U.S. are presented in Table 4. Imazapyr rates are indicated for the control of the various weed species listed.

2. Latin America

In Latin America, the recommended rate of imazapyr for noncrop uses is from 0.5 to 2.0 kg a.e./ha, with the most common rate being about 0.75 kg a.e./ha. In Brazil, the recommended rate for application to railways is 2.5 kg a.e./ha.

Postemergence use when weeds are actively growing is the preferred treatment; however, soil activity does control annual weeds which germinate after application. Length of residual activity varies from 2 to 3 months in high-rainfall tropical environments, 3 to 6 months in subtropical climates with moderate rainfall, 6 to 12 months in humid temperate regions, and over 12 months in dry, temperate areas. Excellent control of perennial species has been obtained with applications in the summer or autumn, when the sap translocates from the aerial portions of the plant to the storage organs.

The diversity of environmental conditions throughout Latin America makes it difficult to target imazapyr applications to a particular species or group of species. However, *Paspalum* spp. in Central America and *Cyperus* spp. throughout all Latin America are high on the list of principal target weeds. Weeds controlled by imazapyr in Latin America are listed in Table 5.

TABLE 4
Weeds Controlled by Imazapyr in the U.S.

Species	Growth habit[a]	Species	Growth habit[a]
0.56 to 0.84 kg a.e./ha[b]		**0.56 to 0.84 kg a.e./ha[b]**	
Grasses		Broadleaved weeds	
Agropyron repens	P	*Amaranthus* spp.	A
Avena fatua	A	*Ambrosia elatior*	A
Brachiaria mutica	P	*Ambrosia psilostachya*	P
Brachiaria platyphylla	A	*Arctium* spp.	B
Bromus inermis	P	*Brassica campestris*	B
Bromus tectorum	A	*Brassica juncea*	A
Cenchrus spp.	A	*Chenopodium album*	A
Dactylis glomerata	P	*Chenopodium murale*	A
Eragrostis spp.	A/P	*Chrysanthemum leucanthemum*	P
Festuca spp.	A/P	*Conyza canadensis*	A
Lolium multiflorum	A	*Daucus carota*	B
Panicum capillare	A	*Descurainia pinnata*	A
Paspalum urvillei	P	*Erigeron* spp.	A
Poa annua	A	*Erodium* spp.	A
Poa compressa	P	*Eupatorium capillifolium*	A
Poa pratensis	P	*Franseria tomentosa*	P
Setaria spp.	A	*Geranium carolinianum*	A
Sorghum halepense	P	*Helianthus* spp.	A
Sporobolus cryptandrus	A	*Heterotheca subaxillaris*	P
		Lactuca spp.	A/B
0.84 to 1.12 kg a.e./ha[b]		*Lepidium* spp.	A
		Lespedeza spp.	P
		Melilotus spp.	A/B
Andropogon spp.	P	*Mollugo verticillata*	A
Aristida oligantha	P	*Montia perfoliata*	A
Bromus secalinus	A	*Oxalis stricta*	P
Digitaria spp.	A	*Pastinaca sativa*	B
Eleusine indica	A	*Plantago* spp.	P
Hordeum spp.	A	*Polygonum* spp.	A/P
Panicum dichotomiflorum	A	*Rumex* spp.	P
Panicum repens	P	*Salsola kali*	A
Phalaris arundinacea	P	*Stellaria media*	A
		Taraxacum officinale	P
1.12 to 1.68 kg a.e./ha[b]		*Tribulus terrestris*	A
		Trifolium spp.	A/P
		Verbascum spp.	B
Andropogon gerardii	P	*Verbena stricta*	P
Cynodon dactylon	P		
Distichlis stricta	P	**0.84 to 1.12 kg a.e./ha[b]**	
Imperata cylindrica	P		
Muhlenbergia frondosa	P		
Panicum maximum	P	*Alhagi pseudalhagi*	P
Paspalum dilatatum	P	*Atriplex* spp.	A
Paspalum notatum	P	*Centaurea diffusa*	A
Pennisetum villosum	P	*Centaurea solstitialis*	A
Phleum pratense	P	*Chondrilla juncea[c]*	B
Phragmites australis	P	*Cirsium vulgare*	B
Spartina patens	P	*Gutierrezia sarothrae[c]*	P
Sporobolus cryptandrus	P	*Lythrum salicaria[c]*	P
Typha spp.	P	*Phytolacca americana*	P

TABLE 4 (continued)
Weeds Controlled by Imazapyr in the U.S.

Species	Growth habit[a]	Species	Growth habit[a]
Portulaca spp.	A	\multicolumn{2}{c}{1.12 to 1.68 kg a.e./ha[b]}	
Rumex spp.	P		
Solidago spp.	P	*Campsis radicans*	P
Urtica dioica[c]	P	*Parthenocissus quinquefolia*	P
Xanthium strumarium	A	*Pueraria lobata*[d]	P
		Rubus spp.[e]	P
\multicolumn{2}{c}{1.12 to 1.68 kg a.e./ha}		*Vitis* spp.	P
Ambrosia trifida	A		
Asclepias spp.	P	\multicolumn{2}{c}{1.12 to 1.68 kg a.e./ha[b]}	
Centaurea repens	P		
Chrysothamnus nauseosus	P	Brush species	
Cirsium arvense	P	*Acer macrophyllum*	P
Cirsium texanum	P	*Acer negundo*	P
Malva parviflora	B	*Acer rubrum*	P
Oenothera kunthiana	P	*Acer* spp.	P
Pluchea sericea	A	*Alnus rubra*	P
Polygonum cuspidatum	P	*Carya* spp.	P
Solanum elaeagnifolium	P	*Cornus* spp.	P
Sonchus spp.	A	*Crataegus* spp.	P
		Diospyros virginiana	P
\multicolumn{2}{c}{0.28 kg a.e./ha[b]}		*Epaeagnus angustifolia*	P
		Fagus grandifolia	P
Vines and brambles		*Fraxinus* spp.	P
Calystegia sepium	A	*Ligustrum vulgare*	P
Convolvulus arvensis	P	*Liquidambar styraciflua*	P
		Liriodendron tulipifera	P
\multicolumn{2}{c}{0.56 to 0.84 kg a.e./ha[b]}		*Melia azedarach*	P
		Morus spp.	P
Polygonum convolvulus	P	*Nyssa sylvatica*	P
		Oxydendrum arboreum	P
\multicolumn{2}{c}{0.84 to 1.12 kg a.e./ha[b]}		*Populus* spp.	P
		Prunus spp.	P
Brunnichia cirrhosa	P	*Quercus* spp.	P
Ipomoea spp.	A/P	*Rhus* spp.	P
Lonicera spp.	P	*Salix* spp.	P
Rhus radicans	P	*Sapium sebiferum*	P
Rosa bracteata	P	*Sassafras albidum*	P
Rosa multiflora	P	*Tamarix ramosissima*	P
Rosa spp.	P	*Taxodium distichum*	P
Smilax spp.	P		

[a] Growth habit: A = Annual, B = Biennial, P = Perennial
[b] The higher rates should be used where heavy or well established infestations occur.
[c] For best results, early postemergence applications are required.
[d] A minimum of 700 l/ha of spray solution should be used. Control of established stands may require repeat applications.
[e] The degree of control is species dependent. Some *Rubus* species may not be completely controlled.

3. Europe and Southeast Asia

For Europe and Southeast Asia, imazapyr is recommended for noncrop uses at rates ranging from 0.5 kg a.e./ha up to 2.0 kg a.e./ha, depending on the weeds to be controlled and the timing of application. The most frequently used rate is 0.75 kg a.e./ha. Postemergence application to actively growing weeds is the most common and preferred treatment. A clear

TABLE 5
Key Weeds Controlled by Imazapyr in Latin America

Amaranthus sp.	Hydrocotyle sp.
Ammi majus	Hyparrhenia rufa
Andropogon bicornis	Ibicella lutea
Anoda cristata	Imperata brasiliensis
Anthemis cotula	Ipomoea spp.
Baccharis dracunculifolia	Ixophorus unisetus
Bidens pilosa	Lantana camara
Borreria alata	Leptochloa uninervia
Bowlesia incana	Melampodium divaricatum
Brachiaria purpurascens	Panicum fasciculatum
Brassica campestris	Panicum maximum[a]
Carduus nutans	Panicum muticum
Cenchrus spp.	Parthenium hysterophorus
Chloris sp.	Paspalum conjugatum
Cirsium vulgare	Paspalum dilatatum
Clidemia sp.	Paspalum fasciculatum
Cleome viscosa	Paspalum urvillei
Commelina diffusa	Paspalum virgatum
Conyza bonariensis	Pennisetum clandestinum
Coronopus didymus	Polygonum aviculare
Cynodon dactylon	Polygonum convolvulus
Cyperus esculentus	Pothomorphe umbellata
Cyperus ferax	Pueraria phaseoloides
Cyperus rotundus	Raphanus sativus
Datura ferox	Rottboellia exaltata
Digitaria horizontalis	Rumex crispus
Digitaria sanguinalis	Scoparia dulcis
Diolea lasiocarpa	Sisymbrium sp.
Echinochloa colonum	Solidago microglossa
Echinochloa crus-galli	Sorghum halepense
Eichhornia crassipes	Tagetes minuta
Eleusine indica	Typha domingensis
Euphorbia heterophylla	Verbesina encelioides
Euphorbia peplus	Wedelia paludosa
Fumaria sp.	

[a] Adequate control of this weed requires large plants to be chopped or cut, then treated within 15 d after cutting. Under average conditions, only suppression is achieved.

advantage for imazapyr use is the residual activity resulting from postemergence applications. Key weeds controlled by imazapyr in Europe and Southeast Asia are listed in Table 6.

B. CONTROL OF WEEDS UNDER PAVED SURFACES

In the U.S., imazapyr herbicide, at a rate of 1.68 kg a.e./ha, can be used under asphalt, pond liners, and other paved areas, but only in industrial sites or where the pavement has a suitable barrier along the perimeter that prevents encroachment of roots of desirable plants.

Imazapyr should be used only where the area to be treated has been prepared according to good construction practices. If rhizomes, stolons, tubers, or other vegetative plant parts are present in the site, they should be removed by scalping with a grader blade to a depth sufficient to ensure their complete removal. The soil should not be moved following application of the herbicide. Imazapyr needs to be activated by moisture or mechanical soil incorporation before the area is paved.

TABLE 6
Weeds Controlled by Imazapyr in Europe and
Southeast Asia

Europe	Southeast Asia

Grasses and sedges

Europe	Southeast Asia
Agrostis spp.	*Brachiaria mutica*
Avena fatua	*Cynodon dactylon*
Cynodon dactylon	*Cyperus* spp.
Dactylis glomerata	*Digitaria adscendens*
Elymus repens	*Imperata cylindrica*
Lolium perenne	*Ischaemum muticum*
Phragmites spp.	*Ottochloa nodosa*
Poa annua	*Panicum repens*
Poa pratensis	*Paspalum commersonii*
Sorghum halepense	*P. conjugatum*
	Pennisetum spp.

Broadleaved weeds

Europe	Southeast Asia
Capsella bursa-pastoris	*Ageratum conyzoides*
Cirsium arvense	*Amaranthus spinosus*
Convolvulus arvensis	*Asystasia intrusa*
C. sepium	*Borreria laevis*
Plantago lanceolata	*B. latifolia*
P. major	*Cleome rutidosperma*
Senecio jacobaea	*Clidemia hirta*
S. vulgaris	*Commelina communis*
Solidago canadensis	*Eupatorium odoratum*
Taraxacum officinale	*Hedyotis* spp.
Trifolium repens	*Ipomoea hardwickii*
Urtica dioica	*Melastoma malabathricum*
Viola arvensis	*Mikania cordata*
Vicia sativa	*M. micrantha*

Ferns

Nephrolepis biserrata
Stenochlaena palustris

V. FORESTRY MANAGEMENT

Imazapyr fits well in a forestry management program since many conifer species are tolerant to the compound. Forestry uses for imazapyr include site preparation, conifer release, and control of undesirable woody vegetation with stump, cut-stem, or basal-bark treatments.

A. SITE-PREPARATION TREATMENTS
1. United States

Imazapyr is used to control grasses, broadleaved weeds, vines and brambles, and woody brush and trees on forest sites before regeneration (see Table 4). Dead herbaceous grass and broadleaved weeds may provide fuel for a site-preparation burn, if desired, to control residual pines. Imazapyr can be used for site preparation before regeneration with the conifer species listed in Table 7.

The higher label rates of imazapyr and higher spray volumes are more effective when

TABLE 7
Recommended Rates of Imazapyr for Site
Preparation in the U.S.

Species to be planted	Rate (kg a.e./ha)
Pinus taeda	0.84—1.4
Pinus virginiana	0.84—1.4
Pinus elliottii	0.84—1.4
Pinus echinata	0.84—1.4
Pinus taeda × *Pinus rigida* hybrid	0.84—1.4
Pinus resinosa	0.56—0.84
Pinus banksiana	0.43—1.12
Pinus strobus	0.43—1.12
Pinus ponderosa	0.43—1.12
Pseudotsuga menziesii	0.43—1.12
Picea glauca	0.43—1.12
Picea rubens	0.43—1.12
Picea mariana	0.43—1.12

controlling particularly dense or multilayered canopies of hardwood stands or difficult-to-control species.

A 3-month interval is required between application and planting of seedlings on sites that have been broadcast treated or zones that have been treated by spot- or banded-application methods. A 6-month waiting period between application and planting is required for *Pinus resinosa*.

Occasionally, tank mixes may be necessary to control residual conifers and other tolerant species.

2. Europe

Site preparation involves application of imazapyr several months prior to conifer transplanting. Site-preparation treatments provide excellent control of most herbaceous and brush species with good residual activity to eliminate competition and allow the conifers to grow quickly.

Imazapyr herbicide site-preparation treatments range from 0.375 to 1 kg a.e./ha. Optimum results are obtained when imazapyr is applied in late summer to early autumn[1,2] and the conifers transplanted the following spring. This 6- to 9-month delay in transplanting allows ample time for the residues to dissipate.

Brownout of existing vegetation may take several months as imazapyr is a slow-acting herbicide.[3] In areas where quicker kill is desired, imazapyr may be tank mixed with another, faster-acting herbicide.

Most conifers can successfully be transplanted in imazapyr-treated sites 6 months after application. For example, in Ireland, *Picea sitchensis*, *Pinus contorta*, and *Pinus nigra* have all been shown to be tolerant of imazapyr preplant applications.

B. CONIFER RELEASE FROM HERBACEOUS WEEDS

In the U.S., imazapyr is used to control herbaceous annual and perennial grasses and broadleaved weeds in the following conifer species:

Species	Rate (kg a.e./ha)
Pinus taeda	0.21—0.34
Pinus virginiana	0.21—0.34
Pinus taeda × *Pinus rigida* hybrid	0.21—0.34
Pinus elliottii	0.14—0.34

Imazapyr herbicide, applied by helicopter or ground equipment as a broadcast or direct-spray treatment, releases young pines from herbaceous weeds. The height of conifers may be inhibited when release treatments are made during periods of active conifer growth. A suitable nonionic surfactant may be added at a rate not to exceed 0.25% by volume in water. For best results, applications should be made to newly emerged weeds. For difficult-to-control weeds, the higher rates of imazapyr should be used.

C. CONIFER RELEASE FROM COMPETING WOODY VEGETATION

In the U.S., imazapyr is used to release the following conifers from competing woody vegetation:

Species	Rate (kg a.e./ha)
Pinus taeda	0.56—1.12
Pinus virginiana	0.56—1.12
Pinus taeda × *Pinus rigida* hybrid	0.56—1.12
Pinus strobus	0.28—0.56

Studies of conifer tolerance are being conducted to identify appropriate timing and rates for other conifer species.

Imazapyr may be applied as a broadcast or direct-spray treatment using helicopter or ground-spray equipment to control the weed species listed on the product label. As mentioned above, some minor growth inhibition may be observed when release treatments are made during periods of active conifer growth. To minimize injury, imazapyr should not be applied to conifer stands before resting buds have formed at the end of the second growing season.

Low-volume, direct-spray applications may be made in conifer stands of all ages and species by targeting the unwanted vegetation and avoiding direct application to the conifer. Higher rates of imazapyr are required to control particularly dense stands or difficult-to-control species. Injury may occur to nontarget or desirable hardwoods or conifers if they extend from the same root system, if their root systems are grafted to those of the treated tree or if their roots extend into the treated zone.

D. STUMP AND CUT-STEM TREATMENTS

Imazapyr may be used to control undesirable woody vegetation in forest management by applying the compound either to freshly cut stem surfaces or to cuts on the stem of the target woody vegetation. Imazapyr is absorbed by the woody tissue and translocated throughout the undesired plant. Injury may occur to nontarget or desirable woody plants if they extend from the same root system or if their root systems are grafted to those of the treated tree.

1. United States

In the U.S., for stump or cut-stem treatments, Arsenal Applicators Concentrate or Chopper formulations may be used during any season of the year, except during periods of heavy sap flow in the spring.

Arsenal Applicators Concentrate may be mixed as either a concentrated or dilute solution for stump and cut-stem treatments. For cut-stump treatments, dilute solutions of Arsenal Applicators Concentrate or Chopper are sprayed or brushed onto a freshly cut-stump surface so that the entire area is thoroughly wet. With Chopper RTU, undiluted solutions may be used for cut-stump treatments. In frill or girdle treatments, the solutions are sprayed or brushed into cuts made through the bark at intervals of 5 cm or less around the tree. Dilute solutions can also be injected into similarly spaced cuts on the tree. Concentrated imazapyr solutions may be used only for application to cuts on the stem allowing fewer cuts on the

TABLE 8
Species Controlled with Imazapyr by Cut-Stump
and Injection Application Methods

South Africa	Europe
Chromolaena odorata	Acer spp.
Eucalyptus spp.	Alnus spp.
Lantana camara	Betula spp.
	Corylus avelana
	Fraxinus spp.
	Populus spp.
	Prunus padus
	Quercus robur
	Salix caprea
	Sorbus aucuparia

plant. For tree injection and frill or girdle treatments, the appropriate cuts are made at the ratio of one for every 10 cm of diameter at breast height (dbh) and are spaced evenly around the tree.

2. Europe and South Africa

Selective treatments include cut-surface and injection applications. Equipment used for cut-surface application includes spot sprayers, dosing guns, circular brush cutting saws equipped with a device which applies herbicide to the underside of the blade (Enso stump treater manufactured by Marttinen ky, Finland), and chain saws with Chopper mixed in the oil.[4] Cut-stump application rates vary widely, however. The most common Chopper concentrations are 1% to 5%.

Injection treatments involve applying Chopper to hatchet cuts made to the trunk of the deciduous tree. The cuts must be made at regular intervals completely around the tree with no more than 2.5 cm between cuts.

Injection methods require a dose of 1 to 2 ml per cut of a 1% to 5% Chopper solution to kill susceptible deciduous species and prevent root or stump shoot regrowth. When the Enso brush cutter or chain saws are used to apply imazapyr solutions, higher concentrations may be required.[4] Optimum results are usually obtained with late-summer or early-autumn treatments.

Species which have been successfully controlled by these methods are listed in Table 8.

E. BASAL-BARK TREATMENTS

In the U.S., a special formulation of imazapyr, sold under the trademark Chopper, is required for basal-bark applications. Chopper is miscible in both diesel oil and commercially available penetrating oils. The oils help carry the herbicide through the bark of the target trees and permit application in freezing weather. A premixed, ready-to-use formulation of Chopper, called Chopper RTU, can also be used for basal-bark applications. For low-volume, basal-bark treatments, a Chopper/oil mixture solution or undiluted Chopper RTU can be sprayed on the lower 30 to 45 cm of the plant stem. Brush species should not be treated if larger than 20 cm (tree diameter at breast height).

Chopper and Chopper RTU control the species listed in Table 9.

VI. INDUSTRIAL TURF MANAGEMENT

In the U.S., imazapyr is used on unimproved Cynodon dactylon and Paspalum notatum turf, such as roadsides, utility rights-of-way, and other noncrop industrial sites. Imazapyr

TABLE 9
Species Controlled with Chopper Herbicide

Acer macrophyllum	*Populus* spp.
A. negundo	*Populus tremuloides*
A. rubrum	*Prunus* spp.
Acer spp.	*Prunus serotina*
Alnus rubra	*Quercus alba*
Carya spp.	*Quercus laevis*
Cornus spp.	*Quercus laurifolia*
Crataegus spp.	*Quercus nigra*
Diospyros virginiana	*Quercus rubra*
Elaeagnus angustifolia	*Quercus virginianna*
Elaeagnus umbellata	*Rhus* spp.
Fraxinus spp.	*Rosa arkansana*
Ligustrum vulgare	*Rosa multiflora*
Liquidambar stryaciflua	*Salix* spp.
Liriodendron tulipifera	*Sassafras albidum*
Lithocarpus densiflorus	*Tamarix* spp.
Morus spp.	*Tamarix ramosissima*

TABLE 10
Weeds Controlled by Imazapyr in
C. dactylon* and *P. notatum

Daucus carota	*Ptilimnium capillaceum*
Festuca spp.	*Ranunculus parviflorus*
Galium spp.	*Setaria* spp.
Geranium carolinianum	*Sorghum halepense*
Hordeum pusillum	*Trifolium repens*
Oxalis stricta	

controls many broadleaved and grass weeds (Table 10) in established common and coastal *C. dactylon* and *P. notatum* by eliminating these weeds and releasing the grasses. *C. dactylon* treated with imazapyr results in a compacted growth habit and seedhead inhibition.

In early spring, when the grass is still dormant, imazapyr should be applied at 0.1 to 0.21 kg a.e./ha on *C. dactylon* and at 0.07 to 0.13 kg a.e./ha for *P. notatum*. Later, when the grass in the stand has initiated new growth but 25% of the grass in the stand has not yet turned green, imazapyr can be applied at 0.1 to 0.13 kg a.e./ha for *C. dactylon* and at 0.07 to 0.13 kg a.e./ha for *P. notatum*.

For applications to turf, a surfactant should not be added to imazapyr spray solution. Imazapyr treatments should not be made to turf during its first growing season or to turf that is under stress. Temporary yellowing of turfgrass may occur when treatment is made after growth commences.

VII. USE IN RUBBER PLANTATIONS

A. SOUTHEAST ASIA

In rubber plantations in Southeast Asia, imazapyr is effective for the control of spot and sheet *Imperata cylindrica* (also known as lalang, alang-alang, and cogon grass) as well as a range of annual and perennial grass and broadleaved weeds, sedges, and woody species in rubber plantations. In addition, imazapyr controls newly germinating weeds because of its soil-residual activity.

1. Crop Tolerance

In Southeast Asia, crop-tolerance trials showed that postplant, directed applications of imazapyr around the base of rubber trees or between the rows are selective at the recommended dosages. Broadcast applications to the foliage are not selective, and care must be taken not to contact the foliage or young bark of rubber trees with imazapyr.

The degree of tolerance to imazapyr is directly related to dosage and the age of the trees at application. Minor symptoms, such as transient leaf elongation and chlorosis of new leaves, have occasionally occurred with postplant treatments of 0.375 kg a.e./ha or more around 6- to 24-month-old trees; however the affected trees have recovered quickly. These symptoms have not been observed on rubber trees older than 2 years. Results indicate that imazapyr used at recommended dosages has no effect on tree girth and height or latex flow and yield even when minor symptoms are observed.

Trials conducted in Malaysia, Thailand, Indonesia, and the Philippines on trees 3 years of age or older showed good tolerance to postplant, directed applications of imazapyr herbicide at dosages ranging from 1 to 3 kg a.e./ha. These applications were selective at dosages up to 1 kg a.e./ha in 2-year-old rubber trees and at lower dosages of 0.25 to 0.375 kg a.e./ha in rubber treated 6 months after transplanting. Rubber seedlings were also tolerant of preplant (site preparation) treatments of imazapyr at dosages up to 2 kg a.e./ha applied 6 weeks prior to transplanting.

Crop tolerance has been excellent with repeat applications around trees which are 2 years of age or older. In Malaysia, the girth of 2-year-old trees was not affected when imazapyr herbicide at 0.75 to 1.5 kg a.e./ha was applied six times at 90-day intervals.[5] In the Philippines, no symptoms were observed after six directed applications of 0.125 to 2 kg a.e./ha to bare soil around 10-year-old rubber trees at 90-day intervals.[6] This trial continued for a total of 30 months in which 10 soil applications were made; these applications had no effect on height, girth, and leaf development or monthly and cumulative latex yields.

As a result of the crop tolerance data collected, the product labels in Southeast Asia do not permit postplant, directed applications to rubber trees less than 2 years old. In addition, rubber trees cannot be planted for 6 weeks following a site-preparation treatment with imazapyr.

2. Weed Spectrum

In trials conducted in Southeast Asia (Malaysia, Indonesia, Thailand, and the Philippines), imazapyr provided 85% or better control of *I. cylindrica*, the primary weed problem in rubber plantations, and other weeds for 3 to 6 months longer than standard herbicides such as glyphosate, dalapon, and paraquat. Imazapyr is rapidly absorbed by *I. cylindrica* and interferes immediately with meristematic growth; however, apparent signs of death are slow to appear.[7] As a result of the rapid foliar absorption, imazapyr activity is not affected by rainfall occurring more than 2 to 3 h after application.

Chlorotic effects may be observed 15 to 30 d after treatment, and maximum control is usually achieved 90 to 120 d after treatment. The collapse of *I. cylindrica* sheet stands generally occurs 120 d after treatment or later. Optimum control is obtained when imazapyr is applied to actively growing weeds under normal soil moisture conditions.

Imazapyr at 0.5 to 1 kg a.e./ha kills *I. cylindrica* with little or no regrowth when the herbicide is applied under good growing conditions. The higher dosage is recommended for use on mature *I. cylindrica* or in areas with dense stands of the weed.

In comparative tests, imazapyr has provided better control of *I. cylindrica* for a longer period than any of the standard herbicides. In trials conducted in Thailand, imazapyr at 0.5 to 1 kg a.e./ha provided 95% to 97% control at 175 d after treatment, while glyphosate at 1.90 to 2.56 kg a.e./ha provided only 30% to 85% control. Regrowth at 298 d after treatment was greater than 40% in the plots treated with glyphosate compared to less than 7% for the imazapyr treatments.[5]

When used for general weed control in rubber, most weed species are controlled by dosages of imazapyr ranging from 0.125 to 0.5 kg a.e./ha. Perennial grass weeds, such as *Pasapalum conjugatum*, are very susceptible and are well controlled with 0.125 kg a.e./ha. *Ottochloa nodosa* is effectively controlled with 0.25 kg a.e./ha of imazapyr. Less susceptible species, such as *Ischaemum muticum, Mikania cordata*, and *Cyperus* spp. require higher dosages of 0.5 to 1 kg a.e./ha for complete kill. Woody shrubs, including *Eupatorium odoratum* and *Melastoma malabathricum*, are well controlled with a single application of imazapyr at 0.125 to 0.375 kg a.e./ha.

The length of control ranges from 3 to 5 months, depending on the predominant weed species, weed density, dosage applied, climatic conditions, and the age of the plantation. *Paspalum* spp., and *Ottochloa nodosa* have been effectively controlled for 14 to 20 weeks, while regrowth of *Eleusine indica, Commelina communis*, and *Cyperus* spp. has been observed within 40 to 50 d after treatment. Trial results indicate that the duration of control and suppression of regrowth are better in older, shaded plantations than in young, unshaded ones.

3. Methods of Application and Rates

Imazapyr is applied in broadcast, site-preparation treatments in Southeast Asia for the control of *I. cylindrica* at rates ranging from 0.5 to 1 kg a.e./ha. Following these applications, rubber trees cannot be planted for 6 weeks.

Postplant, directed applications of imazapyr are also made to rubber trees 2 or more years old, using dosages of 0.5 to 1 kg a.e./ha. When used for general weed control, dosages from 0.25 to 0.5 kg a.e./ha can be used in rubber trees 1 or more years old. When making these applications, care must be taken to avoid contacting the foliage or green bark of the trees, or injury may result.

B. LATIN AMERICA

1. Crop Tolerance

In Central America and Brazil, $1^1/_2$- to 2-year-old rubber trees have shown good tolerance to inter-row applications of imazapyr at up to 1.0 kg a.e./ha. In nursery trials in Brazil, leaf elongation and premature shedding were observed when seedlings were planted 3 months after the soil was treated with 0.125 to 0.75 kg a.e./ha. However, seedling development and survival were not significantly affected. Results of Brazilian trials also confirm that imazapyr at the recommended rate and timing has no detrimental effect upon latex production and dry rubber quality.

2. Weed Spectrum

Many of the weeds cited in Table 5 frequently appear in rubber plantations in Latin America and are controlled by imazapyr. In particular, *Paspalum fasciculatum*, a problematic weed in Guatemala, is adequately controlled with treatments of 0.5 to 0.75 kg/ha of imazapyr, for a 3- to 4-month period. *Hyparrenhia rufa* is another important weed controlled by imazapyr, but it is not well controlled by other commercial herbicides. In Brazil, the following weeds are controlled by imazapyr in rubber plantations: *Digitaria horizontalis, Imperata brasiliensis, Paspalum conjugatum, Paspalum virgatum, Borreria capitata, Clidemia* sp., *Hydrocotyle* sp., and *Wedelia paludosa*. The length of residual control varies from 3 to 4 months in young unshaded plantations to 6 to 8 months in 7-year-old shaded plantations.

3. Methods of Application and Rates

Arsenal should be applied between the rows of rubber trees, avoiding direct contact with the trees. The recommended rate varies between 0.5 to 1 kg/ha.

VIII. USE IN OIL PALM PLANTATIONS

A. SOUTHEAST ASIA

1. Crop Tolerance

In trials in Malaysia and Indonesia, oil palms were tolerant of preplant (site-preparation) treatments of imazapyr at up to 2 kg a.e./ha applied 6 weeks before transplanting. In Malaysia, Indonesia, and the Philippines, postplant, directed applications of 0.5 to 1 kg a.e./ha were selective to 4-year-old palms. Some localized symptoms, such as chlorotic and twisted young fronds, were noted when the herbicide spray came into contact with the foliage. Subsequent growth was normal.

Postplant, directed applications to young palms (less than 27 months old) have resulted in moderate to severe injury, such as chlorosis and necrotic spotting of young fronds and occasional twisting or breaking of mature fronds. Oil palms have fully recovered except when palms of 1 year or less have been treated with dosages greater than 1 kg a.e./ha.

In trials in Malaysia and the Philippines, imazapyr at dosages up to 1 kg a.e./ha had no effect on fruit development and production.

Based on these results, oil palms should be planted 12 weeks or more following a site-preparation application of imazapyr. When used in postplant, directed applications, oil palm must be 4 or more years old. Care must be taken when applying postplant treatments to avoid contact with the foliage or young bark as unacceptable crop injury may result.

2. Weed Spectrum

The weed spectrum controlled by imazapyr in oil palm plantations is identical to those species controlled in rubber plantations.

3. Methods of Application and Rates

For site-preparation treatments, broadcast applications of 0.5 to 1 kg a.e./ha of imazapyr are required. The higher rate is needed for control of *I. cylindrica*.

When used as a postplant, directed treatment, imazapyr dosages range from 0.18 to 0.56 kg a.e./ha.

B. LATIN AMERICA

1. Crop Tolerance

In Central America, imazapyr at 0.5 to 0.75 kg/ha is selective in oil palm trees 2 or more years old when applied between the rows. Spray contact with the foliage results in chlorotic and twisted fronds. Fruit development and production is similar in treated and untreated palms.

2. Weed Spectrum

A number of weeds mentioned in Table 5 are controlled by imazapyr in oil palm plantations. For acceptable control of *Panicum maximum*, the weeds must be chopped or cut 15 d before application of the maximum rate of 0.75 kg/ha.

3. Methods of Application and Rates

In Central America, imazapyr should be applied postemergence to the weeds between the rows of oil palm at a minimum distance of 1.5 m from the base of young trees (less than 2 years old) or over the whole plantation if the trees are older than 2 years. The spray should not come in contact with the fronds. The recommended rate is between 0.5 and 0.75 kg/ha.

TABLE 11
Weeds Controlled by Imazapyr in Sugar Cane
in Latin America

Grasses	Broadleaves	Sedges
Digitaria sanguinalis	*Amaranthus* sp.	*Cyperus* spp.
Echinochloa colonum	*Clitoria ternatea*	
Echinochloa crus-galli	*Caperonia* sp.	
Leptochloa spp.	*Colocasia esculenta*	
Panicum fasciculatum	*Eclipta alba*	
Rottboellia exaltata	*Euphorbia* spp.	
Setaria sp.	*Ipomoea* spp.	
Sorghum arundinaceum	*Melampodium divaricatum*	
Sorghum halepense	*Mucuna pruriens*	
	Phyllanthus sp.	
	Portulaca oleracea	
	Richardia sp.	

IX. USE IN SUGAR CANE PLANTATIONS

A. CROP TOLERANCE

Numerous trials have been conducted in Argentina, Brazil, Central America, Colombia, the Dominican Republic, and Venezuela. In these trials, sugar cane has shown tolerance to imazapyr when the compound is applied preemergence to plant cane at rates up to 0.25 kg/ha. Application at the "spike" stage of plant cane frequently results in severe phytotoxicity to the crop. Ratoon cane is not tolerant to imazapyr, and directed sprays have not always resulted in adequate selectivity. Rates above 0.25 kg/ha show significant phytotoxic symptoms especially in situations involving light soils and heavy rains after application.

B. WEED SPECTRUM

Imazapyr has shown good activity on grasses, broadleaves, and sedges in sugar cane. *Cyperus* spp., a particularly noxious weed in sugar cane, is suppressed by imazapyr. *Rottboellia exaltata* is also well controlled by the herbicide; however, under a high level of infestation, a combination treatment with pendimethalin has given superior results. A list of weeds controlled by imazapyr in sugar cane is shown in Table 11.

C. METHODS OF APPLICATION AND RATES

To ensure adequate sugar cane selectivity, imazapyr should be applied preemergence to plant cane, generally within a week of planting. The postemergence activity of the compound will adequately control the weeds that have emerged. In Venezuela, the recommended rate is 150 to 200 g a.e./ha.

REFERENCES

1. **Lund-Høie, K. and Rognstad, A.,** The effect of foliage applied imazapyr and glyphosate on common forest weed species and Norway spruce, *Crop Protec.*, 9, 52, 1990.
2. **Winfield, R. J. and Bannister, C. J.,** Imazapyr for broad spectrum weed control in forestry, in *Aspects of Appl. Biol.*, 16, Association for Applied Biology, Warwick, England, 1988, 79.
3. **Christensen, P.,** Danish results with a new herbicide, imazapyr, in forestry, in *Aspects of Appl. Biol.*, 16, Association for Applied Biology, Warwick, England, 1988, 105.

4. **Lund-Høie, K. and Rognstad, A.,** Stump treatment with imazapyr and glyphosate after cutting of hardwoods to prevent regrowth of suckers, *Crop Protec.,* 9, 59, 1990.

5. **Boonsrirat, C., Lee, S. C., and Chee, K. S.,** AC 252,925, a new herbicide for use in rubber plantations, *Proc. 10th Conf. Asian-Pacific Weed Sci. Soc.,* 1985, 99.

6. **Manimtim, M. B., Lapada, B. E., and Calora, F. B.,** Evaluation of some factors affecting the efficacy of AC 252,925 on "cogon grass", *Imperata cylindrica* and selectivity to rubber, *Hevea brasiliensis, Proc. 1st Trop. Weed Sci. Soc. Conf.,* 1984, 281.

7. **Shaner, D. L.,** Absorption and translocation of imazapyr in *Imperata cylindrica* (L.) Raeuschel and effects on growth and water usage, *Trop. Pest Manage.,* 34, 388, 1988.

Chapter 18

IMAZAMETHABENZ-METHYL HERBICIDE

Barnett Bernstein, Kenneth Kirkland, and Cletus Youmans

TABLE OF CONTENTS

I. INTRODUCTION

Imazamethabenz-methyl is a selective, postmergence herbicide used to control weeds in cereal crops. Imazamethabenz-methyl controls many of the economically important weeds which infest wheat, barley, and rye, such as *Avena* spp., *Apera spica-venti, Poa trivialis, Raphanus raphanistrum, and Brassica* spp. In addition, imazamethabenz-methyl suppresses but does not usually kill *Alopecurus myosuroides, Galium aparine, Polygonum convolvulus,* and *Fagopyrum tartaricum.* In Europe, commercial coformulations of imazamethabenz-methyl with other herbicides, such as isoproturon, pendimethalin, difenzoquat, and mecoprop expand the weed-control spectrum to include *Alopecurus myosuroides, Galium aparine, Polygonum* spp., *Matricaria* spp., *Stellaria media, Veronica hederaefolia,* and other weeds which are either suppressed or not effectively controlled with the use of imazamethabenz-methyl alone.

Imazamethabenz-methyl is readily absorbed through the foliage and roots of sensitive weeds and is slowly translocated to the meristematic regions, where the herbicidally active acid accumulates. The amount of herbicide absorbed through either the foliage or roots varies according to weed species. For example, in greenhouse experiments, *Avena* spp. absorbed approximately 60% through the foliage and 40% through the roots, whereas *A. myosuroides* absorbed approximately 40% through the foliage and 60% through the roots. Cell division is halted at interphase 24 h after treatment. The first sign of herbicidal activity is chlorosis of the youngest leaves. Later in the season, the whole plant becomes necrotic or remains as a small, noncompetitive weed at the base of the crop.

Imazamethabenz-methyl also has soil residual activity; thus, susceptible weeds that germinate after application can be controlled.

II. REGULATORY STATUS

A. IMAZAMETHABENZ-METHYL ALONE

Imazamethzbenz-methyl is registered in Belgium, Bulgaria, Canada, Czechoslovakia, Greece, Hungary, Ireland, Italy, Luxembourg, Morocco, Netherlands, Norway, Poland, Portugal, South Africa, Spain, the U.K., and the U.S. In the U.K., the tradename for imazamethabenz-methyl is Dagger. In the other countries, the tradename is Assert.

B. COMBINATION FORMULATIONS

Coformulations of imazamethabenz-methyl and other herbicides are registered in the countries listed in Table 1.

III. USE IN WHEAT, BARLEY, AND RYE

A. CROP TOLERANCE

In several years of testing, most major varieties of spring and winter wheat, barley, and rye, as well as some varieties of winter triticale, have shown excellent tolerance to post-emergence applications of imazamethabenz-methyl herbicide at twice the normal recommended dosages. Most varieties of durum wheat show acceptable levels of tolerance. Imazamethabenz-methyl can be safely applied to these crops after two leaves have fully expanded.

B. WEED SPECTRUM/APPLICATION TIMING
1. Europe

In Europe, the recommended dosages for imazamethabenz-methyl range from 0.4 to 0.9 kg a.i/ha, depending upon the growth stage of *Avena* spp. and the geographic region.

TABLE 1
Registration Status of Imazamethabenz-methyl
Coformulations

Coformulation partner	Country
Pendimethalin	France, Italy, Spain
Isoproturon	Belgium, France, Spain, U.K.
CMPP	France, Italy
Difenzoquat	France
Trifluralin	Italy
Chlortoluron	Italy

TABLE 2
Grass Weeds Susceptible to Imazamethabenz-methyl
(Europe)

Northern Europe	Southern Europe
Alopecurus myosuroides[a]	*Alopecurus myosuroides*[a]
Apera spica-venti	*Apera spica-venti*
Arrhenatherum elatius	*Avena fatua*
Avena fatua	*A. ludoviciana*
A. ludoviciana	*A. sterilis (macrocarpa)*
Poa trivialis	*Milium scabrum*
	Poa trivialis

[a]   Under favorable conditions.

TABLE 3
Broadleaved Weeds Susceptible to
Imazamethabenz-methyl (Europe)

Northern Europe	Southern Europe
Brassica kaber	*Brassica kaber*
Polygonum convolvulus	*Brassica nigra*
Raphanus raphanistrum	*Capsella bursa-pastoris*
Spergula arvensis	*Cardamine hirsuta*
	Polygonum convolvulus
	Ranunculus arvensis

Imazamethabenz-methyl herbicide provides excellent control or suppression of many economically important weeds which infest wheat, barley, and rye. The susceptibility of grass and broadleaved weeds to postemergence application of imazamethabenz-methyl at local recommended dosages is presented in Tables 2 and 3, respectively.

For effective control of *Avena* spp., imazamethabenz-methyl must be applied from emergence to the early tillering stage. *Avena* spp. past the early tillering stage may not be effectively controlled. The soil residual activity of autumn or winter treatments may control *Avena* spp. germinating the following spring. Where vigorous spring flushes of *Avena* spp. emerge under dry conditions, a sequential treatment with difenzoquat (Avenge wild oat herbicide) may be required in the spring.

Under favorable conditions, imazamethabenz-methyl will control *A. myosuroides* at dosages recommended for *Avena* spp. control; otherwise, the product will provide excellent suppression of this weed. Where heavy populations of *A. myosuroides* are present, a

<table>
<tr><td colspan="2">**TABLE 4**
Weeds Susceptible to
Imazamethabenz-methyl (Canada)</td><td colspan="2">**TABLE 5**
Weeds Susceptible to
Imazamethabenz-methyl (U.S.)</td></tr>
</table>

Broadleaved weeds	Grass weeds	Broadleaved weeds	Grass weeds
Brassica kaber	*Avena fatua*	*Brassica kaber*	*Avena fatua*
Thlaspi arvense		*Descurainia pinnata*	*Poa trivialis*
		D. sophia	
		Sisymbrium irio	
		Thlaspi arvense	

sequential treatment or tank mixture with a herbicide specific for *A. myosuroides* is recommended.

Good control of broadleaved weeds such as *Raphanus* spp. and *Brassica* spp. is obtained when imazamethabenz-methyl is applied to plants up to 10 cm tall. Under favorable conditions, control of *Galium* spp. may also be achieved, particularly when applications are made prior to the two true-leaf stage.

2. Canada

In Canada, imazamethabenz-methyl is used either alone or in tank mixtures with other herbicides to control *Avena fatua* and broadleaved weeds in spring wheat and spring barley. A list of susceptible weeds is presented in Table 4. Dosage and timing of application depends on the growth stage of the weed. A dosage of 0.5 kg a.i./ha should be used when *A. fatua* plants are in the one- to four-leaf stage. A dosage of 0.4 kg a.i./ha can be used when the majority of *A. fatua* plants are in the one- to three-leaf stage.

When applied from the cotyledon to the six-leaf stage of the weeds, imazamethabenz-methyl used alone will control or suppress several broadleaved weeds in the Cruciferae family, such as *Thlaspi arvense*.

Polygonum convolvulus and *Fagopyrum tataricum* are suppressed by imazamethabenz-methyl treatments.

3. United States

Imazamethabenz-methyl is primarily used in the northern areas of the U.S. (west of and including Minnesota), along the western coastal states, and in several Rocky Mountain states. Imazamethabenz-methyl is applied postemergence in wheat and barley in the autumn (Washington and Idaho), spring, or early summer. Increased control is frequently noticed when temperatures are warm although very high temperatures can stress weeds and decrease compound uptake, therefore limiting weed control.

Imazamethabenz-methyl controls several weeds frequently found in cereals at the recommended rate of 350 to 540 g a.i./ha. These are listed in Table 5.

Fagopyrum tataricum, Galium aparine, and *Polygonum convolvulus* are suppressed by imazamethabenz-methyl treatments.

A 300 g a.i./l liquid concentrate formulation is currently marketed in the U.S. With this formulation, the addition of a surfactant to the spray solution at a rate of 0.25% is required for optimum herbicidal activity. The addition of oil (crop or petroleum), in addition to surfactant, can boost performance under stressed conditions.

C. COMBINATIONS
1. Tank Mixtures

In the U.S. and Canada, for additional control of broadleaved weeds, imazamethabenz-methyl can be tank-mixed with many other herbicides including MCPA ester, 2,4-D ester, chlorsulfuron, metsulfuron methyl, thifensulfuron, and tribenuron. Bromoxynil and clopyralid are also used in the U.S. Imazamethabenz-methyl cannot be tank-mixed with dicamba or salts of 2,4-D, MCPA, and CMPP because of antagonism (reduced control of *Avena* spp.).

In the U.S. (Idaho, Oregon, Utah, and Washington), imazamethabenz-methyl at 263 g a.i./ha can be tank-mixed with difenzoquat at 500 g a.i./ha to control *Avena fatua*. The tank mixture can be applied from the two- to five-leaf stage of *A. fatua*.

In the U.S., imazamethabenz-methyl can be applied with a liquid fertilizer, and a combination of the two can be tank-mixed with additional herbicides such as MCPA ester, 2,4-D ester, or bromoxynil plus 2,4-D ester. When a postemergence application of liquid fertilizer is made to crops, foliage burning may occur.

In Europe, pendimethalin and isoproturon are the most commonly used tank-mix partners with imazamethabenz-methyl.

The product labels for each country should be consulted for specific recommendations.

2. Coformulations

In Europe, the following coformulations have been developed to expand the weed-control spectrum beyond that controlled by imazamethabenz-methyl alone.

a. *Imazamethabenz-methyl Plus Pendimethalin*

Application rates varying from 0.5 to 0.8 plus 0.625 to 1.0 kg a.i./ha imazamethabenz-methyl and pendimethalin can be made beginning at the one-leaf stage of the cereal crop. For optimal broadleaved weed control, applications should be made from preemergence up to the two true-leaf stage for most species. Optimal application timings for grass weeds vary by species and location.

A list of susceptible weeds is given in Table 6.

b. *Imazamethabenz-methyl Plus Isoproturon*

The combination of imazamethabenz-methyl and isoproturon has been developed for use as a postemergence herbicide in winter wheat (excluding durum) and winter barley.

This product can be applied from the one-leaf stage until the first node of the crop at the recommended dose of 500 and 1500 g a.i./ha imazamethabenz-methyl and isoproturon. The weeds controlled by this combination and the maximum growth stage for effective weed control are listed in Table 7.

The combination of imazamethabenz-methyl and isoproturon improves the control of *Alopecurus myosuroides* compared with imazamethabenz-methyl alone and widens the weed spectrum to include important weeds such as *Matricaria* spp.

c. *Imazamethabenz-methyl Plus CMPP*

Imazamethabenz-methyl plus CMPP is a postemergence herbicide combination used to control annual grasses and broadleaved weeds in winter wheat and winter barley. Imazamethabenz-methyl plus CMPP can be applied from the beginning of tillering to the first node. At a dosage of 625 and 2000 g a.i./ha of imazamethabenz-methyl and CMPP the weeds listed in Table 8 are controlled.

d. *Imazamethabenz-methyl Plus Difenzoquat*

Imazamethabenz-methyl plus difenzoquat is a postemergence herbicide combination developed for use in winter and spring cereals for the control of *Alopecurus myosuroides*, *Avena* spp., *Poa trivialis*, and *Apera spica-venti* as well as certain broadleaved weeds such as *Cruciferae* spp. In winter cereals, this product can be applied from the one-leaf stage until the first node, while in spring cereals application must be made before the first node stage. A dosage of 625 and 500 g a.i./ha imazamethabenz-methyl plus difenzoquat is recommended.

TABLE 6
Weeds Susceptible to Imazamethabenz-methyl
Plus Pendimethalin (Europe)

Broadleaved weeds	Grass weeds
Adonis aestivalis	*Alopecurus myosuroides*
Anagallis arvensis	*Apera spica-venti*
Aphanes arvensis[a]	*Arrhenatherum elatius*
Arabidopsis thaliana	*Avena fatua*
Atriplex sp.	*Phalaris* spp.
Capsella bursa-pastoris	*Poa trivalis*
Cardamine hirsuta	
Cerastium arvense	
Chenopodium album	
Euphorbia sp.	
Fallopia convolvulus	
Galium aparine	
Geranium spp.	
Legousia speculum-veneris	
Lamium purpureum	
Linaria sp.	
Matricaria spp.[a]	
Myosotis sp.	
Papaver rhoeas	
Polygonum spp.	
Raphanus raphanistrum	
Spergula arvensis	
Stellaria media[a]	
Sonchus arvensis	
Thlaspi arvense	
Veronica spp.	
Viola tricolor[a]	

[a]   These species are sensitive only to preemergence applications.

TABLE 7
Maximum Growth Stages for Effective Weed Control
by Imazamethabenz-methyl Plus Isoproturon

Weeds controlled	Maximum stage of growth
Alopecurus myosuroides	Mid-tillering
Avena spp.	3 tillers
Poa annua	Early tillering
P. trivialis	Early tillering
Apera spica-venti	Beginning of tillering
Broadleaved weeds	2 true leaves

The correct growth stages for application against specific grass weeds are as follows: *Avena* spp. — one leaf to beginning of tillering; *Alopecurus myosuroides* and *Apera spica-venti* — two leaves to mid-tillering.

IV. ROTATIONAL CROP SAFETY

A. NORTHERN EUROPE

Spring cereals, winter wheat, winter barley, sunflowers, field beans, maize, oilseed rape, sugar beet, and most other major crops can be grown in normal rotation following

TABLE 8
Weeds Controlled by Imazamethabenz-methyl
Plus CMPP (Europe)

Broadleaved weeds	Grass weeds
Anagallis arvensis	*Alopecurus myosuroides*
Arabidopsis thaliana	*Apera spica-venti*
Brassica kaber	*Avena* spp.
Capsella bursa-pastoris	
Cerastium sp.	
Chenopodium album	
Fumaria officinalis	
Galium aparine	
Lapsana communis	
Papaver rhoeas	
Polygonum aviculare	
Raphanus raphanistrum	
Spergula arvensis	
Stellaria media	
Veronica hederaefolia	

cereals treated with recommended dosages of imazamethabenz-methyl or imazamethabenz-methyl combination products.

In the U.K., if imazamethabenz-methyl products are used later than March 31 in a cereal crop, the ground should be moldboard plowed before sowing oilseed rape or sugar beet.

Sugar beet sown the year following a cereal crop treated with imazamethabenz-methyl should be grown according to good agricultural practices. In particular, soil pH must be greater than 6.0 and free calcium should be at an optimum level to ensure good sugar beet growth. Autumn-sown sugar beet should not be planted following a cereal crop treated with imazamethabenz-methyl.

When imazamethabenz-methyl products and products containing either chlorsulfuron or metsulfuron methyl are used on the same crop during the same growing season, only wheat, barley, or triticale should be grown as the following crop.

B. SOUTHERN EUROPE

In Italy, a 12-month interval is required between imazamethabenz-methyl application and planting of sugar beet, oilseed rape, fodder mustard, and sensitive legume crops.

In Spain, sugar beet, oilseed rape, vetch, or lentils cannot be planted as first-year follow crops.

C. CANADA

Normal rotation of follow crops is permitted the year after an application of imazamethabenz-methyl in the brown and dark brown soil zones for spring wheat, durum wheat, barley, and sunflower. In the black and grey wooded soil zones canola, flax, oats, and peas are also permitted. Imazamethabenz-methyl should not be applied to the same field in 2 successive years.

D. UNITED STATES

The following crops can be sown during the planting season following an imazamethabenz-methyl application: barley, corn, edible beans, safflower, soybean, sunflower, and wheat. Sugar beets are restricted to 20 months after an imazamethabenz-methyl application regardless of rainfall and/or irrigation. All other rotational crops are restricted to being planted a minimum of 15 months following an imazamethabenz-methyl application.

Peas can be planted the season following an application of the imazamethabenz-methyl/difenzoquat tank mixture if 45 cm of rainfall has been received during the year of application.

Chapter 19

IMAZAQUIN HERBICIDE

Prithvi Bhalla, Neil Hackett, Ronaldo Hart, and Edward Lignowski

TABLE OF CONTENTS

I. INTRODUCTION

Imazaquin was the first imidazolinone to be widely accepted as a herbicide for broad spectrum weed control. In 1981, field testing began in the U.S. and Brazil. Initial results demonstrated that imazaquin could be applied preplant incorporated, preemergence, or postemergence for the control of many difficult-to-control weed species. In Brazil, excellent control of *Euphorbia heterophylla* was observed. In the U.S., a significant finding was that imazaquin soil-applied treatments provided excellent control of *Xanthium strumarium*, which was not controlled by standard preplant-incorporated or preemergence herbicides. In early testing, it was discovered that imazaquin applied as a postemergence treatment uniquely provided both burndown of susceptible weeds present and residual control of later germinating weeds.

Subsequent testing demonstrated that imazaquin selectively controlled *Cyperus rotundus* and other problem weeds in warm-season turfgrasses.

II. REGULATORY STATUS

Experimental Use Permits were granted by the U.S. Environmental Protection Agency (EPA) for field testing of imazaquin in 1984 and 1985.

Trials in Mississippi demonstrated that imazaquin controlled *Cassia obtusifolia*, a highly competitive soybean weed. In 1985, the EPA permitted growers in five southern states to use imazaquin under a Section 18 Emergency Exemption Clearance for control of *C. obtusifolia* since there were no registered herbicides which controlled this weed.

The first registration for imazaquin was granted in Argentina in 1984 for use in soybeans. In 1986, imazaquin was registered for use in soybeans in the U.S. Table 1 provides a list of countries, along with formulations and use rates, where imazaquin is registered for use in soybeans. Imazaquin has the tradename Scepter in all countries where it is registered for use in soybeans.

Imazaquin is under development in Bolivia, Colombia, Venezuela, and the U.S.S.R.

III. USE IN SOYBEANS

A. CROP TOLERANCE

Soybeans exhibit excellent tolerance to imazaquin when applied either as a soil or postemergence treatment. Once absorbed, soybeans rapidly metabolize imazaquin into inactive components. Growing conditions affect the soybean plant's ability to metabolize imazaquin. Adverse growing conditions, such as cool, wet weather, that slow or temporarily stop soybean growth and development will lengthen the time required to metabolize absorbed imazaquin. Adverse conditions may occasionally cause temporary height reductions in soybeans. Under favorable growing conditions, symptoms rapidly disappear, and yields are not affected.

B. WEED SPECTRUM/APPLICATION TIMING

Table 2 lists the weeds that are controlled by preplant-incorporated or preemergence applications of imazaquin at 140 g a.e./ha in soybeans in the U.S.

The following weeds are suppressed by soil applications of imazaquin: *Desmodium tortuosum, Echinochloa crus-galli, Caperonia castanaefolia, Eleusine indica, Sorghum bicolor, Cassia obtusifolia, Brachiaria platyphylla, Euphorbia maculata,* and *Caperonia palustris.* Preplant-incorporated treatments of imazaquin provide the best activity on *Ipomoea hederacea* and *Pharbitis purpurea* and are required for suppression of *Cyperus esculentus.*

In the U.S., postemergence applications of imazaquin control the weeds listed in Table 3. Application rates range from 70 to 140 g a.e./ha depending on weed species and height.

TABLE 1
Countries Where Imazaquin Is Registered for Use in Soybeans

Country	Formulation	Use rate (g a.e./hectare)
Argentina	200 AS[a]	200
Brazil	150 AS	150
Bulgaria	150 AS	150
Hungary	150 AS	150
Nicaragua	150 AS	150
Nigeria[b]	180 AS	180
Paraguay	150 AS	150
Romania	150 AS	150
U.S.[c]	180 AS	70—140
	70 DG[d]	70—140
Yugoslavia	150 AS	150

[a] Formulations in this table designated AS contain the indicated amount of imazaquin as the ammonium salt in an aqueous solution. For example, 200 AS contains 200 grams of imazaquin per liter.

[b] Imazaquin is also registered for use in cowpeas (*Vigna sinensis*).

[c] Imazaquin is also registered for use in turfgrass.

[d] Dispersible granule which contains 70% imazaquin.

TABLE 2
Weeds Controlled by Imazaquin Applied Preplant Incorporated or Preemergence at 140 g a.e./ha (U.S.)

Broadleaved weeds	Grass weeds
Abutilon theophrasti[a]	*Setaria faberi*
Acanthospermum australe	*S. viridis*
Amaranthus hybridus	*S. lutescens*
A. palmeri	*Sorghum halepense* (seedling)
A. retroflexus	
A. spinosus	
A. tuberculatos	
Ambrosia artemisiifolia	
A. trifida[a]	
Brassica spp.	
Chenopodium album	
Datura stramonium	
Euphorbia heterophylla	
Helianthus annuus	
Hibiscus trionum	
Ipomoea lacunosa	
I. wrightii	
Jacquemontia sandwicensis	
Melochia corchorifolia[a]	
Polygonum pensylvanicum	
P. persicaria	
Richardia scabra	
Sicyos angulatus[a]	
Sida spinosa	
Solanum ptycanthum[a]	
Tribulus terrestris	
Xanthium strumarium	

[a] Preplant-incorporated treatments only.

TABLE 3
Weeds Controlled by Postemergence
Applications of Imazaquin (U.S.)

Weeds controlled	Weed height (cm)	Imazaquin rate (g a.e./ha)
Amaranthus palmeri	Up to 30	140
A. retroflexus	Up to 10	70[a]
	12.5 to 30	140
A. hybridus	Up to 10	70[a]
	12.5 to 30	140
A. spinosa	Up to 30	140
A. tuberculatos	Up to 30	140
Euphorbia heterophylla	Up to 15	140
Helianthus annuus	Up to 10	70[a]
	12.5 to 20	140[a]
Volunteer corn	Up to 20	70[a]
	22.5 to 30	140
Xanthium strumarium	Up to 20	70[a]
	20 to 30	140

[a]   For control of emerged weeds.

TABLE 4
Dicot Weeds Controlled by Soil Applications of
Imazaquin at 150 g a.e./ha (Brazil)

Acanthospermum australe	*Galinsoga parviflora*
A. hispidum	*Ipomoea* sp.
Alternanthera ficoidea	*Nicandra physaloides*
Amaranthus sp.	*Phyllanthus niruri*
Bidens pilosa	*Physalis angulata*
Borreria alata	*Portulaca oleracea*
Chenopodium album	*Raphanus raphanistrum*
Commelina sp.	*Richardia brasiliensis*
Eclipta alba	*Sida* sp.
Emilia sonchifolia	*Solanum* sp.
Euphorbia heterophylla	*Spergula arvensis*
E. pilulifera	

For postemergence applications the use of a nonionic surfactant or crop oil concentrate is required.

In Brazil, imazaquin is used as a soil-applied broadleaf herbicide to control major soybean weeds such as *Euphorbia heterophylla* (the most widely distributed and economically significant weed), *Bidens pilosa, Commelina virginica, Sida rhombifolia, Amaranthus* spp., *Ipomoea* spp., and *Acanthospermum australe* in conventional tillage and no-till systems. Table 4 provides a list of weeds controlled by imazaquin in Brazil.

In Brazil, the main method of application for imazaquin in conventional tillage in soybeans is preplant incorporation. Under average climatic conditions this method has proven to give more consistent results than preemergence application. Shallow (not more than 5 cm) incorporation is recommended; the disc harrow is most commonly used to incorporate imazaquin. "Aplique plante" is a method in which imazaquin is applied before planting and then incorporated at planting time with the use of machinery that mixes the herbicide to a depth of 2 to 3 cm. Imazaquin is applied preemergence to the crop and weeds in zero-tillage systems.

TABLE 5
Weeds Controlled by Soil Applications of
Imazaquin at a Rate of 200 g a.e./ha (Argentina)

Dicots	Monocots
Amaranthus hybridus	Digitaria sanguinalis
A. lividus	Echinochloa crus-galli
Ammi majus	Sorghum halepense (seedling)
Anoda cristata	
Anthemis cotula	
Brassica campestris	
Chenopodium album	
Coronopus didymus	
Datura ferox	
Euphorbia dentata	
Galinsoga parviflora	
Helianthus annuus	
Ipomoea purpurea	
Lamium amplexicaule	
Portulaca oleracea	
Sida sp.	
Tagetes minuta	
Tithonia tubaeformis	
Xanthium spinosum	
X. strumarium	

In Argentina, imazaquin is used in the Pampa region (the area north of Buenos Aires, south of Santa Fe, and southeast of Cordoba) as an early preplant- or preplant-incorporated herbicide. Preemergence application in conventional tillage is recommended in the northwestern soybean area of Argentina. When the rainfall pattern is regular and the product receives rain within 5 to 7 d after application, preemergence performance is excellent. Late preemergence applications, when weeds and crop are emerging from the soil, have also given very good results.

Imazaquin at 200 g a.e./ha controls the major annual monocot and dicot soybean weeds in Argentina, including *Amaranthus hybridus, Chenopodium album, Datura ferox, Anoda cristata, Tagetes minuta, Echinochloa crus-galli, Digitaria sanguinalis,* and seedling *Sorghum halepense.* A list of weeds controlled by imazaquin in Argentina is provided in Table 5.

The following weeds are suppressed by soil applications of imazaquin in Argentina: *Cenchrus echinatus, Cyperus* spp., *Eleusine indica, Sorghum halepense* (rhizome), and *Wedelia glauca.*

In Paraguay, imazaquin is used preplant incorporated at 150 g a.e./ha. The spectrum of weeds controlled is similar to those controlled in Brazil and includes the key weeds *Euphorbia heterophylla* and *Ipomoea nil.*

Imazaquin is used preemergence in Nicaragua at an application rate of 150 g a.e./ha for the control of *Euphorbia heterophylla, Portulaca oleracea, Amaranthus* spp., and *Ipomoea* spp.

In Eastern Europe, imazaquin is applied preplant incorporated, preemergence, and early postemergence at a rate of 150 g a.e./ha in soybeans. Weeds controlled by imazaquin with soil-applied treatments include *Abutilon theophrasti, Amaranthus* spp., *Brassica* spp., *Chenopodium album, Datura stramonium, Helianthus annuus, Setaria* spp., *Solanum nigrum,* seedling *Sorghum halepense,* and *Xanthium strumarium.* Postemergence treatments control *Amaranthus* spp. and *Xanthium* spp.

In Nigeria, imazaquin applied preemergence or early postemergence at 180 g a.e./ha controls the main weeds in soybeans including *Euphorbia heterophylla, Talinum triangulare,*

Commelina benghalensis, and Brachiaria spp. Imazaquin is also registered for preemergence use in cowpeas (*Vigna sinensis*).

C. COMBINATIONS

For soil applications, imazaquin can be tank-mixed with the following grass herbicides: alachlor, metolachlor, pendimethalin, and trifluralin. Imazaquin can also be applied sequentially as a preemergence or postemergence treatment following application of these grass herbicides. No compatibility problems have been observed and these tank-mix combinations provide effective control of a wide spectrum of annual grasses and broadleaved weeds.

In the U.S., imazaquin is a component of three premix formulations, Squadron, Tri-Sept, and Scepter O.T. Squadron is a concentrated emulsion containing 40 g imazaquin and 240 g pendimethalin per liter. Tri-Sept is a concentrated emulsion formulation containing 52 g imazaquin and 308 g of trifluralin per liter. Scepter O.T. is an aqueous formulation containing 60 g imazaquin (sodium salt) and 240 g acifluorfen per liter.

For minimum and no-till uses, imazaquin may be tank-mixed with paraquat or glyphosate for use in applications prior to soybean emergence for increased burndown activity on weeds.

Imazaquin cannot be tank-mixed with postemergence grass herbicides because antagonism may occur. However, imazaquin applications may be followed by postemergence applications of fluazifop-*P*-butyl or sethoxydim.

In the U.S., when heavy infestations of *Abutilon theophrasti* are expected, imazaquin may be soil applied in combination with pendimethalin or trifluralin plus clomozone.

In Argentina, imazaquin can be used in a postemergence tank mixture with bentazon for control of *Datura ferox* and *Amaranthus hybridus*. Bentazon provides excellent control of *Datura* spp., but is weak on *Amaranthus* spp., imazaquin is very active on *Amaranthus* spp. Rates for imazaquin are 60 to 100 g a.e./ha. The higher rate is for "rescue" treatments, when weeds are over 30 cm tall.

D. SEQUENTIAL PROGRAM

In the U.S., the label permits certain sequential imazaquin treatments, i.e., a postemergence imazaquin application following a soil-applied imazaquin application. Sequential applications can be made in Alabama, Arkansas, Florida, Georgia, Louisiana, Mississippi, North Carolina, Oklahoma, South Carolina, Tennessee, and Texas.

Imazaquin may be applied preplant incorporated or preemergence followed by a sequential postemergence application of imazaquin at 70 to 140 g a.e./ha for the control of the following weeds: *Amaranthus* spp., *Euphorbia heterophylla*, *Helianthus annuus*, *Xanthium strumarium*, and volunteer corn. The total amount of active ingredient should not exceed 210 g a.e./ha.

A sequential postemergence application of Scepter O.T. (imazaquin at 70 g a.e./ha and acifluorfen at 280 g a.e./ha) additionally controls *Ipomoea* spp. and *Sesbania exaltata*.

Imazaquin applied postemergence at 140 g a.e./ha, following a soil application of imazaquin at 140 g a.e./ha controls *Desmodium tortuosum*, *Caperonia castanaefolia*, *C. palustris*, and *Cassia obtusifolia* in addition to the weeds controlled with a single application of imazaquin.

E. ROTATIONAL CROP SAFETY
1. United States

The restrictions on rotational crops are dependent on the areas in which imazaquin herbicide has been applied. In the U.S., the areas in which imazaquin is used has been divided into two zones, the Southern use area and the Northern use area.

The Southern use area includes states along the Atlantic and Gulf coasts, those west to southeastern Nebraska, and the eastern portions of Kansas, Oklahoma, and Texas. The

TABLE 6
Rotational Crop Restrictions for Soil-Applied
Imazaquin Herbicide in the Southern Use Area (U.S.)

Crop	Time interval
Barley	11 months
Edible beans	11 months
Grain sorghum	11 months
Oats	11 months
Peanuts	11 months
Tobacco	11 months
Field corn[a]	11 months
Wheat	4 months
Rice	Spring following application
Sugar beets	26 months
All other crops	18 months

[a] Field corn may be planted in the spring (11 months) following soil application of imazaquin unless drought conditions develop or less than 37.5 cm of rainfall or irrigation is received from the date of application through October.

northern border of this area includes all of Missouri, the southern two thirds of Illinois, Indiana, and Ohio, and all of Pennsylvania and New Jersey.

The Northern use area includes the eastern portions of Nebraska and South Dakota, Iowa, Michigan (excluding the peninsula), southern Minnesota, Wisconsin, and the northern third of Illinois, Indiana, and Ohio.

a. Southern Use Area

When imazaquin or products containing imazaquin have been applied to the soil, the rotational crops listed in Table 6 may be planted at specified intervals.

For eastern Oklahoma, Arkansas, the southeast corner of Missouri, Tennessee, North Carolina and states south, rotation to field corn requires 37.5 cm of rainfall or irrigation must be received within 6 months of the date of soil application. A minimum of 25 cm of rainfall or irrigation is needed for postemergence applications if the amount of imazaquin applied does not exceed 70 g a.e./ha. A minimum of 37.5 cm of rainfall or irrigation is needed for postemergence applications exceeding 70 g a.e./ha.

b. Northern Use Area

When imazaquin or products containing imazaquin are applied to the soil, only soybeans, edible beans, grain sorghum, and tobacco can be planted 11 months after application. All other crops can be planted 18 months after application of imazaquin, except for sugar beets, which have a 26-month rotation restriction.

If a postemergence application of 70 g a.e./ha of imazaquin is made and at least 25 cm of rainfall or irrigation has been received from the date of application through October, then field corn, barley, and oats can be planted 11 months after application. Wheat can be planted 4 months after application.

2. Latin America

Common rotational crops in Latin America are soybeans, wheat, barley, oats, corn, rice, sunflowers, cotton, legumes (peanuts, dry beans, peas, lentils), and, in some areas vegetables. The application rate, weather conditions during the soybean growing season (rainfall and temperature), tillage practices, interval between application and follow-crop planting, and sensitivity of the follow crop are factors that affect the potential for carryover effects. As environmental conditions differ substantially from region to region within Latin America, so does the potential for carryover; thus, evaluation of rotational crop behavior has led to label recommendations which are specific for each country and in accordance with local field study results.

3. Other Countries

In Eastern Europe, rotational crop restrictions (the interval after the last imazaquin application) are as follows: wheat — 4 months; barley, oats, edible beans, grain sorghum, tobacco — 11 months; sunflowers — 18 months; and sugar beets and other rotational crops — 26 months. Rice may be planted in the spring of the year following the imazaquin application. Maize can be planted in the spring following imazaquin application if normal or above normal rainfall occurred after application. If drought occurred during the soybean growing season, winter cereals or legumes should be the next follow crops.

In Nigeria, rotational crop restrictions are as follows: small grain, rice — 4 months; maize, edible beans, grain sorghum, peanuts, tobacco, cotton — 11 months; and other rotational crops — 18 months.

IV. USE IN TURFGRASSES

Several warm-season turfgrasses are tolerant to imazaquin postemergence applications which control many problem weeds such as *Cyperus rotundus* and *Allium vineale*. Imazaquin has been developed for use in turfgrass in the U.S. and Japan.

A. UNITED STATES

In the U.S., imazaquin is marketed under the tradename of Image herbicide. The formulation is an aqueous solution (AS) containing 180 g a.e. of imazaquin per liter.

For postemergence applications, a minimum of 25 cm of rainfall or irrigation is needed if the amount of imazaquin applied does not exceed 70 g a.e./ha. A minimum of 37.5 cm of rainfall or irrigation is needed for postemergence applications exceeding 70 g a.e./ha. This restriction applies only to field corn.

Rotation to field corn (following soil applications or postemergence applications exceeding 70 g/ha) requires a minimum of 37.5 cm of rainfall or irrigation received from 2 weeks before the last application through November 15 of the same year. Postemergence applications at the rate of 70 g/ha require a minimum of 25 cm of rainfall during the same period.

Imazaquin can be applied at a rate of 420 to 560 g a.e./ha to the following established turfgrasses: bermudagrass (*Cynodon dactylon*), centipedegrass (*Eremochloa ophiuroides*), St. Augustinegrass (*Stenotaphrum secundatum*), and zoysiagrass (*Zoysia* spp.). These warm-season grasses show good tolerance to imazaquin applied at twice the use rate. Acceptable short-lived yellowing may occur on some varieties.

For postemergence control of summer weeds, imazaquin applications should be made following spring transition of turfgrass and prior to the onset of winter dormancy. This application timing coincides with the period of active turfgrass and summer weed growth. Imazaquin applications should not be made just prior to or during transition of turfgrass or during periods of very slow turf growth because severe discoloration may occur.

Table 7 provides a list of summer weed species controlled by imazaquin.

Imazaquin will provide good suppression of *Hydrocotyle umbellata*.

For the control of winter weeds, imazaquin application should be timed to follow the first killing frost. Table 8 provides a list of winter weeds controlled by imazaquin. Imazaquin may be applied prior to or soon after emergence of the weeds listed in Table 8, with the exception of *Allium vineale* and *A. canadense* which must be emerged when the application is made.

TABLE 7
Summer Weeds Controlled by Imazaquin in Turfgrass (U.S.)

Sedge weeds	Grass weeds
Cyperus esculentus	Cenchrus incertus
C. globosus	
C. iria	
C. rotundus	
Kyllinga brevifolia	

TABLE 8
Winter Weeds Controlled by Imazaquin in Turfgrass (U.S.)

Broadleaved weeds	Grass weed
Alchemilla arvensis	Lolium perenne (overseeded)
Allium canadense	
A. vineale	
Cardamine hirsuta	
Cerastium vulgatum	
Geranium carolinianum	
G. molle	
Lamium amplexicaule	
L. purpureum	
Medicago lupulina	
Oenothera laciniata	
Ranunculus parviflorus	
Rumex acetosella	
Scleranthus annuus	
Soliva pterosperma	
Stellaria media	
Trifolium repens	

B. JAPAN

Imazaquin is being developed for use in zoysiagrass in Japan and will be marketed under the tradename of Tone-Up. The formulation is an aqueous solution containing 200 g a.e./l.

Cyperus rotundus and *C. brevifolius* are key problem weeds controlled by imazaquin applied postemergence, spring-summer or fall-winter, at a rate of 0.6 to 0.8 kg a.e./ha. Postemergence applications have both foliar and soil activity on these sedge species. *Zoysia matrella* and *Z. japonica* show good tolerance to imazaquin treatments.

Imazaquin plus pendimethalin combination treatments are being evaluated in Japan. The addition of imazaquin widens the application timing and weed spectrum for pendimethalin, which is registered for use in zoysiagrass. Imazaquin at 0.2 kg a.e./ha plus pendimethalin at 2.0 kg a.i./ha provides effective control of annual grasses and broadleaved weeds such as *Digitaria* spp., *Poa annua, Ambrosia elatior, Chenopodium album*, and *Euphorbia supina*.

Chapter 20

IMAZETHAPYR HERBICIDE

Ronaldo Hart, Edward Lignowski, and Fred Taylor

TABLE OF CONTENTS

I. INTRODUCTION

Imazethapyr is a selective imidazolinone herbicide used to control a wide spectrum of broadleaved weeds and grasses in soybeans and several other leguminous crops. Imazethapyr is a uniquely flexible herbicide which can be applied early preplant, at planting (preplant incorporated or preemergence), or postemergence. When applied postemergence, imazethapyr provides both burndown of susceptible weeds present and residual control of later germinating weeds. The herbicide is very effective in no-till as well as conventional tillage.

In 1982, field testing of imazethapyr was begun in the U.S. in soybeans. Initial results identified imazethapyr as a unique product, not only because of its flexibility of application, but also because of the wide spectrum of weeds it controlled in the midwestern U.S. These susceptible weeds include *Abutilon theophrasti, Amaranthus* spp., *Helianthus annuus, Solanum* spp., and *Setaria* spp. Residual season-long control was recognized as an important feature for imazethapyr. In general, it was determined that optimum application rates for soil and postemergence applications are the same, and efficacy is not affected by soil type or characteristics.

Subsequent field programs were expanded to the major soybean growing countries, e.g., Brazil and Argentina in 1985. After crop selectivity studies identified several other tolerant leguminous crops, the scope of testing was further enlarged to include most of the agricultural areas in the world where such leguminous crops as peanuts (groundnuts), peas, *Phaseolus* beans, and alfalfa are grown.

II. REGULATORY STATUS

Experimental Use Permits were granted by the U.S. Environmental Protection Agency (EPA) for field testing of imazethapyr in soybeans in 1987 and 1988 under grower use conditions.

University researchers in Minnesota discovered that imazethapyr was highly effective in controlling *Helianthus tuberosus* (up to 25 cm tall). Because there were no labeled herbicides which controlled this weed in soybeans, the EPA permitted Minnesota growers to use imazethapyr under a Section 18 Emergency Exemption Clearance in 1987 and 1988 for the control of this specific weed.

In several other states, university research showed that imazethapyr could fill the void left by dinoseb, which was removed from the marketplace in 1986, for weed control in edible beans and peas. Section 18 use was granted to five states in 1987, to seven states in 1988, and to 15 states in 1989.

The first registration for imazethapyr was granted in Argentina for use in the 1987 to 1988 growing season as an early postemergence herbicide in soybeans. In 1989, imazethapyr was registered for use in soybeans in the U.S. Table 1 provides a list of countries, along with crops and use rates, where imazethapyr is registered.

III. USE IN SOYBEANS

A. CROP TOLERANCE

Soybeans exhibit excellent tolerance to imazethapyr when it is applied either to the soil or postemergence. Under stressful growing conditions (cold, wet) internode shortening may occur, resulting in shortened plants. When heavy rainfall occurs following a postemergence application in coarse or high pH soil, slight chlorosis may occur. When imazethapyr is used postemergence at rates greater than 70 g a.e./ha, occasional crinkling of the upper leaves may occur. Under favorable growing conditions, soybeans quickly outgrow these temporary symptoms and yields are not affected.

TABLE 1
Registration of Imazethapyr Herbicide in Various Countries

Country	Tradename	Formulation	Crop	Use rate (g a.e./ha)
Argentina	Pivot	100 AS[a]	Soybeans	100
			Peanuts	100
Brazil	Pivot	100 AS	Soybeans	100
Bulgaria	Pivot	100 AS	Alfalfa	100
Canada	Pursuit	240 AS	Soybeans	75—100
Chile	Pivot	100 AS	Alfalfa (for seed)	100
Czechoslovakia	Pivot	100 AS	Alfalfa	100
			Peas	100
Hungary	Pivot	100 AS	Soybeans	80
			Horsebeans	80
			Alfalfa	100—120
			Red clover	100—120
Mexico	Pivot	100 AS	Soybeans	100
PRC[b]	Pursuit	50AS	Soybeans	75—100
Paraguay	Pivot	100 AS	Soybeans	100
South Africa	Hammer	100 AS	Dry beans	30—50
			Peanuts (groundnuts)	40—50
			Soybeans	
Thailand	Pursuit	50 AS	Soybeans	100—125
U.S.	Pursuit	240 AS	Soybeans	70
			Peas	50
			Beans	35—70
Uruguay	Pivot	100 AS	Soybeans	100
Yugoslavia	Pivot	100 AS	Soybeans	80
			Alfalfa	80

[a] All formulations in this table contain the indicated amount of imazethapyr as the ammonium salt in an aqueous solution. For example, 100 AS contains 100 g of imazethapyr per l.
[b] People's Republic of China.

B. WEED SPECTRUM/APPICATION TIMING

Table 2 lists the weeds that are controlled in soybeans by soil or postemergence applications of imazethapyr at 70 g a.e./ha in the U.S.

In the U.S., imazethapyr may be applied early preplant (up to 45 d prior to planting), preplant incorporated, preemergence, or postemergence. For optimum early postemergence control of most susceptible weed species, the imazethapyr soybean label recommends that application be made when broadleaved weeds are in the one- to four-leaf stage (2.5 to 7.5 cm). For control of more sensitive weeds, such as *Xanthium strumarium* or *Amaranthus* spp., imazethapyr can be applied up to the 8-leaf stage (20 cm). The control of some weed species, such as *Chenopodium album* and *Ipomoea* spp., requires an earlier imazethapyr application at the one- to two-leaf stage (2.5 to 5.0 cm).

For optimum postemergence activity in all crops, a nonionic surfactant or crop oil concentrate must be added to the imazethapyr spray solution. In the U.S. and Canada, in addition to an adjuvant, liquid fertilizer (e.g., 10-34-0, 28-0-0, or 32-0-0) is recommended in the spray solution to increase the activity of imazethapyr on several weed species, such as *Abutilon theophrasti*, in soybeans.

In Brazil, imazethapyr controls *Euphorbia heterophylla* well and *Brachiaria plantaginea* adequately; in Argentina, the compound is active on the three main soybean weeds, *Datura ferox*, *Amaranthus quitensis*, and *Sorghum halepense*. Table 3 lists weeds controlled in soybeans in Argentina and Brazil by imazethapyr applied early postemergence at 100 g a.e./ha.

TABLE 2
Broadleaved and Grass Weeds Controlled by Soil or Postemergence
Applications of Imazethapyr at 70 g a.e./ha (U.S.)

Weeds controlled	Soil applied	Postemergence
Broadleaved Weeds		
Abutilon theophrasti	X[a]	X
Amaranthus chlorostachys	X	X
A. palmeri	X	X
A. retroflexus	X	X
A. spinosus	X	X
A. tuberculatos		X
Ambrosia elatior		X
A. trifida		X
Brassica spp.	X	X
Chenopodium album	X[a]	X
Datura stramonium	X[a]	X
Euphorbia maculata	X	X
E. prostata	X	X
Galinsoga spp.	X	
Helianthus annuus	X[a]	X
H. tuberosus		X
Ipomoea hederacea var. *integriuscula*		X
I. hederacea		X
I. lacunosa		X
Iva xanthifolia		X
Jacquemontia tamnifolia		X
Kochia scoparia		X
Mollugo verticillata	X	
Pharbitis purpurea		X
Portulaca oleracea	X	
Richardia scabra	X	
Sida spinosa	X[a]	
Solanum nigrum		X
S. ptycanthum		X
S. sarrachoides		X
Tribulus terrestris	X	
Xanthium strumarium	X[b]	X
Grasses		
Digitaria ischaemum		X
D. sanguinalis		X
Echinochloa crus-galli		X
Eriochloa villosa[c]		X
Oryza sativa		X
Setaria faberi	X[a]	X
S. lutescens	X[a]	X
S. viridis var. *major*	X[a]	X
S. viridis var. *robusta-alba*	X[a]	X
S. viridis var. *robusta-purpurea*	X[a]	X
S. viridis	X[a]	X
Sorghum bicolor		X
S. halepense	X[a]	X

[a]   More consistently controlled by preplant-incorporated treatments.
[b]   Light-to-moderate infestation only. Must be preplant incorporated for best results.
[c]   Imazethapyr controls emerged *Eriochloa villosa* only.

TABLE 3
Weeds Controlled by Imazethapyr at 100 g a.e./ha Early Postemergence (Argentina/Brazil)

Dicots	Monocots
Acanthospermum australe	Brachiaria plantaginea[a]
Alternanthera ficoidea	Cenchrus echinatus
Amaranthus quitensis	Commelina virginica
Anoda cristata	Cyperus rotundus[b]
Bidens pilosa	Digitaria sanguinalis
Chenopodium album	Echinochloa colonum
Datura ferox	Echinchloa crus-galli
Euphorbia heterophylla	Sorghum halepense (seed)
Hyptis suaveolens	S. halepense (rhizome)[b]
Nicandra physaloides	
Portulaca oleracea	
Richardia braziliensis	
Solanum sisymbriifolium	
Tagetes minuta	
Xanthium strumarium	

[a] Infestations of not more than 40 plants/m^2.
[b] Suppression only.

Extensive testing has shown imazethapyr to be most effective in Latin America as an early postemergence herbicide. The following weeds, which are well controlled by postemergence applications of imazethapyr, are not well controlled by soil applications: *Sorghum halepense, Brachiaria plantaginea, Euphorbia heterophylla, Datura ferox,* and *Ipomoea* spp. Some weed species require a very early application; *Chenopodium album, Portulaca oleracea,* and *Tagetes minuta* are not well controlled when the application is beyond the second true-leaf stage of the weed.

In Canada, imazethapyr can be applied early preplant, preemergence, or early postemergence in soybeans at 75 to 100 g a.e./ha. In addition to many weeds listed in Table 2, imazethapyr also controls *Brassica kaber* and *Polygonum convolvulus.*

In Thailand, imazethapyr applied at 100 to 125 g a.e./ha preemergence or early postemergence controls the following weed species: *Amaranthus viridis, Boerhaavia erecta, Corchorus* spp., *Chloris barbata, Cleome viscosa, Eleusine indica, Euphorbia geniculata, Dactyloctenium aegyptium, Ipomoea gracilis, Paederia* spp., *Physalis minima, Polygonum pubescens, Setaria geniculata,* and *Trianthema portulacastrum.*

In South Africa, imazethapyr can be applied preemergence or early postemergence to soybeans at a rate of 40 to 50 g a.e./ha.

In Eastern Europe and the People's Republic of China, imazethapyr can be applied preplant incorporated, preemergence, or early postemergence in soybeans. In addition to several weeds species listed in Table 2, in the People's Republic of China, imazethapyr at 75 to 100 g a.e./ha controls *Cyperus iria, Leptochloa chinensis,* and *Polygonum bungeanum.* In Eastern Europe, weeds controlled by imazethapyr include *Abutilon theophrasti, Amaranthus* spp., *Brassica* spp., *Chenopodium album, Datura stramonium, Digitaria* spp., *Echinocloa crus-galli, Helianthus annuus, Setaria* spp., *Solanum nigrum, Sorghum halepense* (from seed), and *Xanthium strumarium.*

C. COMBINATION TREATMENTS

For the control of grasses which are not susceptible to imazethapyr, herbicides such as pendimethalin, trifluralin, alachlor, and metolachlor can be tank-mixed with imazethapyr for soil applications. In the U.S., the premix formulations Pursuit Plus (imazethapyr/pendimethalin) and Passport (imazethapyr/trifluralin) are used for broad spectrum weed control. For both premix formulations, the application rate of imazethapyr is 70 g a.e./ha; the pendimethalin rate for Pursuit Plus is 925 g a.i/ha; and the trifluralin rate for Passport is 825 g a.i./ha.

When postemergence grass herbicides (e.g., fluazifop-*P*-methyl or sethoxydim) are tank-mixed with imazethapyr, a slight reduction in grass control may occur. The grass herbicide can be applied five or more days following the imazethapyr application with no reduction in activity.

Soil-applied tank mixture of imazethapyr with linuron or metribuzin can be used for excellent control of certain problem weeds, such as *Ambrosia elatior*. For conservation tillage uses, paraquat or glyphosate tank mixtures with imazethapyr provide complete burn-down of susceptible weed species before crop emergence.

IV. USE IN PEAS AND *PHASEOLUS* BEANS

A. CROP TOLERANCE

In the U.S., southern peas (cowpeas) (*Vigna sinensis*) and English peas (*Pisum sativum*) are very tolerant to soil and early postemergence applications of imazethapyr, but they are not quite as tolerant as soybeans. Lentils (*Lens culinaris*), dry peas (*P. arvense or P. sativum*), lima beans (*Phaseolus lunatus*), and several *Phaseolus vulgaris* dry bean types (navy, Great Northern, red kidney, black turtle, and cranberry) exhibit good to excellent tolerance to soil applications of imazethapyr; however these crops are not as tolerant to postemergence applications, which may cause a delay in crop maturity. The tolerance of pinto beans (*P. vulgaris*) is acceptable, although not as good as that of the other dry bean types. Of the edible bean types, snap beans (*P. vulgaris*) are the least tolerant of imazethapyr and preplant-incorporated or postemergence applications may cause unacceptable stunting. Snap beans are most tolerant to applications made after planting but before emergence.

Dry pea tolerance to soil and early postemergence treatments of imazethapyr has been good in field development programs conducted in Australia and Canada. In Australia, horsebeans (*Vicia faba*) are tolerant to imazethapyr applied preemergence.

B. WEED SPECTRUM/APPLICATION TIMING

The U.S. label rate for peas and *Phaseolus* beans ranges from 35 to 70 g a.e./ha. The use rate is dependent upon the crop, the region within the U.S., and the targeted weed species. Imazethapyr may be applied preplant incorporated, preemergence, and early post-emergence to Southern peas and English peas. Preplant-incorporated or preemergence treatments can be made to lima, navy, red kidney, Great Northern, black turtle, and cranberry beans. Imazethapyr can be applied to lentils, dry peas, and snap beans after planting but before crop emergence.

In South Africa, imazethapyr can be used preemergence only in dry beans at 30 to 50 g a.e./ha.

In Australia, weeds controlled by imazethapyr applied preemergence (50 to 75 g a.e./ha) include *Amsinckia intermedia*, *Capsella bursa-pastoris*, *Carthamus lanatus*, *Erodium* spp., *Juncus bufonius*, *Raphanus raphanistrum*, *Sisymbrium orientale*, and *Urtica incisa*.

V. USE IN ALFALFA (LUCERNE)

A. CROP TOLERANCE

Both seedling (from second trifoliate leaf stage) and established alfalfa (*Medicago sativa*) have demonstrated excellent tolerance to postemergence applications of imazethapyr. Occasionally, under cool and/or wet conditions, the alfalfa internodes may be shortened, but

TABLE 4
Weeds Controlled by Postemergence Applications of Imazethapyr at 70 to 105 g a.e./ha (U.S.)

Scientific name	Scientific name
Amsinckia spp.	*Lamium amplexicaule*
Calandrinia caulescens	*Malva pusilla*
Capsella bursa-pastoris	*Polygonum aviculare*
Chenopodium murale	*Raphanus raphanistrum*
Claytonia perfoliata	*Rumex obtusifolius*
Coronopus didymus	*Salsola kali*
Descurainia pinnata	*Sisymbrium irio*
Epilobium palustre	*Solanum* spp.
Erodium cicutarium	*Stellaria media*
E. moschatum	*Urtica dioica*

alfalfa soon outgrows the temporary symptoms. In fact, studies have shown that forage quality (total digestible nutrients) is substantially improved as a result of imazethapyr treatments. It has been observed that preplant, soil applications of imazethapyr can cause unacceptable alfalfa injury.

B. WEED SPECTRUM/APPLICATION TIMING

In the U.S., postemergence applications of imazethapyr at 70 to 105 g a.e./ha control weeds commonly found in alfalfa (Table 4) in addition to those weeds listed in Table 2. The 70 g a.e./ha rate is sufficient for many weeds. However, when weeds are larger than 7.5 cm, or species such as *Chenopodium album* are present, 105 g a.e./ha provides better control. Additional research is being conducted to expand the list of weeds controlled by imazethapyr for the initial label.

Field trials in the U.S. indicate that, in seedling alfalfa, imazethapyr should be applied when the alfalfa has at least two fully expanded trifoliate leaves and weeds are less than 7.5 cm tall. Applications to established alfalfa can be made during dormancy, after breaking dormancy, or following cuttings before growth/regrowth exceeds 3 in. (7.5 cm).

In addition to those in the U.S., programs to develop uses for imazethapyr in alfalfa (lucerne) are being conducted in Australia and Canada.

VI. USE IN PEANUTS (GROUNDNUTS)

A. CROP TOLERANCE

Peanuts are very tolerant to imazethapyr; however, heavy rainfall or irrigation following application of imazethapyr and before emergence of peanut plants may cause some stunting. Occasionally, a slight yellowing of the top leaves of treated peanuts may occur after a postemergence application of imazethapyr. These minor effects are transitory and yields are not affected.

B. WEED SPECTRUM/APPLICATION TIMING

In the U.S., imazethapyr can be applied at a rate of 70 g a.e./ha preplant incorporated, preemergence, at cracking (ground cracking at emergence of the peanut plant) or postemergence. Imazethapyr may also be applied in a sequential (split) application — preplant incorporated or preemergence followed by at cracking or postemergence (35 g a.e./ha followed by 35 g a.e./ha).

In the U.S., imazethapyr works well in peanuts because of its activity on *Cyperus esculentus* and *C. rotundus* when applied to the soil at cracking or postemergence. It is also

TABLE 5
Weeds Controlled by Soil and Postemergence Applications
of Imazethapyr in Peanuts (U.S.)

Weeds controlled	Soil/at cracking	Postemergence
Acalypha ostryaefolia	X	
A. virginica	X	
Anoda cristata	X	X
Brachiaria platyphylla	X	
Eclipta alba	X	
Euphorbia heterophylla	X	
Indigofera hirsuta	X	X

very effective on *Euphorbia heterophylla* when soil-applied. Imazethapyr controls the weeds listed in Table 5 in the peanut-growing regions of the U.S.

In South Africa, imazethapyr can be applied preemergence or early postemergence to peanuts (groundnuts) at a rate of 40 to 50 g a.e./ha. Weeds controlled include *Acanthospermum hispidum, Amaranthus deflexus, A. thunbergii, Galinsoga parviflora,* and *Schkuhria pinnata.* Imazethapyr applied preemergence at a minimum rate of 45 g a.e./ha controls the problem weed *Tribulus terrestris.*

In Latin America, the recommended application timing and rate for imazethapyr in peanuts is early postemergence at 100 g a.e./ha. The spectrum of weeds controlled is similar to that of soybeans. In Argentina, imazethapyr provides excellent control of *Anoda cristata,* one of the most important weeds in the peanut-growing area.

C. COMBINATION TREATMENTS

Imazethapyr can be used in tank mixtures with pendimethalin, trifluralin, alachlor, metolachlor, benefin, ethalfluralin, and vernolate for enhanced grass control in peanuts.

In the U.S., the premix formulation Pursuit Plus (imazethapyr/pendimethalin) provides excellent grass and broadleaved weed control in peanuts. The application rates are 70 g a.e./ha of imazethapyr and 925 g a.i./ha of pendimethalin. Imazethapyr enhances control of *Brachiaria platyphylla* and *Digitaria* spp. which may escape control by pendimethalin alone. The addition of 2,4-DB to imazethapyr, when applied at cracking or postemergence, enhances the control of weeds such as *Ipomoea* spp., *Sida spinosa, Chenopodium album,* and *Cassia tora.* The addition of paraquat to at-cracking or postemergence applications of imazethapyr improves the control of *Desmodium tortuosum* and several other broadleaved and grass weeds.

In South Africa, the premix formulation Sweep is being developed for broad spectrum weed control in groundnuts (also for soybeans and dry beans). This formulation contains 30 g a.e. of imazethapyr and 750 g a.e. of metolachlor per liter. Application rates depend on soil texture.

VII. IMIDAZOLINONE-TOLERANT CORN

A. CROP TOLERANCE

Corn is sensitive to imazethapyr and may be injured by applications made directly to the crop. A program was initiated by Cyanamid in collaboration with Molecular Genetics, Inc. (MGI) in 1982 to develop imidazolinone-tolerant corn using callus tissue techniques. By 1984, corn plants had been regenerated from tolerant cell lines and crossed with public inbread lines. The imidazolinone-tolerant corn plants were not produced using genetic-engineering techniques, e.g., gene transformation. Rather, these herbicide-tolerant plants were the result of standard selection procedures utilizing tissue culture. The only difference between this approach and classical breeding selection is that selection was done at the cellular level rather than at the whole-plant level. Cyanamid has granted a license to Pioneer Hi-Bred International, Inc. to develop imazethapyr-tolerant Pioneer corn hybrids.

In the U.S., the imazethapyr-tolerant Pioneer hybrids are expected to be introduced for sale in 1992. They will have a 2X to 4X crop safety factor for both soil and postemergence applications of imazethapyr at 70 g a.e./ha. Label registration for imazethapyr in corn is anticipated in mid-1991.

B. WEED SPECTRUM/APPLICATION TIMING

In the U.S., tests conducted with imazethapyr in corn have demonstrated the same wide spectrum of control of grass and broadleaved weeds as that shown in soybeans. Postemergence application provides control of *Sorghum bicolor, S. halepense, Panicum miliaceum,* and *Erichloa villosa,* none of which are well controlled with currently registered corn herbicides.

Imazethapyr may be used preplant incorporated, preemergence, and early postemergence at 70 g a.e./ha on tolerant corn hybirds. For postemergence applications, both a nonionic surfactant at 0.25% v/v and a fluid fertilizer such as 28-0-0 at 2.8 liters/ha should be added to the spray solution.

C. COMBINATION TREATMENTS

For increased control of grass species, label rates of alachlor, metolachlor, EPTC, or butylate can be tank-mixed with imazethapyr. Imazethapyr can be tank-mixed with atrazine for both soil and postemergence treatments.

VIII. ROTATIONAL CROP SAFETY

A. United States

Corn is quite tolerant to soil residues of imazethapyr when the herbicide is applied the previous season according to the label. Even following the drought of 1988, there were very few effects on corn planted in 1989. Under normal growing conditions, corn will tolerate soil residues up to 30 ppb with little or no effect on the plants. Soils with low binding potentials and high moisture levels following planting may result in high bioavailability of imazethapyr to corn plants. When under stress, corn exposed to soil residues of imazethapyr may exhibit temporary chlorosis and/or stunting as plant metabolism (detoxification) of imazethapyr is reduced. When grown under favorable conditions, corn is able to quickly outgrow the early symptoms. In 95% of extensive follow-crop trials conducted since 1983 in the U.S., there was no noticeable effect on corn when planted following an imazethapyr treatment the previous year. In those instances where an effect was observed, corn plants exhibited slight stunting, chlorosis and/or interveinal reddening. Lateral root growth inhibition was also noted and may result in a "bottlebrush" effect. In all cases, the symptoms were short-lived, and by the time the corn had tasseled, there were no observable effects. In no case had yield been reduced as a result of the imazethapyr treatment.

Winter wheat planted in the fall is tolerant to soil residues of imazethapyr from the previous spring application, and spring wheat is tolerant to imazethapyr residues from the previous season. Barley tolerance is similar to that of wheat, whereas oats are less tolerant than wheat. (The rotational crop restriction for oats is 18 months after imazethapyr application compared with four for wheat and barley.)

Grain sorghum, cotton, and rice are sensitive to imazethapyr soil residues. Follow-crop restrictions of 18 months are required for these crops in order for the residues to degrade to safe levels.

Most nonleguminous vegetable crops are sensitive to imazethapyr residues. This is especially true for members of the Solanaceae family and root crops. Research available to date in the U.S. has shown that the 18-month rotational crop restriction is sufficient to allow vegetable crops to be planted safely. Field research is continuing on these crops.

Sugar beets are extremely sensitive to residues of imazethapyr. Soil residue levels less than 1 ppb may cause severe injury to sugar beets. (The current chemical assay method has a validated sensitivity of 5 ppb of imazethapyr in soil.) Canola (oilseed rape) is also very sensitive to imazethapyr. The follow-crop restriction for sugar beets, canola, and potatoes is 26 months after applying imazethapyr. In addition to the 26-month interval, sugar beet growers in the U.S. are required to conduct a field bioassay with sugar beets. If no injury is observed, sugar beets may be planted the following year.

B. OTHER COUNTRIES

In Argentina and Paraguay, winter wheat, winter oats, corn, and legumes can be planted as follow crops the season after an imazethapyr application. Sunflowers are sensitive to imazethapyr residues and can only be planted the second growing season after application.

Winter cereals (wheat, barley, and oats), annual ryegrass, and legumes can be planted in Brazil in the fall after an imazethapyr application. Corn and legumes can be planted the next spring.

In Mexico, winter cereals (wheat, barley, and oats), corn, and legumes can be planted the season after an imazethapyr application.

Winter wheat, spring wheat, field corn, and spring barley can be planted as follow crops in Canada the season after an imazethapyr application. All other crops can be planted 2 years following application.

In the People's Republic of China, spring wheat and maize can be planted as follow crops the season after applying imazethapyr in soybeans. All other rotational crops can be planted 26 months after imazethapyr application.

In Thailand, transplanted rice, maize, or mungbeans can be planted the season after applying imazethapyr in soybeans. All other follow crops can be planted 12 months after imazethapyr application.

In South Africa, maize and any legume crop can be planted 10 months; wheat can be planted 16 months; and all other crops can be planted 24 months after applying imazethapyr.

In Eastern Europe, winter wheat, wheat barley, maize, and legumes can be planted the season following imazethapyr application. Sunflowers, sorghum, rice, and spring cereals can be planted 18 months; and all other crops (including sugar beets, fodder beets, rape, and vegetables) can be planted the third year after imazethapyr application.

Chapter 21

IMIDAZOLINONES AS PLANT GROWTH REGULATORS

Prithvi Bhalla and S.M. Shehata

TABLE OF CONTENTS

I. INTRODUCTION

Plant growth regulators (PGRs) represent one of the smallest sectors of the agricultural chemicals market. PGRs are compounds that regulate the growth and metabolism of plants in a beneficial manner and include a diverse series of chemical structures which may be synthetic or naturally occurring. The imidazolinones are being extensively studied to determine their PGR characteristics. The imidazolinones augment numerous morphological and physiological PGR-like responses, some of which are axillary branching, tillering, flowering, and yield of some agronomic and horticultural crops. While the imidazolinones can cause a large number of changes in the plant, the role of these compounds as PGRs in agriculture is presently limited to use in wheat and turfgrass. These applications will be discussed here. Research in other areas continues, but the results are preliminary and are not included in this chapter.

II. IMAZAQUIN AND ITS INTERACTION WITH CHLORMEQUAT IN WHEAT

Chlormequat or CCC, 2-chloroethyltrimethylammonium chloride, (chlorocholine chloride), also known as Cycocel, is acknowledged as the most widely used plant growth regulant in agriculture. The PGR properties of chlormequat were first described in 1959. Chlormequat appears to affect plant growth by interfering with the biosynthesis of gibberellins, which leads to a prevention of cell elongation. The product has been registered for use in a range of cereals and several other field, horticultural, and ornamental crops in many countries since the mid-1960s.

In Europe, the prevention of lodging in wheat is the most significant agricultural use of chlormequat. The main responses of wheat to treatment with chlormequat are reduced stem height and shortened and thickened lower internodes, which thus reduce lodging caused by adverse climatic conditions. Treated wheat also remains dwarfed under optimal fertilization. Chlormequat also often causes grain yield increases in some wheat varieties even in the absence of lodging.

In 1987, a coformulation of chlormequat and imazaquin was registered in France as Cycocel CL. Imazaquin, used at very low rates, exhibits PGR activity, particularly when combined with chlormequat. Chlormequat plus imazaquin was registered as GR-90 in Belgium in 1988. Both products are used for control of lodging in winter wheat.

Cycocel CL is applied from the beginning of tillering to the first node stage. The product shortens the internodes as well as thickens the base of the stem, both of which effects provide excellent lodging control. Under optimum growing conditions, an increase in tillering and an improvement of tiller synchrony have been recorded on treated winter wheat. This results in a larger number of uniform-sized ears and consequently an increase in yield. These effects on tillers are even more prominent when Cycocel CL is applied during the tillering stage. In trials where lodging has been severe, yield increases have always been recorded.

The role of imazaquin in the absorption, translocation, and distribution of chlormequat chloride in Cycocel CL has been investigated using ^{14}C-labeled imazaquin foliar applications to winter wheat. Imazaquin was shown to increase the mobility and the pattern of distribution of chlormequat chloride in the plant.

III. IMAZAQUIN AS A PGR IN WARM-SEASON TURFGRASSES

Imazaquin slows grass growth and reduces the frequency for mowing when applied during the post-greenup growth stage. The rate of grass growth is slowed enough to reduce the need for mowing from weekly to 3- to 4-week intervals after application. The length of time between mowing depends on various lawn care practices and environmental influences. Imazaquin is effective on *Stenotaphrum secundatum, Cynodon dactylon,* and *Zoysia* sp., but tends to injure *Paspalum notatum, Dichondra repens,* and *Festuca arundinacea.*

IV. IMAZETHAPYR AND IMAZAPYR AS "CHEMICAL MOWERS" IN COOL-SEASON TURFGRASSES

In the U.S., field research has demonstrated that Event, which is a combination of imazethapyr and imazapyr, suppresses seedhead formation, seedhead height, and vegetative height of several cool-season grasses, such as *Festuca arundinacea, Festuca ovina, Festuca* var. *commutata,* and *Lolium perenne.* In the U.S., Event is formulated as an aqueous solution containing 169 g a.i./l of imazethapyr and 6.2 g a.i./l of imazapyr. The recommended use rate is 0.58 to 0.73 l of formulation per hectare applied after grasses have reached at least 5 cm of vertical growth and are actively growing. A nonionic surfactant can be added. Reductions in grass height have been reported for the following cultivars: *Festuca arundinacea* var. Falcon and var. Jaguar and *Lolium perenne* var. Cowboy and var. Palmer. *Poa pratensis* did not respond adequately to Event. One application of Event can eliminate two to three mowings by reducing the foliar growth of established *F. arundinacea, Lolium perenne, Poa pratensis,* and *Paspalum notatum.* Event also controls seedhead production in *F. arundinacea* and *L. perenne* and controls or suppresses seedhead production in *P. notatum* for 60 to 90 d. Event is ideal for use in limited-care areas, such as roadsides, airports, and fairgrounds, and in limited-wear areas such as industrial, institutional, and cemetery grounds.

V. FORAGE CROPS AND PASTURE GRASSES

Field work has been conducted on the feasibility of using imidazolinones, especially imazethapyr, in forage management. The objective of such research is to regulate the growth of various components of pasture vegetation to prolong the vegetative cycle resulting in increased nutritional value of forage for grazing, hay, and silage. In field trials, imazethapyr increased *in vitro* dry matter digestibility and reduced yield at the first cutting of *Phleum pratense,* and *Lolium perenne* and *L. multiforum.*[2] Total seasonal yield was similar to that of the control. When the first cutting was deferred, imazethapyr treatments minimized quality declines in contrast with those seen in untreated plots. Experiments with a *Dactylis glomerata* and *Medicago sativa* mixture demonstrated the potential for use of imazethapyr in grass/legume pastures.[3]

REFERENCES

1. **Guckert, A., Blouet, A., and Arissian, M.,** Du mode d'action d'une association a base d'imazaquine sur blé tender d'hiver, *Phytoma,* No. 407, 15, 1989.
2. **Fales, S. L.,** Chemical regulation of growth and nutritive values of forage grasses, Proc. 12th Gen. Meet. Eur. Grassland Fed., Dublin, Ireland, 1988.
3. **Fales, S. L.,** Chemical regulation of growth and quality of orchardgrass/alfalfa mixtures, Proc. XVI Int. Grassland Congr., Nice, France, 1989.

APPENDIX A
Product Descriptions

Imazapyr

Chemical names	2-[4,5-dihydro-4-methyl-4-(1-methylethyl)-5-oxo-1*H*-imidazol-2-yl]-3-pyri-dinecarboxylic acid (CA)
	2-(4-isopropyl-4-methyl-5-oxo-2-imidazolin-2-yl)nicotinic acid (IUPAC)
Trademarks	Formulations
	ARSENAL, ASSAULT, CHOPPER, CHOPPER RTU, CONTAIN
	Coformulations
	ARSENAL XL, ARSENAL CDA (imazapyr/atrazine);
	MARBLE, MARBRE (imazapyr/diuron);
	MONARCH (imazapyr/atrazine)
Other designations	AC252,925; CL 252,925; CL 243,997
CAS Registry No.	81334-34-1
EPA Registration No.	241-286 (technical)
	241-273 (formulation)
Molecular formula	$C_{13}H_{15}N_3O_3$
WLN	T6NJ CVQ B-1 ET5MV DN CHJ CY = 1 & 1 Cl && (RS)
Molecular weight	261.3
Physical state and color	White to tan powder
Odor	Slight acetic-acid odor
Melting point	169 to 173°C
Vapor pressure	$<1 \times 10^7$ mmHg at 60°C
pK_a	pK_1 = 1.9; pK_2 = 3.6
Partition coefficient (Kow) (*n*-octanol/water)	1.3 at 22°C (not corrected for degree of dissociation)
pH	3.0 to 3.5 as a 1% solution in water at 25°C
Solubility	Distilled water; 9740 ppm at 15°C; 11,272 ppm at 25°C; 13,470 ppm at 35°C
	Other solvents (g/100 ml)
	Acetone 3.39
	Dimethyl sulfoxide 47.1
	Hexane 0.00095
	Methanol 10.5
	Methylene chloride 8.72
	Toluene 0.180

Imazamethabenz-Methyl

Chemical names	Methyl 2-[4,5-dihydro-4-methyl-4-(1-methylethyl)-5-oxo-1*H*-imidazol-2-yl]-4-methylbenzoate and methyl 2-[4,5-dihydro-4-methyl-4-(1-methylethyl)-5-oxo-1*H*-imidazol-2-yl]-5-methylbenzoate (60:40) (CA — 9th CI)
	Methyl 2-(4-isopropyl-4-methyl-5-oxo-2-imidazolin-2-yl)-*p*-toluate mixed with methyl 6-(4-isopropyl-4-methyl-5-oxo-2-imidazolin-2-yl)-*m*-toluate (3:2) (IUPAC)
Trademarks	Formulations
	ASSERT, DAGGER
	Coformulations
	MEGAPLUS, CHACAL, ASSERT Combi, SAIFLOS (imazamethabenz-methyl/pendimethalin)
	MEGANET, IMIVENGE (imazamethabenz-methyl/difenzoquat)
	ASSERT M, SERTOX, IMIFEN (imazamethabenz-methyl/mecoprop)
	ASSERT PLUS, PINNACLE, SWELL (imazamethabenz-methyl/isoproturon)
	STALK (imazamethabenz-methyl/trifluralin)
	IMITOL (imazamethabenz-methyl/chlortoluron)
Other designations	AC 222,293; Cl 222,293
CAS Registry No.	81405-85-8
EPA Registration No.	241-290 (technical)
	241-285 (formulation)

APPENDIX A (continued)
Product Descriptions

Imazamethabenz-Methyl

Molecular formula	$C_{16}H_{20}N_2O_3$
WLN	*p*-isomer: T5MV DN CHJ CY1&1 C1 ER C1 FVO1
	m-isomer: T5MV DN CHJ CY1&1 C1 ER D1 BVO1
Molecular weight	288.35
Physical state and color	Off-white fine powder with a tendency to form easily friable aggregates
Odor	Slight, musty odor
Melting point	Softening begins at 108 to 117°C; melting starts at 113 to 122°C and is completed at 144 to 153°C
Vapor pressure	1.13×10^{-8} mmHg at 25°C
pK_a	2.9 at 23.5°C ± 1°C
Partition coefficient (*n*-octanol/water)	At 25°C
	p-isomer: 35
	m-isomer: 66
pH	6.05 in water/dioxane solution
Solubility	At 25°C (g/100 ml solvent)

Acetone	23.0	Isopropyl alcohol	18.3
Dimethyl sulfoxide	21.6	Methanol	30.9
n-Heptane	0.06	Methylene chloride	17.2
Toluene	4.5		

At 25°C (ppm)

Distilled water
p-isomer   857
m-isomer   1,370

Imazethapyr

Chemical names	(±)-2-[4,5-dihydro-4-methyl-4-(1-methylethyl)-5-oxo-1*H*-imidazol-2-yl]-5-ethyl-3-pyridinecarboxylic acid (CA)
	(±)-5-ethyl-2-(4-isopropyl-4-methyl-5-oxo-2-imidazolin-2-yl)nicotinic acid (IUPAC)
Trademarks	Formulations
	PURSUIT; PIVOT; HAMMER
	Coformulations
	EVENT (imazethapyr/imazapyr)
	PASSPORT (imazethapyr/trifluralin)
	PURSUIT PLUS (imazethapyr/pendimethalin);
	SWEEP (imazethapyr/metolachlor)
Other designations	AC 263,499; CL 263,499
CAS Registry No.	81 335-77-5
EPA Registration No.	241-309 (technical)
	241-310 (formulation)
Molecular formula	$C_{15}H_{19}N_3O_3$
WLN	T6NJ CVQ E2 B–ET5MV DN CHJ CY1&1 C1 && (RS)
Molecular weight	289.34
Physical state and color	Off-white to tan solid
Odor	Slightly pungent
Melting point	169 to 173°C
Vapor pressure	$<1 \times 10^{-7}$ mmHg at 60°C
pK_a	pK_1 = 2.1, pK_2 = 3.9
Partition coefficient (*n*-octanol/water)	At 25°C
	11 at pH 5
	31 at pH 7
	16 at pH 9
pH	3.0 as a 0.14% solution in water at 21°C

APPENDIX A (continued)
Product Descriptions

Imazethapyr

Solubility
At 25°C (g/100 ml solvent)

Acetone	4.82
Dimethyl sulfoxide	42.25
Heptane	0.09
Methanol	10.50
Methylene chloride	18.48
2-Propanol	1.73
Toluene	0.50
Water (distilled)	0.14

Imazaquin

Chemical names
2-[4,5-dihydro-4-methyl-4-(1-methylethyl)-5-oxo-1H-imidazol-2-yl]-3-quinoline-carboxylic acid (CA)
2-(4-isopropyl-4-methyl-5-oxo-2-imidazolin-2-yl)-3-quinolinecarboxylic acid (IUPAC)

Trademarks
Formulations
SCEPTER; IMAGE
Coformulations
SQUADRON (imazaquin/pendimethalin)
TRI-SCEPT (imazaquin/trifluralin)
GR90, CYCOCEL CL (imazaquin/chlormequat)

Other designations
AC 252,214; CL 252,214

CAS Registry No.
81335-37-7

EPA Registration No.
241-287 (technical)
241-289 (formulation)

Molecular formula
$C_{17}H_{17}N_3O_3$

WLN
T66 BNJ DVQ C– ET5MV DN CHJ CY1&1 C1 && (±)

Molecular weight
311.3

Physical state and color
Tan solid

Odor
Odorless

Melting point
219 to 224°C

Vapor pressure
$<1 \times 10^{-7}$ mmHg at 60°C
$<2 \times 10^{-8}$ mmHg at 45°C

pK_a
$pK_2 = 3.8$

Partition coefficient (n-octanol/water)
2.2 at 22°C (not corrected for degree of dissociation)

pH
3.8 in 1% solution in water at 23°C

Solubility
Water: 60 ppm at 25°C
Dimethyl sulfoxide: 15.9 g/100 ml
Dimethylformamide: 6.8 g/100 ml
Toluene: 0.04 g/100 ml
Methylene chloride: 1.4 g/100 ml

APPENDIX B
List of Scientific and Common Names of Species

The following is a list of species used throughout *Imidazolinone Herbicides*. The species have been arranged by scientific name in alphabetical order with the common name(s) appearing to the right, if available.

Species	Common name
Abutilon theophrasti	Velvetleaf
Acalypha ostryaefolia	Hophornbean copperleaf
Acalypha virginica	Virginia copperleaf
Acanthospermum australe	Prostrate starbur
Acanthospermum hispidum	Bristly starbur
Acer macrophyllum	Bigleaf maple
Acer negundo	Boxelder
Acer rubrum	Red maple
Adonis aestivalis	Summer adonis
Ageratum conyzoides	Tropic ageratum
Agropyron repens	Quackgrass
Alchemilla arvensis	Parsley-piert
Alchemilla vulgaris	Lady's mantle
Alhagi pseudalhagi	Camelthorn
Allium canadense	Wild onions
Allium vineale	Wild garlic
Alnus rubra	Red alder
Alopecurus myosuroides	Blackgrass
Alternanthera ficoidea	Perrotleaf; alligatorweed
Amaranthus chlorostachys	Smooth pigweed
Amaranthus deflexus	Spreading amaranth
Amaranthus hybridus	Smooth pigweed
Amaranthus lividus	Livid amaranth
Amaranthus palmeri	Palmer amaranth
Amaranthus quitensis	
Amaranthus retroflexus	Redroot pigweed
Amaranthus spinosus	Spiny amaranth
Amaranthus thunbergii	Red pigweed
Amaranthus tuberculatos	Tall waterhemp
Amaranthus viridis	Slender amaranth
Ambrosia artemisiifolia	Common ragweed
Ambrosia elatior	Hogweed
Ambrosia psilostachya	Perennial ragweed
Ambrosia trifida	Giant ragweed
Ammi majus	Greater ammi
Amsinckia intermedia	Coast fiddleneck
Anagallis arvensis	Red chickweed
Andropogon bicornis	West Indian foxtailgrass
Andropogon gerardii	Big bluestem
Anoda cristata	Spurred anoda
Anthemis cotula	Mayweed chamomile
Apera spica-venti	Loose silky-bent
Aphanes arvensis	Break-stone
Arabidopsis thaliana	Mouseearcress
Aristida oligantha	Prairie threeawn
Arrhenatherum elatius	Onion couch
Asystasia intrusa	
Avena fatua	Wild oat
Avena ludoviciana	Winter wild oat
Baccharis dracunculifolia	
Bidens pilosa	Hairy beggarticks
Boerhaavia erecta	Erect spiderling

APPENDIX B (continued)
List of Scientific and Common Names of Species

Species	Common name
Borreria alata	Broadleaf buttonweed
Borreria capitala	Poya
Borreria laevis	Buttonplant
Borreria latifolia	Broadleaf buttonweed
Bowlesia incana	Hairy bowlesia
Brachiaria mutica	Paragrass
Brachiaria plantaginea	Alexandergrass
Brachiaria platyphylla	Broadleaf signalgrass
Brachiaria purpurascens	Paragrass
Brassica campestris	Wild turnip
Brassica juncea	Indian mustard
Brassica kaber (see also	Wild mustard
Sinapis arvensis)	
Brassica napus	Rape
Brassica nigra	Black mustard
Brassica tournefortii	Mediterranean mustard
Bromus inermis	Smooth brome
Bromus secalinus	Cheat
Bromus tectorum	Downy brome
Brunnichia cirrhosa	Redvine
Calandrinia caulescens	Redmaids rockpurslane
Calystegia sepium	Hedge bindweed
Campsis radicans	Trumpetcreeper
Caperonia castanaefolia	Mexicanweed
Caperonia palustris	Texasweed
Capsella bursa-pastoris	Shepherd's purse
Cardamine hirsuta	Hairy bittercress
Carduus nutans	Musk thistle
Carthamus lanatus	Woolly safflower
Cassia obtusifolia	Sicklepod
Cassia tora	Sicklepod
Cenchrus echinatus	Southern sandbur
Cenchrus incertus	Field sandbur
Centaurea diffusa	Diffuse knapweed
Centaurea repens	Russian knapweed
Centaurea solstitialis	Yellow starthistle
Cerastium arvense	Field chickweed
Cerastium vulgatum	Mouseear chickweed
Chenopodium album	Common lambsquarter
Chenopodium murale	Nettleleaf goosefoot
Chloris barbata	Swollen fingergrass
Chondrilla juncea	Rush skeletonweed
Chromolaena odorata	Bitterbush
Chrysanthemum leucanthemum	Oxyeye daisy
Chrysothamnus nauseosus	Grey rabbittbrush
Cirsium arvense	Canada thistle
Cirsium texanum	
Cirsium vulgare	Bull thistle
Claytonia perfoliata	Miners lettuce
Cleome rutidosperma	Yellow cleome
Cleome viscosa	Tickweed
Clidemia hirta	Curse
Clitoria ternatea	Conch-flower creeper
Colocasia esculenta	Chinese potato
Commelina benghalensis	Tropical spiderwort
Commelina communis	Common dayflower

APPENDIX B (continued)
List of Scientific and Common Names of Species

Species	Common name
Commelina diffusa	Spreading dayflower
Commelina virginica	Virginian dayflower
Convolvulus arvensis	Field bindweed
Convolvulus sepium, see *Calystegia sepium*	
Conyza bonariensis	Hairy fleabane
Conyza canadensis	Canada horseweed
Coronopus didymus	Swinecress
Corylus avelana	European filbert
Cynodon dactylon	Bermudagrass
Cyperus brevifolius	Green kyllinga
Cyperus esculentus	Yellow nutsedge
Cyperus ferax	Flatsedge
Cyperus globosus	
Cyperus iria	Rice flatsedge
Cyperus rotundus	Purple nutsedge
Dactylis glomerata	Orchardgrass
Dactyloctenium aegyptium	Crowfootgrass
Datura ferox	Large thornapple
Datura stramonium	Jimsonweed
Daucus carota	Wild carrot
Descurainia pinnata	Tansymustard
Descurainia sophia	Flixweed
Desmodium tortuosum	Florida beggarweed
Dichondra repens	Kidneyweed
Digitaria adscendens	Henry crabgrass
Digitaria horizontalis	Jamaica crabgrass
Digitaria ischaemum	Smooth crabgrass
Digitaria sanguinalis	Large crabgrass
Diolea lasiocarpa	
Diospyros virginiana	Common persimmon
Distichlis stricta	Desert saltgrass
Echinochloa colonum	Junglerice
Echinochloa crus-galli	Common barnyardgrass
Eclipta alba	Eclipta
Eichhornia crassipes	Waterhyacinth
Elaeagnus angustifolia	Russian olive
Eleusine indica	Goosegrass
Elymus repens	Quackgrass
Emex australis	Emex
Emilia sonchifolia	Red tassleflower
Epilobium palustre	Marsh willowherb
Eragrostis spp.	Lovegrass
Eremochloa ophiuroides	Centipedegrass
Eriochloa villosa	Wooly cupgrass
Erodium cicutarium	Redstem filaree
Erodium moschatum	Whitestem filaree
Eupatorium capillifolium	Dogfennel
Eupatorium odoratum	Bitterbush
Euphorbia dentata	Toothed spurge
Euphorbia geniculata	Wild poinsettia
Euphorbia heterophylla	Wild poinsettia
Euphorbia maculata	Prostrate spurge
Euphorbia peplus	Petty spurge
Euphorbia pilulifera	Garden spurge
Euphorbia prostrata	Prostrate spurge
Fagus grandifolia	American beech

APPENDIX B (continued)
List of Scientific and Common Names of Species

Species	Common name
Fagopyrum tataricum	Tartary buckwheat
Fallopia convolvulus	Wild buckwheat
Festuca arundinacea	Tall fescue
Festuca commutata	
Festuca duriuscula also *Festuca rubra*	Red fescue
Festuca ovina	Sheep's fescue
Franseria tomentosa	Skeletonleaf bursage
Fumaria officinalis	Common fumitory
Galinsoga parviflora	Smallflower galinsoga
Galium aparine	Catchweed bedstraw
Geranium carolinianum	Carolina geranium
Geranium molle	Dovefoot geranium
Gutierrezia sarothrae	Broom snakeweed
Helianthus annuus	Common sunflower
Helianthus tuberosus	Jerusalem artichoke
Heterotheca subaxillaris	Camphorweed
Hibiscus trionum	Venice mallow
Hordeum leporinum	Wild barley
Hordeum pusillum	Little barley
Hydrocotyle umbellata	Water pennywort
Hyparrhenia rufa	Jaraguagrass
Hyptis suaveolens	Wild spikenard
Ibicella lutea	Devil's claw
Imperata brasiliensis	Brazilian cogongrass
Imperata cylindrica	Cogongrass
Indigofera hirsuta	Hairy indigo
Ipomoea gracilis	
Ipomoea hardwickii	
Ipomoea hederacea	Entireleaf morningglory
Ipomoea lacunosa	White morningglory
Ipomoea nil	Japanese morningglory
Ipomoea purpurea	Tall morningglory
Ipomoea wrightii	Palmleaf morningglory
Ischaemum muticum	Bamboograss
Iva xanthifolia	Common marshelder
Ixophorus unisetus	Flatstem
Jacquemontia sandwicensis	Beach jacquemontia
Jacquemontia tamnifolia	Smallflower morningglory
Juncas bufonius	Toad rush
Kochia scoparia	Kochia
Kyllinga brevifolia	Green kyllinga
Lamium amplexicaule	Henbit
Lamium purpureum	Red deadnettle
Lantana camara	Tickberry
Lapsana communis	Nipplewort
Legousia speculum-veneris	Greater Venus's-lookingglass
Lemna minor	Common duckweed
Lepidium densiflorum	Greenflower pepperweed
Leptochloa chinensis	Chinese sprangletop
Leptochloa uninervia	Mexican sprangletop
Ligustrum vulgare	Common privet
Liquidambar styraciflua	Sweetgum tree
Liriodendron tulipifera	Yellow poplar
Lithocarpus densiflorus	Tanoak
Lolium multiflorum	Italian ryegrass
Lolium perenne	Perennial ryegrass

APPENDIX B (continued)
List of Scientific and Common Names of Species

Species	Common name
Lythrum salicaria	Purple loosestrife
Malva parviflora	Little mallow
Malva pusilla	Dwarf mallow
Matricaria chamomilla	Wild chamomile
Medicago lupulina	Black medick
Melampodium divaricatum	Blackfoot
Melastoma malabathricum	Banks melastoma
Melia azedarach	Chinaberry tree
Melochia corchorifolia	Redweed
Mikania cordata	Mile-a-minute
Mikania micrantha	Hempweed
Milium scabrum	Early millet
Mollugo verticillata	Carpetweed
Montia perfoliata	Miner's lettuce
Mucuna pruriens	Velvetbean
Muhlenbergia frondosa	Wirestem muhly
Nephrolepis biserrata	Fishtail fern
Nicandra physaloides	Apple-of-Peru
Nyssa sylvatica	Blackgum
Oenothera kunthiana	Primrose
Oenothera laciniata	Cutleaf evening primrose
Oryza sativa	Rice
Ottochloa nodosa	Slender panicgrass
Oxalis stricta	Common yellow woodsorrel
Oxydendrum arboreum	Sourwood
Panicum capillare	Witchgrass
Panicum dichtomiflorum	Fall panicum
Panicum fasciculatum	Browntop panicum
Panicum maximum	Guineagrass
Panicum muticum see *Brachiaria mutica*	
Panicum repens	Torpedograss
Panicum texanum	Coloradograss
Papaver hybridium	Round-headed poppy
Papaver rhoeas	Corn poppy
Parthenium hysterophorus	Ragweed parthenium
Parthenocissus quinquefolia	Virginia creeper
Paspalum commersonii	Bull paspalum
Paspalum conjugatum	Sour paspalum
Paspalum dilatatum	Dallisgrass
Paspalum fasciculatum	Bullgrass
Paspalum notatum	Bahiagrass
Paspalum urvillei	Vaseygrass
Paspalum virgatum	Razorgrass
Pastinaca sativa	Wild parsnip
Pennisetum clandestinum	Kikuyugrass
Pennisetum villosum	Feathertop
Phalaris arundinacea	Reed canarygrass
Pharbitis purpurea	Tall morningglory
Phleum pratense	Timothy
Phragmites australis	Common reed
Phyllanthus niruri	Niruri
Physalis angulata	Wild gooseberry
Physalis minima	Sunberry
Phytolacca americana	Common pokeweed
Plantago lanceolata	Buckhorn plantain
Plantago major	Broadleaf plantain

APPENDIX B (continued)
List of Scientific and Common Names of Species

Species	Common name
Pluchea sericea	Arrowwood
Poa annua	Annual bluegrass
Poa compressa	Canada bluegrass
Poa pratensis	Kentucky bluegrass
Poa trivialis	Roughstalk bluegrass
Polygonum aviculare	Prostrate knotweed
Polygonum bungeanum	Prickly smartweed
Polygonum convolvulus	Wild buckwheat
Polygonum cuspidatum	Japanese bamboo
Polygonum pensylvanicum	Pennsylvania smartweed
Polygonum persicaria	Lady's-thumb
Polygonum pubescens	
Populus tremuloides	Quaking aspen
Portulaca oleracea	Common purslane
Pothomorphe umbellata also *Pothomorphe peltata*	Largeleaf fingertree
Prunus padus	Hagberry tree
Prunus serotina	Black cherrytree
Ptilimnium capillaceum	Mock bishopweed
Pueraria lobata	Kudzu
Pueraria phaseoloides	Tropical kudzu
Quercus alba	White oak
Quercus laevis	Turkey oak
Quercus laurifolia	Laurel oak
Quercus nigra	Water oak
Quercus robur	English oak
Quercus virginiana	Live oak
Ranunculus arvensis	Corn crowfoot
Ranunculus parviflorus	Small-flowered buttercup
Raphanus raphanistrum	Wild radish
Raphanus sativus	Garden radish
Rhus radicans	Poison ivy
Richardia brasiliensis	Brazil callalily
Richardia scabra	Florida pusley
Rosa arkansana	Arkansas rose
Rosa bracteata	Macartney rose
Rosa multiflora	Multiflora rose
Rottboellia exaltata	Itchgrass
Rumex acetosella	Red sorrel
Rumex crispus	Curly dock
Rumex obtusifolius	Broadleaf dock
Salix caprea	Common swallow
Salsola kali	Prickly saltwort
Sapium sebiferum	Chinese tallowtree
Sassafras albidum	Common sassafras
Schkuhria pinnata	Dwarf marigold
Scleranthus annuus	German knotweed
Scoparia dulcis	Sweet broomweed
Senecio jacobaea	Tansy ragwort
Senecio vulgaris	Common groundsel
Sesbania exaltata	Coffeebean
Setaria faberi	Giant foxtail
Setaria geniculata	Knotroot foxtail
Setaria lutescens	Yellow foxtail
Setaria viridis	Green foxtail
Sicyos angulatus	Burcucumber
Sida rhombifolia	Arrowleaf sida

APPENDIX B (continued)
List of Scientific and Common Names of Species

Species	Common name
Sida spinsoa	Prickly sida
Sinapsis arvensis	Wild mustard
Sisymbrium irio	London rocket
Sisymbrium orientale	Eastern rocket
Solanum elaeagnifolium	Silverleaf nightshade
Solanum nigrum	Black nightshade
Solanum ptycanthum	Eastern black nightshade
Solanum sarrachoides	Hairy nightshade
Solanum sisymbriifolium	Sticky nightshade
Solidago canadensis	Canadian goldenrod
Solidago microglossa	
Soliva pterosperma	Carpet burweed
Sonchus arvensis	Field sowthistle
Sorbus aucuparia	Witchwood
Sorghum arundinaceum	Common wild sorghum
Sorghum bicolor	Shattercane
Sorghum halepense	Johnsongrass
Spartina patens	Prairie cordgrass
Spartina pectinata	Corn spurry
Spergula arvensis	Field sowthistle
Sporobulus cryptandrus	Sand dropseed
Stachys arvensis	Fieldnettle betony
Stellaria media	Common chickweed
Stenochlaena palustris	Climbing fern
Stenotaphrum secundatum	St. Augustinegrass
Tagetes minuta	Wild marigold
Talinum triangulare	Pink purslane
Tamarix ramosissima	Salt cedar
Taraxacum officinale	Common dandelion
Taxodium distichum	Bald cypress
Thlaspi arvense	Fanweed
Tithonia tubaeformis also *Tithonia diversifolia*	Tree marigold
Trianthema portulacastrum	Horse purslane
Tribulus terrestris	Common caltrop
Trifolium repens	White clover
Triodanis perfoliata	Clasping bellwort
Typha domingensis	Southern cattail
Urtica dioica	Stinging nettle
Urtica incisa	
Verbena stricta	Hoary vervain
Vebesina encelioides	Golden crownbeard
Veronica hederaefolia	Ivyleaf speedwell
Vicia sativa	Common vetch
Vigna sinensis	Cowpea
Viola arvensis	Field violet
Viola tricolor	Wild violet
Wedelia glauca	Pascalia
Wedelia paludosa	
Xanthium spinosum	Spiny cocklebur
Xanthium strumarium	Common cocklebur
Zoysia matrella	Manilagrass
Zoysia japonica	Japanese lawngrass

GENERAL BIBLIOGRAPHY

IMAZAQUIN

Adcock, T. E., Banks, P. A., and Bridges, D. C., Effects of preemergence herbicides on soybean (*Glycine max*) weed competition, *Weed Sci.*, 38, 108, 1990.

Akinyemiju, O. A., Evbuomwan, F. O., and Akobundu, I. O., Imazaquin residue and persistence in a humid tropical soil, in *Br. Crop Prot. Conf. — Weeds*, Vol. 1, British Crop Protection Council Publications, Farnham, England, 1989, 251.

Anderson, P. C. and Georgeson, M., Herbicide-tolerant mutants of corn. *Genome*, 31, 944, 1989.

Barnes, C. J., Goetz, A. J., and Lavy, T. L., Effects of imazaquin residues on cotton (*Gossypium hirsutum*) *Weed Sci.*, 37, 820, 1989.

Barrentine, W. L., Minimum effective rate of chlorimuron and imazaquin applied to common cocklebur (*Xanthium strumarium*), *Weed Technol.*, 3, 126, 1989.

Barrett, M., Reduction of imazaquin injury to corn (*Zea mays*) and sorghum (*Sorghum bicolor*) with antidotes, *Weed Sci.*, 37, 34, 1989.

Barros, A. C. de, Controle de plantas daninhas, dicotiledoneas, atraves de herbicidas preemergentes na cultura da soja, Communicado Tecnico — *Empresa Goiana de Pesqui. Agropecu.*, No. 16, 1989.

Basham, G. W. and Lavy, T. L., Microbial and photolytic dissipation of imazaquin in soil, *Weed Sci.*, 35, 865, 1987.

Basham, G., Lavy, T. L., Oliver, L. R., and Scott, H. D., Imazaquin persistence and mobility in three Arkansas soils, *Weed Sci.*, 35, 576, 1987.

Birk, J. H., The Mechanism of Imazaquin Activity in Sunflower (*Helianthus annuus* L.), Ph.D. thesis, Cornell University, Ithaca, 1986.

Bodson, B., Haquenne, W., and Maddens, K., Influence of application of growth regulators on the quality of winter wheat, *Meded. Fac. Landbouwwet. Rijksuniv. Gent*, 54, 409, 1989.

Brinen, J. S., Los, M., Kelland, D., and Wallach, E. R., Direct observation of pesticides on plants using LAMMS: imazaquin on soybean leaves, *Surf. Interface Anal.*, 11, 559, 1988.

Buhler, D. D. and Werling, V. L., Weed control from imazaquin and metolachlor in no-till soybeans (*Glycine max*), *Weed Sci.*, 37, 392, 1989.

Callaway, B. B., Phatak, S. C., and Wells, H. D., Interactions of *Puccinia canaliculata* (Schw.) Lagerh. with herbicides on tuber production and growth of *Cyperus esculentus* L., *Trop. Pest Manage.*, 33, 22, 1987.

Cantwell, J. R., Liebl, R. A., and Slife, F. W., Biodegradation characteristics of imazaquin and imazethapyr, *Weed Sci.*, 37, 815, 1989.

Coats, G. E., Munoz, R. F., Anderson, D. H., Heering, D. C., and Scruggs, J. W., Purple nutsedge (*Cyperus rotundus*) control with imazaquin in warm-season turfgrasses, *Weed Sci.*, 35, 691, 1987.

Congleton, W. F., VanCantfort, A. M., and Lignowski, E. M., Imazaquin (SCEPTER): a new soybean herbicide, *Weed Technol.*, 1, 186, 1987.

Croon, K. A., Effects of Bentazon, Imazaquin, and Chlorimuron on Haloxyfop and Fluazifop-P Efficacy, Ph.D. thesis, Texas A. and M. University, College Station, 1987.

Croon, K. A., Ketchersid, M. L., and Markle, M. G., Effect of bentazon, imazaquin, and chlorimuron on the absorption and translocation of the methyl ester of haloxyfop, *Weed Sci.*, 37, 645, 1989.

Croon, K. A. and Merkle, M. G., Effects of bentazon, imazaquin, or chlorimuron on haloxyfop or fluazifop-P efficacy, *Weed Technol.*, 2, 36, 1988.

Defelice, M. S., Brown, W. B., Aldrich, R. J., Sims, B. D., Judy, D. T., and Guethle, D. R., Weed control in soybeans (*Glycine max*) with reduced rates of postemergence herbicides, *Weed Sci.*, 37, 365, 1989.

Dickens, R., Sharpe, S. S., and Turner, D. L., Herbicide effects on tensile strength and rooting of zoysiagrass sod, *J. Prod. Agric.*, 2, 369, 1989.

Durner, J. and Boger, P., Inhibition of purified acetolactate synthase from barley (*Hordeum vulgare* L.) by chlorsulfuron and imazaquin, in *Prospects for Amino Acid Biosynthesis Inhibitors in Crop Protection and Pharamaceutical Chemistry*, Monogr. No. 42, Copping, L. G., Dalziel, I., and Dodge, A. D., Eds., British Crop Protection Council, Farnham, England, 1989, 85.

Edmund, R. M., Jr. and York, A. C., Factors affecting postemergence control of sicklepod (*Cassia obtusifolia*) with imazaquin and DPX-F6025: spray volume, growth stage, and soil-applied alachlor and vernolate, *Weed Sci.*, 35, 216, 1987.

Edmund, R. M., Jr. and York, A. C., Effects of rainfall and temperature on postemergence control of sicklepod (*Cassia obtusifolia*) with imazaquin and DPX-F6025, *Weed Sci.*, 35, 231, 1987.

Endres, C. S. and Longer, D. E., Herbicide selectivity among grain and weedy Amaranthus species, *Agron. J.*, 79, 824, 1987.

Everitt, J. H., Pettit, R. D., and Alaniz, M. A., Remote sensing of broom snakeweed (*Gutierrezia sarothrae*) and spiny aster (*Aster spinosus*), *Weed Sci.*, 35, 295, 1978.

Gebard, J. M., Charest, P. J., Iyer, V. N., and Miki, B. L., Cross-resistance to short residual sulfonylurea herbicides in transgenic tobacco plants, *Plant Physiol.*, 91, 574, 1989.

Garcia-Torres, L., Lopez-Granados, F., Saavedra, M., and Mesa-Garcia, J., Selection of herbicides for the control of broomrape (*Orobanche* spp.) in faba bean (*Vicia faba* L.), *FABIS*, 24, 32, 1989.

Goetz, A. J., Wehtje, G., and Hajek, B., Soil solution and mobility characterization of imazaquin, *Weed Sci.*, 34, 788, 1986.

Griffin, J. L. and Habetz, R. J., Soybean (*Glycine max*) tolerance to preemergence and postemergence herbicides, *Weed Technol.*, 3, 459, 1989.

Hagood, E. S., Jr. and Komm, D. A., Effect of rate and timing of imazaquin application on the growth and yield of flue-cured tobacco, *Tob. Sci.*, 30, 1, 1987.

Helms, R. S., Tripp, T. N., Smith, R. J., Jr., Baldwin, F. L., and Hackworth, M., Rice (*Oryza sativa*) response to imazaquin residues in a soybean (*Glycine max*) and rice rotation, *Weed Technol.*, 3, 513, 1989.

Holshouser, D. L. and Coble, H. D., Compatibility of sethoxydim with five postemergence broadleaf herbicides, *Weed Technol.*, 4, 128, 1990.

Isaacs, M. A., Murdock, E. C., Toler, J. E., and Wallace, S. U., Effects of late-season herbicide applications on sicklepod (*Cassia obtusifolia*) seed production and viability, *Weed Sci.*, 37, 761, 1989.

Johnson, B. J., Carrow, R. N., and Murphy, T. R., Foliar-applied iron enhances Bermuda grass tolerance to herbicides, *J. Am. Soc. Hortic. Sci.*, 115, 422, 1990.

Kalinova, S. and Ampova, G., Study on the effect of some new herbicides for use on tobacco on the microbiological properties of the soil, *Pochvoan. Agrokhim.*, 22, 41, 1987.

Kalinova, S., Georgieva, E., and Chepilova, D., Effect of some herbicides on the assimilation of nitrate nitrogen and on the productivity of virginia tobacco, *Fiziol. Rast.*, 15, 36, 1989.

Keintz, R. C., Cross, B., and Kovacs, G., Development of SCEPTER 1.5 ASU formulation by evaluation of liquid handling properties, in *Pestic. Formulations Appl. Syst.* — 7th Vol., American Society for Testing and Materials, Philadelphia, 1987, 75.

Khodayani, K., Smith, R. J., Jr., and Black, H. L., Red rice (*Oryza sativa*) control with herbicide treatments in soybeans (*Glycine max*), *Weed Sci.*, 35, 127, 1987.

Kollmer, C. W., The Nature of the Antagonistic Interaction between Imazaquin and Postemergence Graminicides, Ph.D. thesis, North Carolina State University, Raleigh, 1988.

Lolas, P. C., Cinmethylin, imazaquin, and metazachlor performance for weed control in tobacco, in *Br. Crop Prot. Conf. — Weeds*, Vol. 3, British Crop Protection Council Publications, Croydon, England, 1985, 841.

Lolas, P. C. and Galapoulos, A., Soil bioactivity, persistence, and leaching of cinmethylin, imazaquin, and metazachlor, *Zizaniologia*, 1, 221, 1985.

Loux, M. M., Soil Persistence and Sorption Characteristics of Imazaquin, Imazethapyr, and Clomazone, Ph.D. thesis, University of Illinois, Urbana, 1988.

Loux, M. M., Liebl, R. A., and Slife, F. W., Availability and persistence of imazaquin, imazethapyr, and clomazone in soil, *Weed Sci.*, 37, 259, 1989.

Loux, M. M., Liebl, R. A., and Slife, F. W., Absorption of imazaquin and imazethapyr on soils, sediments, and selected absorbents, *Weed Sci.*, 37, 712, 1989.

Malefyt, T., Robson, P., and Shaner, D. L., The effects of temperature on AC 252,214 in *Glycine max*, *Abutilon theophrasti*, and *Digitaria sanguinalis*, in *Aspects of Applied Biology*, Association of Applied Biologists, Warwick, England, 1983, 265.

McKinnon, E. J., The Behavior and Bioactivity of Imazaquinin in Soils, Ph.D. thesis, North Carolina State University, Raleigh, 1989.

Mills, J. A., Effects of Soybean Tillage Systems on the Efficacy, Phytotoxicity, and Persistence of Imazaquin, Imazethapyr, and Clomazone, Ph.D. thesis, University of Kentucky, Lexington, 1988.

Mills, J. A. and Witt, W. W., Effect of tillage systems on the efficacy and phytotoxicity of imazaquin and imazethapyr in soybean (*Glycine max*), *Weed Sci.*, 37, 233, 1989.

Mills, J. A. and Witt, W. W., Efficacy, phytotoxicity, and persistence of imazaquin, imazethapyr, and clomazone in no-till double-crop soybeans (*Glycine max*), *Weed Sci.*, 37, 353, 1989.

Minton, B. W., Kurtz, M. E., and Shaw, D. R., Barnyardgrass (*Echinochloa crus-galli*) control with grass and broadleaf weed herbicide combinations, *Weed Sci.*, 37, 223, 1989.

Minton, B. W., Shaw, D. R., and Kurtz, M. E., Postemergence grass and broadleaf herbicide interactions for red rice (*Oryza sativa*) control in soybeans (*Glycine max*), *Weed. Technol.*, 3, 329, 1989.

Moore, B. A., Larson, R. A., and Skroch, W. A., Herbicide treatment of container-grown Gloria azaleas and Merritt Supreme hydrangeas, *J. Am. Soc. Hort. Sci.*, 114, 73, 1989.

Moseley, C. M. and Hagood, E. S., Jr., Effect of simulated carryover or misapplication of chlorimuron or imazaquin on flue-cured tobacco, *Tob. Int.*, 192, 29, 1990.

Moseley, C. M. and Hagood, E. S., Jr., Reducing herbicide inputs when establishing no-till soybeans (*Glycine max*), *Weed Technol.*, 4, 14, 1990.

Nandihalli, U. B., Toxicity, Absorption, and Translocation of Soil and Foliar Applied Imazaquin in Yellow (*Cyperus esculentus*) and Purple (*C. rotundus*) Nutsedge, Ph.D. thesis, Ohio State University, Columbus, 1986.

Nandihalli, U. B. and Bendixen, L. E., Absorption, translocation, and toxicity of foliar-applied imazaquin in yellow and purple nutsedge (*Cyperus esculentus* and *C. rotundus*), *Weed Sci.*, 36, 313, 1988.

Nandihalli, U. B. and Bendixen, L. E., Toxicity and site of uptake of soil-applied imazaquin in yellow and purple nutsedge (*Cyperus esculentus* and *C. rotundus*, *Weed Sci.*, 36, 411, 1988.

Nastasi, P. and Smith, R. J., Jr., Red rice (*Oryza sativa*) control in soybeans (*Glycine max*), *Weed Technol.*, 3, 389, 1989.

Renner, K. A., Factors Affecting Imazaquin and AC 263,499 Persistence in Soil and Subsequent Corn Cultivar Response, Ph.D. thesis, Michigan State University, East Lansing, 1986.

Renner, K. A., Meggitt, W. F., and Leavitt, R. A., Influence of rate, method of application, and tillage on imazaquin persistence in soil, *Weed Sci.*, 36, 90, 1988.

Renner, K. A., Meggitt, W. F., and Penner, D., Effect of soil pH on imazaquin and imazethapyr adsorption to soil and phytotoxicity to corn (*Zea mays*), *Weed Sci.*, 36, 78, 1988.

Renner, K. A., Meggitt, W. F., and Penner, D., Response of corn (*Zea mays*) cultivars to imazaquin, *Weed Sci.*, 36, 625, 1988.

Riley, D. G. and Shaw, D. R., Influence of imazapyr on the control of pitted morningglory (*Ipomoea lacunosa*) and johnsongrass (*Sorghum halepense*) with chlorimuron, imazaquin, and imazethapyr, *Weed Sci.*, 36, 663, 1988.

Riley, D. G. and Shaw, D. R., Johnsongrass (*Sorghum halepense*) and pitted morningglory (*Ipomoea lacunosa*) control with imazaquin and imazethapyr, *Weed Technol.*, 3, 95, 1989.

Risley, M. A., Imazaquin Herbicidal Activity: Efficacy, Absorption, Translocation, and Ethylene Synthesis, Ph.D. thesis, Universtity of Arkansas, Fayetteville, 1986.

Sander, K. W. and Barrett, M., Differential imazaquin tolerance and behavior in selected corn (*Zea mays*) hybrids, *Weeds Sci.*, 37, 290, 1989.

Sauerborn, J., Saxena, M. C., and Meyer, A., Broomrape control in faba bean (*Vicia faba* L.) with glyphosate and imazaquin, *Weed Res.*, 29, 97, 1989..

Schloss, J. V., Ciskanik, L. M., and Van Dyk, D. E., Origin of the herbicide binding site of acetolactate synthase, *Nature*, 331, 360, 1988.

Shaner, D. L. and Anderson, P. C., Mechanism of action of the imidazolinones and cell culture selection to tolerant maize, in *Biotechnology in Plant Science: Relevance to Agriculture in the Eighties*, 287, Zaitlin, M., Day, P. R., and Hollaender, A., Eds., Academic Press, Orlando, 1985, 287.

Shaner, D. L. and Little, D., Effects of imazaquin and sulfometuron methyl on extractable acetohydroxyacid synthase activity in maize and soybeans, in *Prospects for Amino Acid Biosynthesis Inhibitors in Crop Protection and Pharmaceutical Chemistry*, Monograph No. 42, Copping, L. G., Dalziel, I., and Dodge, A. D., Eds., British Crop Protection Council, Farnham, England, 1989, 197.

Shaner, D. L. and Robson, P. A., Absorption, translocation, and metabolism of AC 252,214 in soybean (*Glycine max*), common cocklebur (*Xanthium strumarium*), and velvetleaf (*Abutilon theophrasti*), *Weed Sci.*, 33, 469, 1985.

Sims, G. R., Wehtje, G., McGuire, J. A., and Patterson, M. G., Weed control in peanuts (*Arachis hypogaea*) with imazaquin, *Weed Sci.*, 35, 682, 1987.

Singh, B. K., Newhouse, K. E., Stidham, M. A., and Shaner, D. L., Acetohydroxyacid synthase-imidazolinone interaction, in *Prospects for Amino Acid Biosynthesis Inhibitors in Crop Protection and Pharmaceutical Chemistry*, Monogr. No. 42, Copping, L. G., Dalziel, I., and Dodge, A. D., Eds., British Crop Protection Council, Farnham, England, 1989, 87.

Souza, I. F. de, Avaliacao preliminar de imazaquin para o controle de plantas daninhas em soja no cerrado, *Pesqui. Agropecu. Bras.*, 23, 575, 1988.

Stougaard, R. N., Efficacy, Soil Behavior, and Persistence of Imazaquin, Imazethapyr, Chlorimuron, and FMC 57020, Ph.D. thesis, University of Nebraska, Lincoln, 1987.

Stougaard, R. N., Shea, P. J., and Martin, A. R., Effect of soil type and pH on adsorption, mobility, and efficacy of imazaquin and imazethapyr, *Weed Sci.*, 38, 67, 1990.

Swanson, E. B., Herrgesell, M. J., Arnoldo, M., Sippell, D. W., and Wong, R. S. C., Microspore mutagenesis and selection: canola plants with field tolerance to the imidazolinones, *Theor. Appl. Genet.*, 78, 525, 1989.

Talasova, N., Evaluation of herbicides for legumes, *Agrochemia*, 25, 149, 1985.

Turner, D. L., Sharpe, S. S., and Dickens, R., Herbicide effects on tensile strength and rooting of centipedegrass sod, *Hort. Sci.*, 25, 541, 1990.

Van Ellis, M. R. and Shaner, D. L., Mechanism of cellular absorption of imidazolinones in soybean (*Glycine max*) leaf disks, *Pestic. Sci.*, 23, 25, 1988.

Vidrine, P. R., Johnsongrass (*Sorghum halepense*) control in soybeans (*Glycine max*) with postemergence herbicides, *Weed Technol.*, 3, 455, 1989.

Vollmer, J. S. L., Combination Effect of ACP 2100, Imazaquin, and Triclopyr on Common Dandelion and Three Kentucky Bluegrass Turf Types, Ph.D. thesis, Virginia Polytechnic Institute and State University, Blacksburg, 1989.

Walls, F. R., Jr., The Potential of Imazaquin for Weed Control in Flue-cured Tobacco (*Nicotiana tabacum*), Ph.D. thesis, North Carolina State University, Raleigh, 1985.

Walls, F. R., Jr., Worsham, A. D., Collins, W. K., Corbin, F. T., and Bradley, J. R., Evaluation of imazaquin for weed control in flue-cured tobacco (*Nicotiana tabacum*), *Weed, Sci.*, 35, 824, 1987.

Warmund, M. R., Postemergence control of an oat cover crop and broadleaf weeds in direct-seeded nursery beds, *Hort. Sci.,* 22, 603, 1987

Wesley, R. A., Jr., Shaw, D. R., and Barrentine, W. L., Incorporation depths of imazaquin, metribuzin, and chlorimuron for common cocklebur (*Xanthium strumarium*) control in soybeans (*Glycine max*) *Weed Sci.,* 37, 596, 1989.

Wesley, R. A., Jr., Shaw, D. R., and Barrentine, W. L., Application timing of metribuzin, chlorimuron, and imazaquin for common cocklebur (*Xanthium strumarium*) control, *Weed Technol.,* 3, 364, 1989.

Westberg, D. E., Oliver, L. R., and Frans, R. E., Weed control with clomazone alone and with other herbicides, *Weed Technol.,* 3, 678, 1989.

Wilcut, J. W., Wehtje, G. R., Patterson, M. G., and Cole, T. A., Absorption, translocation, and metabolism of foliar-applied imazaquin in soybeans (*Glycine max*), peanuts (*Arachis hypogaea*), and associated weeds, *Weed Sci.,* 36, 5, 1988.

Wills, G. D. and McWhorter, C. G., Influence of inorganic salts and imazapyr on control of pitted morningglory (*Ipomoea lacunosa*) with imazaquin and imazethapyr, *Weed Technol.,* 1, 328, 1987.

Winder, T. and Spalding, M. H., Imazaquin and chlorsulfuron resistance and cross resistance in mutants of *Chlamydomonas reinhardtii, Mol. Gen. Genet.,* 213, 394, 1988.

Wolt, J. D., Rhodes, G. N., Jr., Graveel, J. G., Glosauer, E. M., Amin, M. K., and Church, P. L., Activity of imazaquin in soil solution as affected by incorporated wheat (*Triticum aestivum*) straw, *Weed Sci.,* 37, 254, 1989.

IMIDAZOLINONES

Anderson, P. C. and Georgeson, M., Herbicide-tolerant mutants of corn, *Genome,* 31, 994, 1989.

Anderson, P. C. and Hibberd, K. A., Evidence for the interaction of an imidazolinone herbicide with leucine, valine and isoleucine metabolism, *Weed Sci.,* 33, 479, 1985.

Brown, M. A., Chiu, T. Y., and Miller, P., Hydrolytic activation versus oxidative degradation of ASSERT herbicide, an imidazolinone aryl-carboxylate, in susceptible wild oat versus tolerant corn and wheat, *Pestic. Biochem. Physiol.,* 27, 24, 1987.

Cantwell, J. R., Liebl, R. A., and Slife, F. W., Biodegradation characteristics of imazaquin and imazethapyr, *Weed Sci.,* 37, 815, 1989.

Cardaciotto, S. J., Mowery, P. C., Thomson, M. L., and Wayne, R. S., Distinguishing between *N*-methylated imidazolinones using mass spectrometry, *Org. Mass Spectrom.,* 22, 342, 1987.

Guaciaro, M. A., Los, M., Russell, R. K., Wepplo, P. J., Lences, B. L., Lauro, P. C., Orwick, P. L., Umeda, K., and Marc, P. A., *o*-(5-Thiono-2-imidazolin-2-yl)aryl carboxylates. Synthesis and herbicidal activity, in *Synthesis and Chemistry of Agrocehmicals,* Baker, D. R., Fenyes, J. G., Moberg, W. K., and Cross, B., Eds., ACS Symp. Ser. No. 355, American Chemical Society, Washington, D.C., 1987, 87.

Haughn, G. W. and Somerville, C. R., A mutation causing imidazolinone resistance maps to the *Csr1* locus of *Arabidopsis thaliana, Plant Physiol.,* 92, 1981, 1990.

Iler, S., Pauls, K. P., and Swanton, C. J., Selection for imidazolinone tolerance in tomato (*Lycopersicon esculentum*) tissue, callus and cell suspension culture, *Plant Physiol.,* 89, 12, 1989.

Los, M., Synthesis and biology of the imidazolinone herbicides, in *Pesticide Science and Biotechnology,* Proc. 6th Int. Congr. Pestic. Chem. (IUPAC), Greenhalgh, R. and Roberts, T. R., Eds., Blackwell, London, 1987, 35.

Muhitch, M. J., Shaner, D. L., and Stidham, M. A., Imidazolinones and acetohydroxyacid synthase from higher plants. Properties of the enzyme from maize suspension culture cells and evidence for the binding of imazapyr to acetohydroxyacid synthase *in vivo, Plant Physiol.,* 83, 451, 1987.

Nalewaja, J. D., Seed oils with herbicides, *Meded. Fac. Landbouwwet. Rijksuniv. Gent,* 51, 301, 1986.

Pandey, J., Efficacy of some herbicides for weed control in gram, *Indian J. Agron.,* 34, 110, 1989.

Primiani, M. M., Cotterman, J. C., and Saari, L. L., Resistance of kochia (*Kochia scoparia*) to sulfonylurea and imidazolinone herbicides, *Weed Technol.,* 4, 169, 1990.

Reed, W. T., Saladini, J. L., Cotterman, J. C., Primiani, M. M., and Saari, L. L., Resistance in weeds to sulfonylurea herbicides, in *Br. Crop Prot. Conf.— Weeds,* Vol. 1, British Crop Protection Council Publications, Croydon, England, 1985, 295.

Sathasivan, K., Haughn, G. W., and Murai, N., Isolation and sequencing of acetolactate synthase gene from imazapyr-resistant *Arabidopsis thaliana* (L.), *Plant Physiol.,* 86, 136, 1988.

Sathasivan, K., Haughn, G. W., and Murai, N., Nucleotide sequence of a mutant acetolactate synthase gene from an imidazolinone-resistant *Arabidopsis thaliana* var. Columbia, *Nucleic Acids Res.,* 18, 2188, 1990.

Saxena, P. K. and King, J., Herbicide resistance in *Datura innoxia*. Cross-resistance of sulfonylurea-resistant cell lines to imidazolinones, *Plant Physiol.,* 86, 863, 1988.

Schloss, J. V., Ciskanik, L. M., and Van Dyk, D. E., Origin of the herbicide binding site of acetolactate synthase, *Nature,* 331, 360, 1988.

Schneider, I. and Klemme, J. H., Sensitivity of a phototrophic bacterium to the herbicide sulfometuron methyl, an inhibitor of branched chain amino acid biosynthesis, *Zeit. Naturforsch. Teil C,* 41, 1037, 1986.

Shaner, D. L. and Anderson, P. C., Mechanism of action of the imidazolinones and cell culture selection of tolerant maize, in *Biotechnology in Plant Science: Relevance to Agriculture in the Eighties*, Zaitlin, M., Day, P. R., and Hollaender, A., Eds., Academic Press, Orlando, 1985, 287.

Shaner, D. L., Anderson, P. C., and Stidham, M. A., Imidazolinones. Potent inhibitors of acetohydroxy acid synthase, *Plant Physiol.*, 76, 545, 1984.

Shaner, D. L., Singh, B. K., and Stidham, M. A., Interaction of imidazolinones with plant acetohydroxyacid synthase: evidence for *in vivo* binding and competition with sulfometuron methyl, *J. Agric. Food Chem.*, 38, 1279, 1990.

Shaner, D., Stidham, M., Muhitch, M., Reider, M., Robson, P., and Anderson, P., Mode of action of the imidazolinones, in *Br. Crop. Prot. Conf. — Weeds*, Vol. 1, British Crop Protection Council Publications, Croydon, England, 1985, 147.

Singh, B. K., Newhouse, K. E., Stidham, M. A., and Shaner, D. L., Acetohydroxyacid synthase-imidazolinone interaction, in *Prospects for Amino Acid Biosynthesis Inhibition in Crop Protection and Pharmaceutical Chemistry*, Monogr. No. 42, Copping, L. G., Dalziel, I., and Dodge, A. D., Eds., British Crop Protection Council, Farnham, England, 1989, 87.

Singh, B. K. and Schmitt, G. K., Flavin adenine dinucleotide causes oligomerization of acetohydroxyacid synthase from Black Mexican Sweet corn cells, *FEBS Letters*, 258, 113, 1989.

Smith, J. K., Schloss, J. V., and Mazur, B. J., Functional expression of plant acetolactate synthase genes in *Escherichia coli*, *Proc. Natl. Acad. Sci. U.S.A.*, 86, 4179, 1989.

Swanson, E. B., Herrgesell, M. J., Arnoldo, M., Sippell, D. W., and Wong, R. S. C., Microscope mutagenesis and selection: canola plants with field tolerance to the imidazolinones, *Theor. Appl. Genet.*, 78, 525, 1989.

Van Ellis, M. R. and Shaner, D. L., Mechanism of cellular absorption of imidazolinones in soybean (*Glycine max*) leaf disks, *Pestic. Sci.*, 23, 25, 1988.

IMAZAMETHABENZ-METHYL

Allen, R. and Caseley, J. C., The persistence and mobility of AC 222,293 in cropped and fallow soils, in *Br. Crop Prot. Conf. — Weeds*, Vol. 3, British Crop Protection Council Publications, Thornton Heath, England, 1987, 569.

Blackshaw, R. E., Derksen, D. A., and Muendel, H. H., Herbicides for weed control in safflower (*Carthamus tinctorius*), *Can. J. Plant. Sci.*, 70, 237, 1990.

Brown, M. A., Chiu, T. Y., and Miller, P., Hydrolytic activation versus oxidative degradation of ASSERT herbicide, an imidazolinone aryl-carboxylate, in susceptible wild oat versus tolerant corn and wheat, *Pestic. Biochem. Physiol.*, 27, 24, 1987.

Cardaciotto, S. J., Mowery, P. C., Thomson, M. L., and Wayne, R. S., Distinguishing between *N*-methylated imidazolinones using mass spectrometry, *Org. Mass Spectrom.*, 22, 342, 1987.

Chow, P. N. P., Broadening weed control with imazamethabenz combinations, *Crop Protec.*, 8, 447, 1989.

Eberlein, C. V., Miller, T. L., and Wiersma, J. V., Influence of thiameturon and DPX-L5300 on wild oats (*Avena fatua*) control with barban, diclofop, AC 222,293, and difenzoquat, *Weed Sci.*, 36, 792, 1988.

Efthimiadis, P. and Skorda, E. A., The efficacy of AC 222,293 applied post-emergence at various growth stages and rates against wild oat and other weeds in cereals under Greek conditions, *Meded. Fac. Landbouwwet. Rijksuniv. Gent*, 50, 257, 1985.

Fellows, G. M., Fay, P. K., Carlson, G. R., and Stewart, V. R., Effect of AC 222,293 soil residues on rotational crops, *Weed Technol.*, 4, 48, 1990.

Friesen, G. H., Wild mustard control in sunflower with AC 222,293, *Can. J. Plant Sci.*, 68, 1159, 1988.

Gibiat, J., La destruction des Rumex en prairie., *Phytoma*, No. 402, 56, 1988.

Graph, S. and Kleifeld, Y., New herbicides for selective postemergence weed control in wheat, *Phytoparasitica*, 16, 382, 1988.

Guaciaro, M. A., Los, M., Russell, R, K., Wepplo, P. J., Lences, B. L., Lauro, P. C., Orwick, P. L., Umeda, K., and Marc, P. A., *o*-(5-Thiono-2-imidazolin-2-yl)aryl carboxylates. Synthesis and herbicidal activity, in *Synthesis and Chemistry of Agrochemicals*, Baker, D. R., Fenyes, J. G., Moberg, W. K., and Cross, B., Eds., ACS Symp. Ser. No. 355, American Chemical Society, Washington, D.C., 1987, 87.

Gussin, E. J., The control of *Arrhenatherum elatius* in cereals by imazamethabenz-methyl, in *Br. Crop Prot. Conf. — Weeds*, Vol. 1, British Crop Protection Council Publications, Farnham, England, 1989, 199.

Hudson, A. A. and Townsend, S. C. E., AC 222,293 for the control of grass weeds in winter cereals in the UK: field studies of efficacy and crop tolerance., in *Br. Crop Prot. Conf. — Weeds*, Vol. 3, British Crop Protection Council Publications, Croydon, England, 923, 1985.

Manthey, F. A., Nalewaja, J. D., and Szelezniak, E. F., Esterified seed oils with herbicides in *Adjuvants Agrochem.*, Vol. 2, Chow, P. N. P., Ed., CRC Press, Boca Raton, 1989, 139.

Miller, S. D. and Alley, H. P., Weed control and rotational crop response with AC 222,293, *Weed Technol.*, 1, 29, 1987.

Moss, S. R., Herbicide resistance in black-grass (*Alopecurus myosuroides*), in *Br. Crop Prot. Conf. — Weeds*, Vol. 3, British Crop Protection Council Publications, Thornton Heath, England, 879, 1987.

Pillmoor, J. B., Influence of Temperature on the Activity of AC 222,293 against *Avena fatua* and *Alopecurus myosuroides*, *Weed Res.*, 25, 433, 1985.

Pillmoore, J. B. and Caseley, J. C., The influence of growth stage and foliage or soil application on the activity of AC 222,293 against *Alopecurus myosuroides* and *Avena fatua*, *Ann. Appl. Biol.*, 105, 517, 1985.

Pillmoore, J. B. and Caseley, J. C., The biochemical and physiological effects and mode of action of AC 222,293 against *Alopecurus myosuroides* Huds. and *Avena fatua* L., *Pestic. Biochem. Physiol.*, 27, 340, 1987.

Swanson, E. B., Herrgesell, M. J., Arnoldo, M., Sippell, D. W., and Wong, R. S. C., Microspore mutagenesis and selection: canola plants with field tolerance to the imidazolinones, *Theor. Appl. Genet.*, 78, 525, 1989.

Swanton, C. J. and Chandler, K., Control of wild mustard in canola with postemergence herbicides, *Can. J. Plant Sci.*, 69, 889, 1989.

Yaduraju, N. T. and Ahuja, K. N., Performance of AC 222,293 and chlorsulfuron for weed control in wheat and their persistence in soil, *Indian J. Agron.*, 33, 312, 1988.

IMAZAPYR

Anderson, P. C. and Hibberd, K. A., Evidence for the interaction of an imidazolinone herbicide with leucine, valine and isoleucine metabolism, *Weed Sci.*, 33, 479, 1985.

Bouchet, F. and Lescar, L., Cereales: l'apparition d'un arsenal impressionant de techniques phytosanitaires pour un secteur de pres de huit millions d'hectares, *Phytoma*, No. 400, 78, 1988.

Bucsbaum, H., Horowitz, M., Kleifeld, Y., Herzlinger, G., and Bargutti, A., Experiments on the control of *Phragmites* in drainage canals, *Phytoparasitica* 13, 249, 1985.

Christensen, P., Danish results with a new herbicide, imazapyr, in forestry, in *Aspects of Applied Biology*, No. 16, Association of Applied Biologists, Warwick, England, 105, 1988.

Christensen, P., Danske resultater med ARSENAL (imazapyr), et nyt herbicid i skovbruget, *Dan. Skovforen. Tidssk.*, 73, 103, 1988.

Clay, D. V. and West, T. M., The response of triazine-resistant and susceptible biotypes of *Erigeron canadensis* to 23 herbicides, *Meded. Fac. Landbouwwet. Rijksuniv. Gent*, 52, 1195, 1987.

Dickens, R., Sharpe, S. S., and Turner, D. L., Herbicide effects on tensile strength and rooting of zoysiagrass sod, *J. Prod. Agric.*, 2, 369, 1989.

Erasmus, D. J. and Noel, D. J. R., Chemical control of *Chromolaena oderata* (L.) K. & R. efficacy of herbicides applied to stumps, *Appl. Plant Sci.*, 3, 18, 1989.

Erasmus, D. J. and Staden, J. V., Screening of candidate herbicides in field trials for chemical control of *Chromolaena odorata*, *S. Afr. J. Plant Soil*, 3, 66, 1986.

Gabard, J. M., Charest, P. J., Iyer, V. N., and Miki, B. L., Cross-resistance to short residual sulfonylurea herbicides in transgenic tobacco plants, *Plant Physiol.*, 91, 574, 1989.

Guaciaro, M. A., Los, M., Russell, R. K., Wepplo, P. J., Lences, B. L., Lauro, P. C., Orwick, P. L., Umeda, K., and Marc, P. A., *o*-(5-Thiono-2-imidazolin-2-yl)aryl carboxylates. Synthesis and herbicidal activity, in *Synthesis and Chemistry of Agrochemicals*, Baker, D. R., Fenyes, J. G., Moberg, W. K., and Cross, B., Eds., ACS Symp. Ser. No. 355, American Chemical Society, Washington, D.C., 1987, 87.

Haughn, G. W. and Somerville, C. R., A mutation causing imidazolinone resistance maps to the *Csr1* locus of *Arabidopsis thaliana*, *Plant Physiol.*, 92, 1081, 1990.

Hawkes, T. R., Studies of herbicides which inhibit branched-chain amino acid biosynthesis, in *Prospects for Amino Acid Biosynthesis Inhibitors in Crop Protection and Pharmaceutical Chemistry*, Monogr. 42, Copping, L. G., Dalziel, I., and Dodge, A. D., Eds., British Crop Protection Council, Farnham, England, 1989, 131.

IsraelAgan Chemical Manufacturers Ltd., Imazapyr — a new herbicide for weed control on noncultivated land, *Phytoparasitica*, 13, 238, 1985.

Johansson, T., Preventing stump regrowth with a herbicide-applying tree cutter, *Weed Res.*, 28, 353, 1988.

Lawrie, J. and Clay, D. V., Tolerance of forestry and biomass broad-leaved tree species to soil-acting herbicides, in *Br. Crop Prot. Conf. — Weeds*, Vol. 1, British Crop Protection Council Publications, Croydon, England, 1989, 347.

Lund-Høie, K. and Rognstad, K., Effect of foliage-applied imazapyr and glyphosate on common forest weed species and Norway spruce, *Crop Protec.*, 9, 52, 1990.

Lund-Høie, K. and Rognstad, K., Stump treatment with imazapyr and glyphosate after cutting hardwoods to prevent regrowth of suckers, *Crop Protec.*, 9, 59, 1990.

Muhitch, M. J., Acetolactate synthase activity in developing maize (*Zea mays* L.) kernels, *Plant Physiol.*, 86, 23, 1988.

Muhitch, M. J., Shaner, D. L., and Stidham, M. A., Imidazolinones and acetohydroxyacid synthase from higher plants. Properties of the enzyme from maize suspension culture cells and evidence for the binding of imazapyr to acetohydroxyacid synthase *in vivo*, *Plant Physiol.*, 83, 451, 1987.

Nir, A., Combined control of both weeds and mosquitoes in drainage canals, *Phytoparasitica*, 16, 385, 1988.

Nir, A. and Raz, A., Herbicide rotation at roadsides as a means of preventing triazine resistance in annual weeds, *Phytoparasitica*, 13, 248, 1985.

Nir, A. and Raz, A., New herbicides for control of perennial weeds at roadsides and in drainage canals, *Phytoparasitica,* 13, 249, 1985.

Nir, A. and Raz, A., Combination of new herbicides for weed control along roadsides, *Phytoparasitica,* 16, 386, 1988.

Peoples, T. R., Today's herbicide: ARSENAL herbicide, *Weeds Today,* 15, 5, 1984.

Poples, T. R., New herbicide "Imazapyr", *Shokubutsu no Kagaku Chosetsu,* 22, 47, 1987.

Pereira, R. C., Imazapyr, a new herbicide for use in rubber, *Pesqui. Agropecu. Bras.,* 22, 469, 1987.

Primiani, M. M., Cotterman, J. C., and Saari, L. L., Resistance of kochia (*Kochia scoparia*) to sulfonylurea and imidazolinone herbicides, *Weed Technol.,* 4, 169, 1990.

Riley, D. G. and Shaw, D. R., Influence of imazapyr on the control of pitted morningglory (*Ipomoea lacunosa*) and johnsongrass (*Sorghum halepense*) with chlorimuron, imazaquin, and imazethapyr, *Weed Sci.,* 36, 663, 1988.

Sathasivan, K., Haughn, G. W., and Murai, N., Isolation and sequencing of acetolactate synthase gene from imazapyr-resistant *Arabidopsis thaliana* (L.), *Plant Physiol.,* 86, 136, 1988.

Sathasivan, K., Haughn, G. W., and Murai, N., Nucleotide sequence of a mutant acetolactate synthase gene from an imidazolinone-resistant *Arabidopsis thaliana* var. Columbia, *Nucleic Acids Res.,* 18, 2188, 1990.

Schneider, I. and Klemme, J. H., Sensitivity of a phototrophic bacterium to the herbicide sulfometuron methyl, an inhibitor of branched-chain amino acid biosynthesis, *Z. Naturforsch. Teil C,* 41, 1037, 1986.

Shaner, D. L., Absorption and translocation of imazapyr in *Imperata cylindrica* (L.), Raeuschel and effects of growth and water usage, *Trop. Pest Manage.,* 34, 388, 1988.

Shaner, D. L. and Anderson, P. C., Mechanism of action of the imidazolinones and cell culture selection of tolerant maize, in *Biotechnology in Plant Science: Relevance to Agriculture in the Eighties,* Zaitlin M., Day, P. R., and Hollaender, A., Eds., Academic Press, Orlando, 1985, 287.

Shaner, D. L. and Reider, M. L., Physiological response of corn (*Zea mays*) to AC 243,997 in combination with valine, leucine and isoleucine, *Pestic. Biochem. Physiol.,* 25, 248, 1986.

Shaner, D., Stidham, M., Muhitch, M., Reider, M., Robson, P., and Anderson, P., Mode of action of the imidazolinones, in *Br. Crop Prot. Conf. — Weeds,* Vol. 1, British Crop Protection Council Publications, Croydon, England, 1985, 147.

Sharpe, S. S., Dickens, R., and Turner, D. L., Herbicide effects of tensile strength and rooting of bermudagrass (*Cynodon dactylon*) sod, *Weed Technol.,* 3, 353, 1989.

Singh, B. K., Newhouse, K. E., Stidham, M. A., and Shaner, D. L., Acetohydroxyacid synthase-imidazolinone interaction, in *Prospects for Amino Acid Biosynthesis Inhibitors in Crop Protection and Pharmaceutical Chemistry,* Monogr. No. 42, Copping, L. G., Dalziel, I., and Dodge, A. D., Eds., British Crop Protection Council, Farnham, England, 1989, 87.

Singh, B. K. and Schmitt, G. K., Flavin adenine dinucleotide causes oligomerization of acetohydroxyacid synthase from Black Mexican Sweet corn cells, *FEBS Letters,* 258, 113, 1989.

Standell, C. J. and West, T. M., Response of several pasture grass species to potential new herbicides for bracken control, in *Br. Crop Prot. Conf. — Weeds,* Vol. 1, British Crop Protection Council Publications, Croydon, England, 1985, 903.

Subagyo, T., Studies on Imazapyr, a Herbicide for Minimum Tillage Purposes in Areas Infested with Lalang (*Imperata cylindrica* (L.) Raeuschel), Ph.D. thesis, University of Reading, Reading, UK, 1989.

Supamataya, K., Lueang-a-Papong, P., and Phromkunthong, W., Phytotoxicity of herbicides in water. I. Acute toxicity of imazapyr on Nile tilapia (*Sarotherodon niloticus*) and silver barb (*Puntius gonionotus*), *Warasan Songkhla Nakkharin,* 9, 309, 1987.

Thomas, J. M., Skalski, J. R., Cline, J. F., McShane, M. C., Simpson, J. C., Miller, W. E., Peterson, S. A., Callahan, C. A., and Greene, J. C., Characterization of chemical waste site contamination and determination of its extent using bioassays, *Environ. Toxicol. Chem.,* 5, 487, 1986.

Tjitrosemito, S. and Suwinarno, D., The performance of soybean (cv. Americana) established by zero tillage technique in *Imperata* field controlled by herbicides, *Biotropia,* 2, 12, 1988/1989.

Tjitrosemito, S., Utomo, I. H., and Sastroutomo, S. S., The efficacy of imazapyr to control alang-alang (*Imperata cylindrica*), *Biotrop. Tech. Bull.,* 1, 1, 1986.

Townson, J. K. and Price, C. E., Tropical weed control: an approach to optimising herbicide performance with very low volume applications, in *Aspects of Applied Biology,* No. 14, Association of Applied Biologists, Warwick, England, 1987, 305.

Townson, J. K. and Price, C. E., Tropical weed control: an approach to optimising herbicide performance with very low volume applications, in *Aspects of Applied Biology,* No. 14, Association Applied Biologists, Warwick, England, 1987, 305.

Turner, D. J., Richardson, W. G., and Tabbush, P. M., The use of additives to improve the activity of herbicides used for bracken and heather control, in *Aspects Appl. Biol.,* No. 16, Association of Applied Biologists, Warwick, England, 1988, 271.

Turner, D. L., Sharpe, S. S., and Dickens, R., Herbicide effects on tensile strength and rooting of centipedegrass sod, *Hortic. Sci.,* 25, 541, 1990.

Van Ellis, M. R. and Shaner, D. L., Mechanism of cellular absorption of imidazolinones in soybean (*Glycine max*) leaf discs, *Pestic. Sci.,* 23, 25, 1988.

Wehtje, G., Dickens, R., Wilcut, J. W., and Hajek, B. F., Sorption and mobility of sulfometuron and imazapyr in five Alabama soils, *Weed Sci.*, 35, 858, 1987.

Wells, M. J. M. and Michael, J. L., Reversed-phase solid-phase extraction for aqueous environmental sample preparation in herbicide residue analysis, *J. Chromatogr. Sci.*, 25, 345, 1987.

West, T. M., The activity of new herbicide on bracken and grass species, in *Br. Crop Prot. Conf. — Weeds*, Vol. 3, British Crop Protection Council Publications, Thornton Heath, England, 1987, 751.

Wills, G. D. and McWhorter, C. G., Influence of inorganic salts and imazapyr on control of pitted morningglory (*Ipomoea lacunosa*) with imazaquin and imazethapyr, *Weed Technol.*, 1, 328, 1987.

Wilson, A. K., Two pot experiments to assess the postemergence activity of various herbicides for the control of *Pennisetum setosum*. *Ann. Appl. Biol.*, 114, 112, 1989.

Winfield, R. J., Imazapyr for the control of bracken (*Pteridium aquilinum*), *Aspects Appl. Biol.*, No. 16, Association of Applied Biologists, Warwick, England, 1988, 281.

IMAZETHAPYR

Anderson, P. C. and Georgeson, M., Herbicide-tolerant mutants of corn, *Genome*, 31, 994, 1989.

Barrett, M., Protection of corn (*Zea mays*) and sorghum (*Sorghum bicolor*) from imazethapyr toxicity with antidotes, *Weed Sci.*, 37, 296, 1989.

Cantwell, J. R., Weed Control Efficacy and Biodegradation Characteristics of Imazethapyr, Ph.D. thesis, University of Illinois, Urbana, 1989.

Cantwell, J. R., Liebl, R. A., and Slife, F. W., Biodegradation characteristics of imazaquin and imazethapyr, *Weed Sci.*, 37, 815, 1989.

Cantwell, J. R., Liebl, R. A., and Slife, F. W., Imazethapyr for weed control in soybean (*Glycine max*), *Weed Technol.*, 3, 596, 1989.

Carrow, R. N. and Johnson, B. J., Response of centipedegrass to plant growth regulator and iron treatment combinations, *Appl. Agric. Res.*, 5,21, 1990.

Cole, T. A., Wehtje, G. R., Wilcut, J. W., and Hicks, T. V., Behavior of imazethapyr in soybeans (*Glycine max*), peanuts (*Arachis hypogaea*), and selected weeds, *Weed Sci*, 37, 639, 1989.

Fales, S. L. and Hoover, R. J., Chemical regulation of alfalfa/grass mixtures with imazethapyr, *Agron. J.*, 82, 5, 1990.

Fales, S. L. and Hoover, R. J., Manipulating seasonal growth distribution and nutritive quality of timothy and orchardgrass using the growth regulator imazethapyr, *Can. J Plant Sci.*, 70, 501, 1990.

Fales, S. L., Hill, R. R., and Hoover, R. J., Chemical regulation of growth and forage quality of cool-season grasses with imazethapyr, *Agron. J.*, 82, 9, 1990.

Gabard, J. M., Charest, P. J., Iyer, V. N., and Miki, B. L., Cross-resistance to short residual sulfonylurea herbicides in transgenic tobacco plants, *Plant Physiol.*, 91, 574, 1989.

Garcia-Torres, L., Lopez-Granados, F., Saavedra, M., and Mesa-Garcia, J., Selection of herbicides for the control of broomrape (*Orobanche* spp.) in faba bean (*Vicia faba* L.), *FABIS*, 24, 32, 1989.

Guaciaro, M. A., Los, M., Russell, R. K., Wepplo, P. J., Lences, B. L., Lauro, P. C., Orwick, P. L., Umeda, K., and Marc, P. A., *o*-(5-Thiono-2-imidazolin-2-yl)aryl carboxylates. Synthesis and herbicidal activity, in *Synthesis and Chemistry of Agrochemicals*, Baker, D. R., Fenyes, J. G., Moberg, W. K., and Cross, B., Eds., ACS Symp. Ser. No. 355, American Chemical Society, Washington, D.C., 1987, 87.

Herrick, R. M., Factors Affecting the Herbicidal Activity of Imazethapyr and its Control of Spurred Anoda (*Anoda cristata* L. Schlecht.), Ph.D. thesis, Rutgers, New Brunswick, 1987.

Iler, S., Pauls, K. P., and Swanton, C. J., Selection for imidazolinone tolerance in tomato (*Lycopersicon esculentum*) tissue, callus, and cell suspension cultures, *Plant Physiol.*, 89, 12, 1989.

Johnson, B. J., Response of bermudagrass (*Cynodon* spp.) to plant growth regulators, *Weed Technol.*, 3, 440, 1989.

Johnson, B. J., Response of centipedegrass (*Eremochloa ophiuroides*) to plant growth regulators and frequency of mowing, *Weed Technol.*, 3, 48, 1989.

Johnson, B. J., Influence of frequency and dates of plant growth regulator applications to centipedegrass on seedhead formation and turf quality, *J. Am. Soc. Hortic. Sci.*, 115, 412, 1990.

Knott, C. M. and Eke, K. R., Imazethapyr/pendimethalin a new herbicide mixture for weed control in legumes, in *Br. Crop Protect. Conf. — Weeds*, Vol. 3, British Crop Protection Council Publications, Farnham, England, 1989, 829.

Loux, M. M., Soil Persistence and Sorption Characteristics of Imazaquin, Imazethapyr, and Clomazone, Ph.D. thesis, University of Illinois, Urbana, 1988.

Loux, M. M., Liebl, R. A., and Slife, F. W., Availability and persistence of imazaquin, imazethapyr, and clomazone in soil, *Weed Sci.*, 37, 259, 1989.

Loux, M. M., Liebl, R. A., and Slife, F. W., Absorption of imazaquin and imazethapyr on soils, sediments, and selected absorbents, *Weed Sci.*, 37, 712, 1989.

Mills, J. A., Effects of Soybean Tillage Systems on the Efficacy, Phytotoxicity, and Persistence of Imazaquin, Imazethapyr, and Clomazone, Ph.D. thesis, University of Kentucky, Lexington, 1988.

Mills, J. A. and Witt, W. W., Effect of tillage systems on the efficacy and phytotoxicity of imazaquin and imazethapyr in soybean (*Glycine max*), *Weed Sci.*, 37, 233, 1989.

Mills, J. A. and Witt, W. W., Efficacy, phytotoxicity, and persistence of imazaquin, imazethapyr, and clomazone in no-till double-crop soybeans (*Glycine max*), *Weed Sci.*, 37, 353, 1989.

Nastasi, P. and Smith, R. J., Jr., Red rice (*Oryza sativa*) control in soybeans (*Glycine max*), *Weed Technol.*, 3, 389, 1989.

Peoples, T. R., Wang, T., Fine, R. R., Orwick, P. L., Graham, S. E., and Kirkland, K., AC 263,499: a new broad-spectrum herbicide for use in soybeans and other legumes, in *Br. Crop. Conf. — Weeds*, Vol. 1, British Crop Protection Council Publications, Croydon, England, 1985, 99.

Renner, K. A., Factors Affecting Imazaquin and AC-263,499 Persistence in Soil and Subsequent Corn Cultivar Response, Ph.D. thesis, Michigan State University, East Lansing, 1986.

Renner, K. A., Meggitt, W. F., and Penner, D., Effect of soil pH on imazaquin and imazethapyr adsorption to soil and phytotoxicity to corn (*Zea mays*), *Weed Sci.*, 36, 78, 1988.

Riley, D. G. and Shaw, D. R., Influence of imazapyr on the control of pitted morningglory (*Ipomoea lacunosa*) and johnsongrass (*Sorghum halepense*) with chlorimuron, imazaquin, and imazethapyr, *Weed Sci.*, 36, 663, 1988.

Riley, D. G. and Shaw, D. R., Johnsongrass (*Sorghum halepense*) and pitted morningglory (*Ipomoea lacunosa*) control with imazaquin and imazethapyr, *Weed Technol.*, 3, 95, 1989.

Schweizer, E. E., Zimdahl, R. L., and Mickelson, R. H., Weed control in corn (*Zea mays*) as affected by till-plant systems and herbicides, *Weed Technol.*, 3, 162, 1989.

Singh, B. K., Newhouse, K. E., Stidham, M. A., and Shaner, D. L., Acetohydroxyacid synthase-imidazolinone interaction, in *Prospects of Amino Acid Biosynthesis Inhibitors in Crop Protection and Pharmaceutical Chemistry*, Monogr. No. 42, Copping, L. G., Dalziel, I., and Dodge, A. D., Eds., British Crop Protection Council, Farnham, England, 1989, 87.

Stougaard, R. N., Efficacy, Soil Behavior, and Persistence of Imazaquin, Imazethapyr, Chlorimuron, and FMC 57020, Ph.D. thesis, University of Nebraska, Lincoln, 1987.

Stougaard, R. N., Shea, P. J., and Martin, A. R., Effect of soil type and pH on adsorption, mobility, and efficacy of imazaquin and imazethapyr, *Weed Sci.*, 38, 67, 1990.

Swanson, E. B., Herrgesell, M. J., Arnoldo, M., Sippell, D. W., and Wong, R. S. C., Microspore mutagenesis and selection: canola plants with field tolerance to the imidazolinones, *Theor. Appl. Genet.*, 78, 525, 1989.

Van Ellis, M. R. and Shaner, D. L., Mechanism of cellular absorption of imidazolinones in soybean (*Glycine max*) leaf discs, *Pestic. Sci.*, 23, 25, 1988.

Vencill, W. K., Wilson, H. P., Hines, T. E., and Hatzios, K. K., Common lambsquarters (*Chenopodium album*) and rotational crop response to imazethapyr in pea (*Pisum sativum*) and snap bean (*Phaseolus vulgaris*), *Weed Technol.*, 4, 39, 1990.

Wills, G. D. and McWhorter, C. G., Influence of inorganic salts and imazapyr on control of pitted morningglory (*Ipomoea lacunosa*) with imazaquin and imazethapyr, *Weed Technol.*, 1, 328, 1987.

Wilson, A. K., Two pot experiments to assess the post-emergence activity of various herbicides for the control of *Pennisetum setosum.*, *Ann. Appl. Biol.*, 114, 112, 1989.

Wilson, R. G., New herbicides for weed control in established alfalfa (*Medicago sativa*), *Weed Technol.*, 3, 523, 1989.

INDEX

Milton Keynes UK
Ingram Content Group UK Ltd.
UKHW051932141024
449569UK00027B/1464

9 781138 562257